Activated Carbon
Progress and Applications

Activated Carbon
Progress and Applications

Edited by

Chandrabhan Verma
King Fahd University Petroleum and Minerals, Saudi Arabia
Email: Chandraverma.rs.apc@itbhu.ac.in

and

Mumtaz A. Quraishi
King Fahd University of Petroleum and Minerals, Saudi Arabia
Email: mumtaz.quraishi@kfupm.edu.sa

ROYAL SOCIETY
OF **CHEMISTRY**

Print ISBN: 978-1-83916-780-5
PDF ISBN: 978-1-83916-986-1
EPUB ISBN: 978-1-83916-987-8

A catalogue record for this book is available from the British Library

The Royal Society of Chemistry is a charity, registered in England and Wales, Number 207890, and a company incorporated in England by Royal Charter (Registered No. RC000524), registered office: Burlington House, Piccadilly, London W1J 0BA, UK, Telephone: +44 (0) 20 7437 8656.

Visit our website at www.rsc.org/books

Printed in the United Kingdom by CPI Group (UK) Ltd, Croydon, CR0 4YY, UK

Preface

Activated carbon, commonly referred to as charcoal or activated charcoal, is a refined powder form of carbon that is widely used for a variety of purposes, including the filtration of toxins from water and air. Activated carbon has a large surface area (32 000 sq. ft $\approx$ 3000 m^2) due to the presence of tiny, low-volume holes in it. This characteristic gives activated carbon great potential for chemical reactions and adsorption. Activated carbon often comes in the form of tiny granules or powders with diameters of 0.15 to 0.25 mm and sizes of not more than 1.0 mm. By attaining a high surface area by activation and/or by altering its surface characteristics and functionalization through chemical treatment, activated carbon's possible uses may be enhanced. Coconut husks and carbon-based wastes from paper mills are two typical sources of activated carbon. Activated coal and activated coke are the names for the activated carbons made from coal and coke, respectively. Numerous industries, including those related to the environment, fuel storage, analytical chemistry, gas adsorption, chemical purification, agriculture, gas storage, skincare, the purification of alcoholic beverages, mercury scrubbing, food additives, *etc.*, could benefit from the use of activated carbon.

This book aims to describe the various aspects of activated carbon including its synthesis, characterization, properties, and applications. This is a reference book for chemists, chemical engineers, students, and high-level professionals working in both R&D and academia who want to learn more about the synthesis, properties, and application of activated carbon. Overall, this book contains thirteen chapters. The first chapter describes the fundamental aspects of activated carbon and its classification and properties. The remaining chapters deal with specific applications of activated carbon including industrial, medical, analytical, environmental, catalytic, beverage, agricultural, fuel storage, food additive, and anticorrosive applications.

Activated Carbon: Progress and Applications
Edited by Chandrabhan Verma and Mumtaz A. Quraishi
© The Royal Society of Chemistry 2023
Published by the Royal Society of Chemistry, www.rsc.org

The different chapters of the book are written by highly recognized scholars working in academia and industry, working in the fields of activated carbon and materials science, chemistry, and engineering.

We editors, Drs Chandrabhan Verma and Mumtaz A. Quraishi, would like to thank all contributors for their great efforts. On behalf of the Royal Society of Chemistry (RSC), we are very thankful to the authors of all chapters for their amazing and passionate efforts in the making of this book. Special thanks to Dr Merlin Fox (Commissioning Editor) and Dr Amina Headley (Editorial Assistant) at the Royal Society of Chemistry (RSC), for their dedicated support and help during this project. In the end, all thanks to the Royal Society of Chemistry (RSC) for publishing the book.

Chandrabhan Verma
Mumtaz A. Quraishi

Editor Biographies

Chandrabhan Verma works at the Interdisciplinary Center for Research in Advanced Materials at King Fahd University of Petroleum and Minerals (KFUPM), Saudi Arabia. He obtained his PhD in Materials Science/Chemistry at the Indian Institute of Technology (Banaras Hindu University) Varanasi, India. He is a member of the American Chemical Society (ACS). He serves as a reviewer and editorial board member of various internationally recognized publishers such as ACS, Royal Society of Chemistry, Elsevier, Wiley and Springer. Dr Verma is the Associate Editor-in-Chief of the Organic Chemistry Plus Journal. He is the author of several research and review articles published by ACS, Elsevier, Royal Society of Chemistry, Wiley and Springer. Dr Verma has also edited a few books for ACS, Elsevier, Royal Society of Chemistry, Springer and Wiley. Dr Verma has received several awards for his academic achievements.

Activated Carbon: Progress and Applications
Edited by Chandrabhan Verma and Mumtaz A. Quraishi
© The Royal Society of Chemistry 2023
Published by the Royal Society of Chemistry, www.rsc.org

Prof. **Mumtaz A. Quraishi** is a Chair Professor at the Interdisciplinary Center for Research in Advanced Materials at King Fahd University of Petroleum and Minerals (KFUPM), Saudi Arabia. He obtained his PhD in synthetic organic chemistry in 1986 from Kurukshetra University and was awarded a DSc. in 2004 by Aligarh Muslim University Aligarh in the field of Corrosion Inhibition of Industrial Metals and Alloys. Before joining KFUPM he was an Institute Professor at IIT BHU Varanasi, India. He also served as the Head (Chairman) of the Department of Chemistry at IIT BHU. He has teaching experience of more than 35 years. He has received several national and international awards. Dr Quraishi is an Associate Editor of Current Materials Science Bentham and a member of the editorial board of more than 30 international journals. Dr Quraishi is a fellow of the Royal Society of Chemistry and a member of the American Chemical Society. He has published more than 450 papers in peer reviewed journals having an H-index of 93 and citations of more than 29 500. His global status is one in terms of the H-index in the field of corrosion inhibitors. He authored *"Heterocyclic Organic Corrosion Inhibitors Principles and Applications"*, Elsevier, 2020.

Contents

Activated Carbon: Progress and Applications
Edited by Chandrabhan Verma and Mumtaz A. Quraishi
© The Royal Society of Chemistry 2023
Published by the Royal Society of Chemistry, www.rsc.org

Chapter 2   Industrial Applications of Activated Carbon   23
Muhammad Sajid

Chapter 3   Medical Applications of Activated Carbon   42
*Payal B. Joshi, Murthy Chavali, Gagan Kant Tripati and
Surabhi Tondwalkar*

CHAPTER 1

Activated Carbon: Fundamentals, Classification, and Properties

RICHIKA GANJOO,[a] SHVETA SHARMA,[a] ASHISH KUMAR*[b] AND
M. M. ARÊMOU DAOUDA*[c]

[a] Department of Chemistry, School of Chemical Engineering and Physical
Sciences, Lovely Professional University, Punjab, India; [b] NCE, Department
of Science and Technology, Government of Bihar, India; [c] International
Chair in Mathematical Physics and Applications, (ICMPA-UNESCO Chair),
Université d'Abomey-Calavi, Benin
*Emails: drashishchemlpu@gmail.com; mouharrab@yahoo.fr

1.1 Introduction

Activated carbon (AC), also known as activated charcoal, is a rough, imperfectly structured kind of graphite. It has a wide spectrum of pores of varying sizes, from obvious fractures and fissures to molecular dimensions. Because of its significant surface area, AC is frequently utilized for a variety of purposes, including removing impurities from air and water. Small, low-volume pores that are present in AC enhance the surface area that is accessible for chemical reactions such as adsorption (which is different from absorption). Before activation, charcoal has a specific surface area of 2.0 to 5.0 $m^2 g^{-1}$, which increases to 1000 $m^2 g^{-1}$ once activated. Gas adsorption analysis suggests that one gram of AC has more than 3000 m^2 surface area because of its high level of microporosity that can deliver a high activation level for practical application. Adsorption of molecules is reinforced on AC

Activated Carbon: Progress and Applications
Edited by Chandrabhan Verma and Mumtaz A. Quraishi
© The Royal Society of Chemistry 2023
Published by the Royal Society of Chemistry, www.rsc.org

by London dispersion forces (van der Waals forces), gas-phase adsorption and liquid-phase adsorption. Descriptions of the forces that occur between molecules are as follows:

(a) London dispersion forces: these have an extremely small range and are sensitive to the distance between the adsorbate molecule and the surface of the carbon. Additionally, they are additive, which means that the adsorption force is the sum of all atomic interactions. Due to the short-range and cumulative nature of these forces, AC has the strongest physical adsorption forces among substances known to mankind.

(b) Gas-phase adsorption: air, natural gas, chemicals, and petrochemicals are typically purified or separated on a large scale using gas-phase adsorption. When gas is exposed to an adsorbent, it draws molecules to its surface where they concentrate and are drawn away from the gas phase. This condensation process causes the bulk-phase molecules to condense in the pores of the AC. The ratio of the compound's partial pressure to vapour pressure controls the adsorption process.

(c) Liquid-phase adsorption: the adhesion of ions, molecules, or atoms from a liquid to a surface is known as liquid-phase adsorption. An adsorbate layer is formed on the adsorbent's surface as a result of this action. The molecules move from the bulk phase and get adsorbed in the pores in a semi-liquid state. The relationship between a compound's concentration and solubility is what drives adsorption.

On AC, all substances can adsorb to some degree. In actual use, AC is utilized to adsorb mostly organic molecules as well as some inorganic substances with higher molecular weights, such as iodine and mercury. An increase in a compound's adsorbability is typically correlated with its molecular weight, the functional groups attached, like halogens or double bonds, and its polarizability.

Adsorption characteristics are frequently improved by additional chemical treatment. Typically, waste materials like coconut husks,[1–3] paper mill waste,[4–6] coal (bituminous coal, anthracite coal, lignite/brown coal, sub-bituminous coal, *etc.*),[7–10] phenolic resins,[11–13] rayon,[14] wood,[15] acrylonitrile,[16] coal tar pitch,[17] petroleum pitch,[18] sawdust, grass ash (peat),[19] bamboo,[20–23] willow peat,[24] and coir[25] are used to manufacture AC (Figure 1.1). Before being "activated," these raw materials are turned into charcoal. Activated coal is the name given to material that is derived from coal.

1.1.1 Structure of AC

The structure of AC has a significant impact on its ability to absorb substances. AC and pure graphite have fundamental chemical structures that are very similar. The layers of fused hexagons that make up the graphite crystal are kept unified by weak van der Waals forces. The structure of AC differs from that of graphite in terms of the distance between layers. The interlayer spacing is 0.34 to 0.35 nm in AC, while it is 0.33 nm in graphite. ACs are divided into

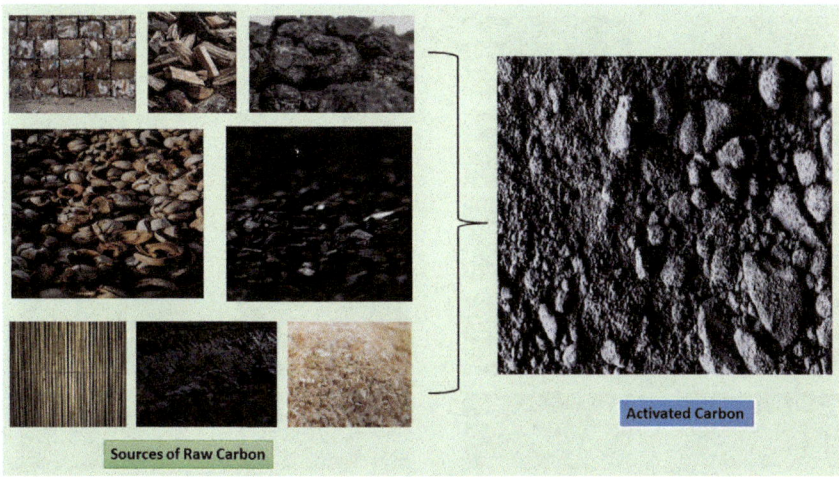

Figure 1.1 Raw sources of AC.

graphitizing and non-graphitizing varieties based on their ability to form graphite. There are several graphene layers in the graphitizing carbon that are positioned parallel to one another. Because of the weak cross-linking between the nearby microcrystallites and the underdeveloped porous structure, the carbon produced is sensitive. The strong cross-linking between crystallites makes the non-graphitizing carbons tough, and they exhibit a well-developed microporous structure.[26,27] When linked oxygen is present or when the amount of hydrogen in the initial raw material is insufficient, non-graphitizing structures with strong crosslinks can form more readily. Carbon–carbon bonds hold the layers together. The AC's ultimate features, which include its pore structure, are largely influenced by the raw material and manufacturing process. It has long been controversial to identify the structure of AC. Harry Marsh and Francisco Rodrguez–Reinoso examined more than 15 models for the structure which is mentioned in their book released in 2006, but they were unable to determine which was the most accurate.[28,29] The structure of ACs consists of heptagonal and pentagonal rings which are quite similar to the fullerene structure, according to recent research utilizing aberration-corrected transmission electron microscopy.[29]

The raw material's structure also affects the porous structure and useful properties of carbonaceous adsorbents. As a result, selecting an appropriate material is just as crucial as choosing an appropriate production technology and figuring out the ideal processing conditions. As a result, efforts have been made to find novel raw resources that could be used in the creation of carbonaceous adsorbents. In particular, biomass waste from wood, food, and other sources as well as agriculture has attracted attention in this respect.

The manufacturing of biomass waste-derived ACs could boost economic returns and lower pollution because wood processing, carpentry, and other

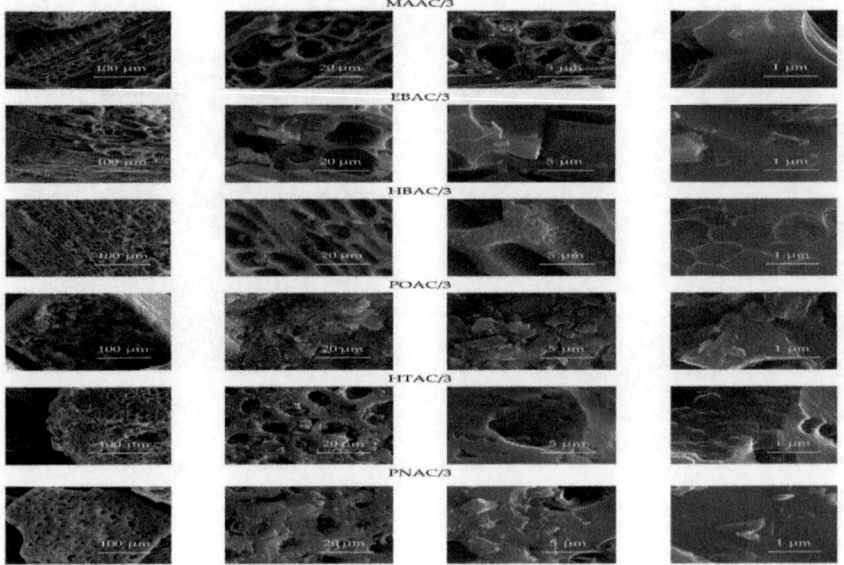

Figure 1.2   SEM micrographs of the AC samples that were obtained from various biomass materials.[30] Reproduced from ref. 30, https://doi.org/10.3390/ma14154121, under the terms of the CC BY 4.0 license https://creativecommons.org/licenses/by/4.0/.

associated industries have huge raw material supply potential. Kwiatkowski *et al.* investigated the production of ACs from numerous biomass-derived resources by activation with KOH. They also carried out a thorough computer analysis of the porous structure and adsorption characteristics of the carbons based on benzene (C_6H_6) adsorption isotherms. SEM imaging results showed that the topological structures of the ACs made from ebony (EBAC/3), mahogany (MAAC/3), and hornbeam wood (HBAC/3) were strikingly comparable (Figure 1.2). The examination of the ACs' porous structure, therefore, revealed the substantial potential for producing ACs from waste biomass with a very high adsorption capacity and significant specific surface area by activation with KOH.[30]

1.1.2   Activation Reaction or Activation of Carbon

Raw carbon is transformed into a porous substance by activation, increasing its surface area. Chemical and physical activation are the two methods of activation, described in the following sections:

1.1.2.1   *Physical Activation*

Hot gases are used to generate AC from source materials.[31,32] The gases are subsequently burned out with the help of air, resulting in a filtered, polished, and brushed form of AC (Figure 1.3). One or more of the following

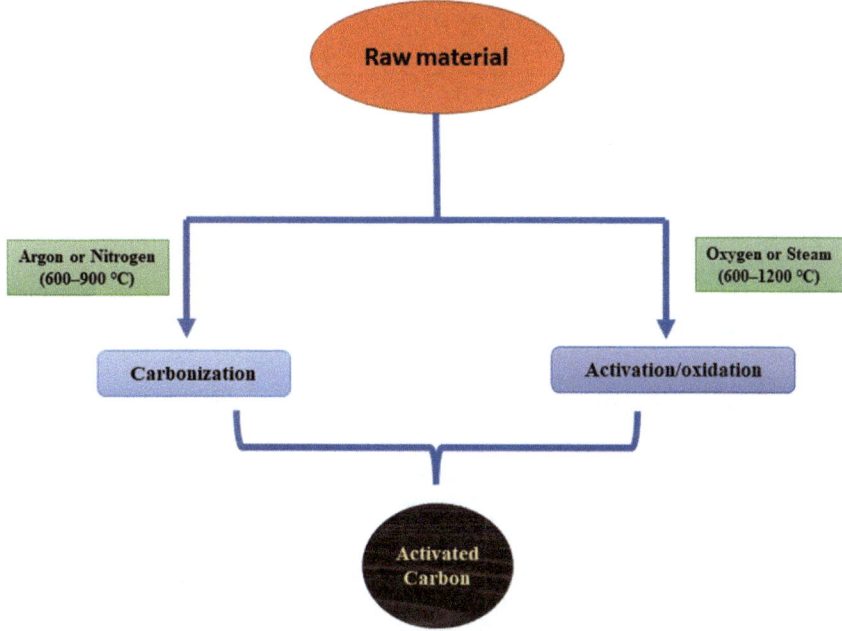

Figure 1.3   Illustration of the physical activation process of carbon.

procedures described in the following list are typically used to accomplish this:

(i) Carbonization: in the presence of gases like nitrogen and argon (inert environment) and at temperatures ranging between 600 and 900 °C, material containing carbon is pyrolyzed.

(ii) Activation/oxidation: at temperatures ranging between 600 and 1200 °C, raw material is exposed to an oxidizing atmosphere (steam or oxygen). The sample is heated for 1 hour at 450 °C in an air-filled muffle furnace to activate it.

1.1.2.2 *Chemical Activation*

Various compounds are infused into the carbon material. Typically, the substance infused is one of the following: phosphoric acid (25%), potassium hydroxide (5%), sodium hydroxide (5%), calcium chloride (25%), and zinc chloride (25%). After blocking the air, carbon is heated to high temperatures (500–700 °C) to prevent the generation of tar and AC (Figure 1.4). The carbon is thought to be activated at this point by the temperature, which causes the substance to crack open and develop more small pores. Because of the lower temperatures, improved uniformity in quality, and smaller activation times required for the material, chemical activation is favoured over physical activation. To prevent the production of tar, wood-based products (such as sawdust and wood chips) are heated to a high temperature. Chemical agents that

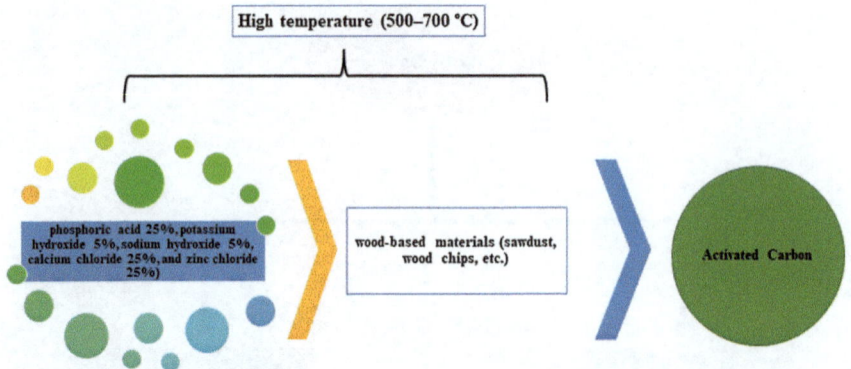

Figure 1.4 Illustration of the chemical activation process of carbon.

degrade the fibres of wood-based materials are applied and are allowed to enter, such as zinc chloride and phosphoric acid. K_2CO_3, KOH, $ZnCl_2$, and H_3PO_4 are the chemical activators that are most frequently utilized.[33–35] Complex processes like condensation, dewatering, carbonization, polymerization, and oxidation are all part of this reaction.

1.2   Categorization of ACs

ACs are intricate products that are categorized based on their behaviour, surface properties, and other essential properties. Some generic classification based on their processing techniques, size, and industrial applications is described in the following sections.[36]

1.2.1   Powdered AC (PAC)

PAC or pulverized AC is the term used by the ASTM, formerly known as the American Society for Testing and Materials, to describe particles that pass through an 80-mesh sieve (0.180 mm). PAC has small AC particles made by milling or pulverizing AC, with an average size that is typically <0.075 mm and a mean particle diameter (MPD) of <0.045 mm. It has a low diffusion distance and a high surface area-to-volume ratio. 95–100% of the carbon particles that make up PAC will pass through a specific mesh sieve after being crushed or powdered. PAC is typically applied to other processing components directly like fast mix basins, raw water intakes, gravity filters, and clarifiers. PAC is used in various processes, including the purification of hydrolyzed vegetable protein, and the control of taste and odour in drinking water.

1.2.2   Granular AC (GAC)

The external surface of GAC is smaller than that of PAC because of its bigger particle size. The size range of the irregularly shaped particles that make up

GAC is 0.2 to 5 mm. The basic material is either activated directly or after being agglomerated to create GACs. Therefore, the adsorbate's diffusion is a significant factor. These carbons are appropriate for gas and vapour adsorption owing to the rapid diffusion of gaseous substances. GACs are utilized for water and air purification, substance segregation, and generalized deodorization in rapid mix basins and flow systems. Coconut shells, coal, peat, and other basic organic materials high in carbon are used to make GAC. As water passes through a filter made of GAC, heat is used to activate the carbon's surface area, eliminating some compounds that have been dissolved in water. Because of its porous characteristics, GAC absorbs the chemical compounds. On the inside surface of the AC, the adsorption takes place. Adsorption is the process of removing substances from liquids or gases by diffusing them to the surface of the adsorbent through the porous structure of AC. Attractive forces then retain the substances on the surface of the adsorbent.

1.2.3   Extruded AC (EAC)

Pulverized anthracite or charcoal and an appropriate binder are combined to create extruded ACs (pressed pellets), which are then extruded at high pressure into cylindrical forms with diameters ranging from 0.8 to 130 mm. To achieve a particular pore shape, activation catalysts like potassium hydroxide are occasionally used before extrusion. Because of their excellent mechanical strength, minimal dust content, and low-pressure loss, these are mostly employed for gas-phase applications.

1.2.4   Bead AC (BAC)

BAC comes in diameters ranging from around 0.35 to 0.80 mm and is manufactured from petroleum pitch. Similar to EAC, although with lower grain size, it is likewise renowned for its low pollution concentration, high mechanical strength, and very little pressure drop. It is favoured for water filtration due to its spherical shape.

1.2.5   Impregnated AC

When chemicals and metal particles are evenly distributed on the inner surfaces of AC's pores, surface impregnation chemically alters the material. The synergism between the chemicals and the carbon significantly increases the adsorptive ability of the carbon. Additionally, it offers a practical means of removing contaminants from gas streams that would not otherwise be possible. Iodine and silver are only two examples of the several inorganic component types found in porous carbons. Aluminium, manganese, zinc, iron, lithium, and calcium are only a few of the cations that have been used for specific applications in the reduction of air pollution, particularly in both exhibitions and museums. AC with silver is utilized as an adsorbent for the filtration of household water because of its antibacterial and antiseptic properties. Natural water can be converted into drinkable water by treating it with a solution of AC and the flocculant $Al(OH)_3$. Hydrogen sulphide (H_2S) and thiols can also be adsorbed on impregnated carbons. There have been reports of H_2S adsorption rates up to 50% by weight. Silver-infused carbon is a powerful adsorbent for purification in earth-based homes and other water systems due to its antimicrobial/antiseptic properties. Flue gases in coal-fired power plants and other air pollution regulator applications are treated using impregnated AC. Aldehydes, radioactive iodine and mercury, and in-organic gases like arsine and phosphine can all be removed using AC by selectively impregnating it with these substances. Inorganic gases such as HCN, H_2S, phosphine, and arsine are the targets of metal-oxide-impregnated carbon.

1.2.6   Polymer-coated Carbon

To create a smooth, permeable covering without clogging the pores, a biocompatible polymer can be applied to a porous carbon using this method. Hemoperfusion can benefit from the resultant carbon. Hemo-perfusion is a medical procedure that removes hazardous compounds from the blood by passing a patient's blood over an adsorbent material.

1.2.7   Woven Carbon

It's feasible to produce AC cloth for carbon filtration using technical rayon fibres. According to the BET theory, activated cloth has an increased cap-ability for absorption than activated charcoal (surface area: 500–1500 $m^2\,g^{-1}$, pore volume: 0.3–0.8 $cm^3\,g^{-1}$).

1.3   Properties of AC

Physically activated carbon uses van der Waals forces or London dispersion forces to bond materials. Alcohols, diols, strong bases and acids, metals, and the majority of inorganic substances, like sodium, lithium, lead, iron, fluorine, arsenic, and boric acid, do not bond well to AC. Iodine is particu-larly well adsorbed by AC.

AC is not very effective at adsorbing carbon monoxide. The fact that the material is used in respiratory filters, fume hoods, or other gas regulator equipment should be of particular concern to those who use them because the gas is hazardous to metabolism and the nervous system and is im-perceptible to the human senses. To increase the adsorptive ability for some inorganic (and hazardous organic) molecules like ammonia (NH_3), H_2S, mercury (Hg), formaldehyde (HCOH), and radioactive iodine-131 (I-131), AC can be employed as a medium for the adsorption of diverse chemicals. In the following sections the various characteristics of ACs are described.

1.3.1 Iodine Number

A method used to assess the adsorption potential of ACs is the iodine number. The quantity of iodine absorbed by 1 g of carbon at the mg level is the iodine number, which measures the porosity of the AC.[37] The usable surface area in $m^2 g^{-1}$ of virgin carbon is indicated by the iodine number (or "iodine value") (ASTM D4607). The iodine number, which is associated with the "activity" of AC and is frequently used as a quality control (QC) parameter in the production and reactivation of AC, does not always offer a measure of the carbon's capacity to adsorb other chemicals. The iodine number is typically 50 to 100 $mg g^{-1}$ lower than the BET surface area. It is widely known that the carbon surface oxygen complexes, pH, and ash components all have an impact on the iodine adsorption from aqueous solutions. A virgin AC's iodine number will decrease by 20 to 50 points after being moistened with water and dried because the surface oxygen complex has changed. In an AC filter, QC samples should be collected dry rather than wet. Many carbons adsorb small compounds selectively. The most essential factor used to describe the performance of AC is the iodine number. It is a measurement of the level of activity (a greater number indicates a higher degree of activation) and is frequently expressed in $mg g^{-1}$ (usual range: 500–1200 $mg g^{-1}$). By iodine adsorption from solution, the micropore content of AC (0–20, or up to 2 nm) is measured. It corresponds to a carbon surface area of 900–1100 $m^2 g^{-1}$. It serves as the benchmark in liquid-phase applications. The iodine values for carbons used in water treatment typically range from 600 to 1100. This metric is frequently used to assess how much carbon is used up. Chemical interactions between the adsorbent and iodine may change the adsorption of iodine if this method is employed carelessly and produce inaccurate results. In light of this, it is only advised to use the iodine number as a gauge of a carbon bed's level of exhaustion if the bed has been demonstrated to be free of chemical interactions with adsorbates and if an experimental association between the iodine number and the degree of exhaustion has been established for the specified application.

1.3.2 Molasses Number

Certain carbons are better at adsorbing heavy compounds. By adsorbing molasses from solution, AC's mesopore content (>2 nm) is measured as the molasses number or molasses efficiency.[38] Increased adsorption of large molecules is indicated by a high molasses number (range 95–600). The decolourizing performance of caramel is comparable to that of molasses. The molasses efficiency is in the range of 40% to 185%, and it corresponds to the molasses number (600 = 185%, 425 = 85%). The decolourization capability of AC is measured using the molasses number (DSTM 16), which also serves as an indicator of the macropore and transport pore structure of the material. A quantity of powdered carbon is combined with a standard molasses solution and heated for a predetermined period. The molasses number describes how much more colour is removed by AC as compared to regular carbon. So, the better the product, the higher the molasses number. The relationship between the European and North American molasses numbers (525–110) is inverse. A standard diluted molasses solution standardized against AC is given a molasses number, which represents the degree of decolourization. The molasses number shows the possible pore capacity that might be accessible for bigger adsorbing entities due to the size of colour bodies.

1.3.3   Apparent Density

A fixed volume of AC's weight is calculated using an apparent density. The solid or skeletal density of ACs normally ranges between 2.0 and 2.1 $g\,cm^{-3}$ (125–130 lbs per ft^3) (*i.e.*, if all pore gaps were removed). Nevertheless, as there is a significant amount of void space between the particles in an AC sample, the actual operational (apparent) density is often lower, ranging from 0.4 to 0.5 $g\,cm^{-3}$ (25–31 lbs per ft^3). When comparing the same source material, greater volume activity and better-activated carbon are typically indicated by higher apparent density. It may also suggest various starting materials (such as bituminous coal *vs.* lignite coal *vs.* wood). ACs typically have a solid or skeletal density of 2000–2100 $kg\,m^{-3}$ (125–130 lbs per cubic foot). The actual or apparent density of an AC sample will be lower, approximately 400 to 500 $kg\,m^{-3}$ (25–31 lbs per cubic foot), because a significant portion of the sample will be made up of air space between the particles. Greater volume activity and better-quality AC is typically associated with higher density. The apparent density of AC is calculated using ASTM D 2854-09 (2014).

1.3.4   Hardness/Abrasion Number

The mechanical strength of AC is one of its fundamental mechanical characteristics. It can be described as a material's resistance to wear and tear while it is being used. There are numerous tests available to assess the mechanical toughness of GAC. These tests measure the number of fines produced or the change in particle size distribution. To quantify the distinct characteristics of strengths, various mechanical strength test methods cannot be mathematically related to one another. The hardness number is the most popular test method, followed by the abrasion number. The hardness number (DSTM 20) gauges how well AC's external integrity holds up to wear along the edges and breaking of small points. After shaking granules under specific circumstances, it is stated as a percentage of loss on a specific sieve. The AWWA B604 abrasion number gauges the structural durability of GAC. It assesses a particle's resistance to shear pressures brought on by particles rubbing against one another or against surfaces like supporting screens or column walls.[39] It is assessed as a percentage decrease in mean particle diameter (MPD) by agitating granules with steel balls in a container under specific conditions. The AC's resistance to attrition is gauged with this parameter. The ability of AC to preserve its physical integrity and withstand frictional forces is a crucial indicator. The raw material and activity levels have a significant impact on the hardness of ACs.

1.3.5 Ash Content

Total ash measures the weight-based mineral oxide concentration of AC. At 800 °C, the mineral components are transformed into their corresponding oxides to determine the value. The amount of ash, which is primarily made up of silica and aluminium, depends on the primary raw material utilized to create the finished product. The typical values for ACs based on wood, coconut shells, and coal are 2–3% w/w, 5% w/w, and 8–15% w/w, respectively. Ash decreases both the total activity of AC and the effectiveness of reactivation; the amount is solely based on the base raw material used to make the AC (*e.g.*, coconut, wood, coal). Discolouration can come from the metal oxides (Fe_2O_3) leaching out of AC. The ash content that is acid/water soluble is more important than the total ash content. For aquarists, the amount of soluble ash can be crucial since ferric oxide can encourage the formation of algae. For marine fish, freshwater fish, and reef tanks, a carbon with a low soluble ash concentration should be used to prevent heavy metal toxicity and excessive plant/algal growth. The ash content of AC is ascertained using the ASTM (D2866 Standard Test Method).

1.3.6 Carbon Tetrachloride Activity

The level of AC activation can be determined using this test procedure. When AC is saturated with carbon tetrachloride (CCl_4) under the circumstances specified in this test method, carbon tetrachloride activity is defined as the weight of CCl_4 absorbed by the AC sample divided by the sample weight (in %).

1.3.7 Particle Size Distribution

An AC's access to surface area and the rate of adsorption kinetics are both improved and accelerated by finer particle size.[40] This must be weighed against pressure drop, which will affect energy costs in systems with vapour phases. Particle size distribution should be carefully taken into account for considerable operational benefits. Nevertheless, the particle size should fall between 3.35 and 1.4 millimetres when utilizing AC for the adsorption of minerals like gold (0.132–0.055 in). It would not be appropriate to elute AC with particles smaller than 1 mm.[41]

1.4   Modification of the Properties and Reactivity of AC

AC is a popular choice for contaminant removal media due to its adaptability and broad variety of uses. The latest studies have concentrated on improving the efficiency of AC by changing its unique features to make it more apt to attract certain pollutants.[42,43] The makeup of the surface functional groups has a significant impact on the properties of acid–base, oxidation–reduction, and particular adsorption processes. Conventional AC has a reactive surface that can be oxidized by ozone, carbon dioxide, and ambient oxygen as well as oxygen plasma steam. Several different chemicals might produce oxidation in the liquid phase (HNO_3, H_2O_2, $KMnO_4$).[44] The surface of oxidized carbon can acquire several basic and acidic groups, which can alter the material's sorption and other characteristics dramatically from its unaltered state.

Natural compounds, polymers, or the use of nitrogenating reagents during the processing of carbon can nitrogenate AC.[45] Fluorine, chlorine, and bromine can all interact with AC.[46,47] Like other carbon materials, the surface of AC can be fluoralkylated *via* a CVD process or treatment with (per)-fluoropolyether peroxide in a liquid phase. Such materials can be employed as electrode materials for supercapacitors because they have excellent electrical and thermal conductivity, high hydrophobicity, and chemical stability.[48] It is believed that the development of ACs from halogenated precursors results in a catalyst that is more efficient because the stability of the remaining halogens is improved. AC with chemically grafted superacid sites is also known in the literature. The presence of the surface-active carbon double bond has been related to a number of the chemical features of AC. AC can get saturated and lose its absorption ability during use. Compared to routinely replacing AC to obtain the desired effect, which wastes resources, AC regeneration has more economic and environmental advantages. Adsorption is a process in which AC, adsorbate, and solvent achieve adsorption equilibrium. To allow AC to desorb and regain its activity, the initial equilibrium conditions must be disrupted on the following grounds:

(i)   External heating to raise the temperature to alter the equilibrium conditions, because the amount of adsorbed mass decreases as the temperature increases, resulting in the desorption of the adsorbed mass.

(ii)   Modifying the adsorbate's chemical characteristics.

(iii)   Adsorbate extraction using a solvent having a strong affinity for the adsorbate.

(iv)   Substitution of the adsorbate with a compound that has a high affinity for the AC, followed by desorption of the replacement material and regeneration of the AC.

(v)   Desorbing an adsorbate involves decreasing the concentration (or pressure) of the solute in the solvent.

(vi)   Decomposing or oxidizing the adsorbed material (organic materials) to remove it.

1.4.1   Regeneration of AC

AC regeneration (*i.e.*, reactivation) is the process of removing adsorbed material from AC by physical or chemical means without altering its original structure, and restoring its adsorption efficiency for reuse. Before selecting the AC regeneration technique, it is essential to have a thorough understanding of the treated object and the extent of the treatment. The largest reactivation facility in the world is located in Feluy, Belgium. There is a reactivation centre for AC also in Roeselare, Belgium. Reactivation or regeneration of ACs entails desorbing impurities from the AC surface to restore the adsorptive ability of saturated AC. There are numerous regeneration methods for AC, including thermal regeneration, biological regeneration, wet oxidation regeneration, electrochemical regeneration, solvent regeneration, and catalytic wet oxidation, as mentioned in the following sections.

1.4.1.1   Thermal Regeneration Method

One of the oldest and most established AC regeneration techniques is thermal regeneration. After treating organic wastewater, AC is typically regenerated in three stages: drying, high-temperature carbonization, and activation depending on how organic matter changes at various temperatures. The volatile elements on the AC are mostly eliminated during the drying process. While some of the organic matter adsorbed on the AC boils, vaporizes, and desorbs in the high-temperature carbonization stage, another organic matter decomposes to produce small molecule hydrocarbons and desorbs, with the remaining organic matter remaining in the pores of the AC to be converted to fixed carbon. The temperature will increase to 800–900 °C at this point. The procedure is typically carried out in a vacuum or an inert environment to prevent oxidation of the AC. The following step of activation involves adding CO_2, CO, H_2, or water vapour to the reactor to purge the AC's micropores and restore its adsorption capacity, which is essential to the whole regeneration process. Despite the broad use and high regeneration effectiveness of the thermal regeneration technique, it needs extra energy for heating during the regeneration process, which raises the cost of both the initial investment and ongoing operations.

1.4.1.2   Biological Regeneration Method

To remove the organic matter that has become adsorbed on AC and further digest and degrade it into H_2O and CO_2, the biological regeneration process uses domesticated bacteria. There are also aerobic and anaerobic approaches, and the biological regeneration method is comparable to the biological wastewater treatment method. It is generally accepted that cell autolysis occurs during the regeneration process, in which case cell enzymes flow to the extracellular space, and AC has an adsorption effect on enzymes, forming an enzymatic centre on the surface of carbon, thereby promoting the decomposition of pollutants and achieving the goal of regeneration. The pore size of AC itself is very small, some only a few nanometers, so micro-organisms cannot enter such pores. The biological process takes a long time and is strongly impacted by water quality and temperature, yet it is straightforward and can be quickly executed with cheap start-up and running expenses.

1.4.1.3   Wet Oxidation Regeneration Method

Under high temperature and pressure conditions, oxygen or air is utilized to oxidize and break down the organic material adsorbed on the AC into tiny molecules. This process is known as wet oxidation regeneration. The following are the ideal regeneration conditions for AC: 230 °C regeneration temperature, 1 h regeneration period, 20.6 MPa oxygenation, 15 g carbon, and 300 mL water. After 5 rounds of regeneration, the regeneration efficiency drops only by 3% from its peak of $45 \pm 5\%$. The primary factor causing the decline in regeneration efficiency is the partial oxidation of AC surface micropores.

1.4.1.4   Solvent Regeneration Method

The solvent regeneration methodology incorporates the phase equilibrium between the AC, solvent, and adsorbed material to disrupt the adsorption equilibrium by adjusting the temperature, solvent pH value, and other variables to desorb the adsorbed material from the AC. For such reversible adsorptions, including the adsorption of organic wastes with high concentrations and low boiling points, the solvent regeneration approach is more suited. It is more focused since a solvent often has a limited range of contaminants that it can desorb, while the water treatment process may deal with a large variety of pollutants.

1.4.1.5   Electrochemical Regeneration Method

A brand-new approach to AC regeneration is electrochemical regeneration. In this procedure, a DC electric field is applied to the electrolyte while AC is sandwiched between the two primary electrodes. A micro-electrolytic cell is created when the electric field causes the AC to become polarized, with one end acting as the anode and the other as the cathode. This technique is simple to use, very effective and uses little energy, and its treatment object is subject to fewer restrictions. If the treatment procedure is flawless, secondary contamination may be prevented.

1.4.1.6   Catalytic Wet Oxidation Method

Regeneration using the conventional wet oxidation approach uses a lot of energy but is only moderately effective. The primary determinant of regeneration efficiency is regeneration temperature, although raising the temperature can also accelerate the surface oxidation of AC, lowering the regeneration efficiency. As a result, it is thought that using a high-efficiency catalyst will allow for the regeneration of AC.

1.5   Applications of AC

The filtration of drinking water, treatment of groundwater and municipal water supplies, reduction of landfill and power plant gas emissions, and recovery of precious metals are just a few of the many industrial and domestic uses for AC. In Figure 1.5, the uses of AC are shown.

In addition to these uses, AC is employed in the production of methane and hydrogen chloride, hydrogen storage, decaffeination, air purification, capacitive deionization, gold extraction, solvent recovery, supercapacitive swing adsorption, water purification, medicine, metal extraction, and sewage treatment.[49] The applications of AC are described in detail in the following sections:[50]

1.5.1   Industrial Applications

Utilizing AC in metal polishing to clean electroplating solutions is a significant industrial use. For instance, it is the primary method for purifying bright nickel-plating solution to eliminate organic contaminants. To improve the deposit attributes of plating solutions and to enhance features like brightness, smoothness, and ductility, a range of organic compounds are added. Organic additives produce undesired breakdown products in solution as a result of the flow of direct current and the electrolytic processes of anodic oxidation and cathodic reduction. They may hurt the physical characteristics of the deposited metal and the plating quality. Such contaminants are eliminated by AC treatment, which also restores the plating performance to the required degree.

Figure 1.5   Applications of AC.

1.5.2   Medications

AC is listed as one of the essential medicines by the World Health Organization. Poisonings and overdoses caused by oral consumption of drugs are treated with AC, but it is not effective in cyanide, lithium, alcohol, and iron poisoning, and AC needs to be administered only in a health care facility. Many nations utilize tablets or capsules of AC to alleviate diarrhoea, indigestion, and gas. Nevertheless, AC has no impact on intestinal gas and diarrhoea and is typically therapeutically worthless when poisoning is caused by ingesting corrosive substances like petroleum products and boric acid. It is also mainly futile when poisoning is caused by strong bases and acids like iron, cyanide, arsenic, lithium, methanol, and ethanol. AC is also used in teeth whitening and oral health. AC can cause side effects such as black stools, black tongue, vomiting, constipation, and gastrointestinal blockage (in severe cases).

1.5.3   Analytical Chemistry

AC can be used in the chromatographic separation of carbohydrates (stationary phase = AC 50% w/w combination with celite, and mobile phase = ethanol solution). Direct oral anticoagulants (DOACs) like apixaban, dabigatran, edoxaban, and rivaroxaban can be removed from blood plasma samples using AC. AC has been manufactured into "minitablets" for this use, with each having 5 mg of AC to treat 1 ml of DOAC samples. Since heparin, most other anticoagulants, and blood clotting factors are unaffected by this AC, a plasma sample may be examined for anomalies that the DOACs would normally influence.

1.5.4   Water Purification

The removal of contaminants in the field of wastewater treatment has been greatly aided by the use of AC made from agricultural waste and products. The price of this adsorbent, however, is highly dependent on the sources of its basic materials. As a result of the financial benefits and environmental benefits, the method of manufacturing AC from agricultural waste is highly recommended. Water filtration systems often employ AC. Numerous field and industrial processes, including groundwater remediation, air purification, spill clean-up, filtration of drinking water, and the capture of volatile organic compounds from dry cleaning, painting, and gasoline dispensing, among others, use carbon adsorption to remove pollutants from air or water streams.

EPA officials created a regulation that advocated mandating the use of GAC in drinking water treatment facilities during the early stages of the Safe Drinking Water Act's adoption in the United States in 1974. The water supply sector, notably the biggest water utilities in California, strongly opposed the so-called GAC regulation because of its high cost throughout the nation. As a result, the agency disregarded the regulation. Due to its versatility, AC filtration is a successful way to purify water. Depending on the pollutants present, distinct kinds of AC filtering techniques and apparatus are recommended.

1.5.4.1 Tannin Removal

Tannins, mixtures of small and big molecules, are a class of naturally occurring organic compounds that are water soluble and created by the metabolism of trees and plants. They are a component of the fulvic acid materials that are formed during the breakdown of vegetation and are resistant to degradation. The tannin adsorption capacity of carbon is measured in parts per million (200–362 ppm). Tannins, which may give water a yellowish to a brown hue, are made of minute, high-molecular-weight colloidal particles that have a very little negative charge. Tannins may be found in practically every water body where there has been a significant breakdown of flora. This often happens in areas with peaty soils, deep woods, or marshlands. Tannins are more prevalent in the Northeast, Southeast, Northwest, and Great Lakes regions of the United States. Typically, only tannins with very high molecular weights may be adsorbed by AC. As a result, it is the choice of treatment that can be used least widely. However, it has the advantage of generally reduced maintenance costs and overall system expenses (as compared to ion exchange).[51]

1.5.4.2 Methylene Blue Adsorption

Dye pollution occurs in wastewater from the textile, cosmetic, printing, dying, food processing, and paper sectors. Because of their harmful and cancer-causing effects on living things, the discharge of coloured effluents poses a significant environmental challenge for developing nations. For the elimination of pigments, dyes, and other inorganic and organic pollutants, AC sorption is very successful.[52]

1.5.4.3 Dechlorination

Dechlorination is a very swift process in which AC takes part, and free chlorine is changed to chloride. In the top 10 cm of the filter bed, the reaction happens quickly. The dechlorination half-life length, which gauges how well-activated carbon removes chlorine, is used to assess certain carbons. The amount of carbon needed to remove chlorine by 50% is known as the dechlorination half-value length. A shorter half-value length indicates higher efficiency.[53]

1.5.5 Agricultural Applications

Organic farmers are permitted to utilize AC (charcoal) in the production of both wine and animals. It serves as a pesticide, an animal feed additive, processing agent, a non-agricultural chemical, and a disinfectant in the production of animals. Utilizing AC as a chemical intermediate to remove brown colour pigments from white grape extracts is permitted in the production of organic wine.

1.5.6 Distilled Alcoholic Beverage Purification

Vodka and whiskey may be cleaned using AC filters to remove organic contaminants that can alter their flavour, colour, and aroma. Vodka with a much higher level of organic purity, as determined by odour and flavour, may be obtained by passing an organically contaminated vodka through an AC filter at the right flow rate.

1.5.7 Fuel Storage

A variety of pore shapes and surface chemistry are available in ACs for the adsorption of gases. The capacity of different ACs to adsorb hydrogen and natural gas is being researched. The porous substance works as a sponge for various gases. van der Waals forces cause the gas to be drawn to the carbon substance. The bonding energies of certain carbons have been seen to range between 5 and 10 kJ mol^{-1}. When exposed to higher temperatures, the gas may then be desorbed and either burned to generate power or used in a hydrogen fuel cell. Compared to pressure tanks, fuel storage in an air conditioner is easier due to the low-mass, low-volume, and low-pressure requirements that may be fulfilled. In the domain of studying and creating nanoporous carbon materials, the US Department of Energy has set out several objectives to be met. Although many organizations, like the ALL-CRAFT programme, are still working in this area, not all of the objectives have yet been met.

1.5.8 Gas Purification

For gas purification GAC is often used. Siloxanes, hydrogen sulphide, carbon dioxide, biomethane, sewage gas, hydrogen gas, landfill gas, and other gases are often removed from the air using filters containing AC in compressed air and gas purification. In the majority of designs, AC is included in the filter medium and there is either a single step of filtration or two stages of filtration. Radioactive gases are kept confined in the air sucked from a nuclear reactor condenser using AC filters. These gases are absorbed and held in place by the massive charcoal beds as they quickly decompose into non-radioactive solid substances. While the air is being filtered, the solids are being held in the charcoal particles.

1.5.9 Chemical Purification

ACs of certain grades are designed to work in acidic chemical processes. To utilize the ACs in acidic settings, some of them are acid washed to eliminate acid soluble ash. Acids like phosphoric acid, HCl, and sulphuric acid are often treated. In the laboratory, AC is often used to clean solutions of organic molecules that include undesirable coloured organic contaminants. AC can be used to treat organic acids such as acetic acid, oxalic acid, lactic acid, malic acid, benzoic acid, and formic acid.

1.5.10 Mercury Scrubbing

AC is very effective in the removal of mercury from flue gas (coal-fired power plants, cement manufacture, industrial boilers, waste incinerators, *etc.*) as well as liquid natural gas. AC is costly to utilize while being effective. The problem with disposing of mercury-filled AC is that it is often not recycled. Federal rules in the United States permit the stabilization of AC (*e.g.*, encasing it in concrete) for landfilling if the mercury content is less than 260 ppm. However, garbage with a mercury content of more than 260 ppm is classified as high-mercury waste and cannot be dumped in landfills (Land-Ban Rule). A 100- ton-per-year accumulation of this material is now occurring in deep abandoned mines and warehouses. The problem of disposing of AC that contains mercury is not exclusive to the United States. In the Netherlands, the AC is disposed of *via* total combustion, producing carbon dioxide (CO_2), and the mercury is mainly recovered.

1.5.11 Food Additive

To give some foods, such as pizza bases, black ice cream, and hotdogs, a smoky flavour, activated charcoal is employed as a food ingredient. Foods and drinks with activated charcoal should be avoided when taking drugs, such as birth control pills and antidepressants, since they lessen the effectiveness of the medication. Activated charcoal may slow down the pace at which the body absorbs hazardous compounds, according to research, but there isn't enough concrete proof to back up its detoxifying abilities.

1.5.12 Skin Care

Activated charcoal is a common ingredient in a variety of skin care treatments because of its absorbent properties. The cleaning power of soap is combined with the absorption power of activated charcoal in products like activated charcoal soaps, face masks, and scrubs.

1.6 Largest Producers of AC in the World

The biggest manufacturer of AC in the world is Norit NV of the Netherlands, a subsidiary of Cabot Corporation. The world's leading providers of AC are as follows:

(i) Shinkwang Chem. Ind. Co., Ltd., South Korea
(ii) Indo German Carbons Limited, Oman

(iii) Haycarb, Indonesia, Sri Lanka, and Thailand
(iv) Japan. Kuraray Co., Ltd, Kuraray Europe
(v) Oxbow Coal SARL, Belgium

1.7   Conclusion

The categorization, characteristics, and uses of AC are briefly discussed in this chapter. A broad variety of carbonized materials with high surface area and porosity are included in AC. In many regions of the globe, it has a variety of uses in the purification of water, analytical chemistry, pharmaceuticals, skin care, fuel storage, industrial wastewater treatment, gas purification, and odour elimination due to its special properties. The production of AC currently involves using a variety of industrial wastes. A combination of physical and chemical processes is used to activate AC. Chemical activation is more cost-effective than physical activation because it has a lower activation temperature, a faster processing time, and higher carbon efficiency.

References

1. I. Tan, A. Ahmad and B. Hameed, *Chem. Eng. J.*, 2008, **137**, 462–470.
2. I. Tan, A. Ahmad and D. B. Hameed, *J. Hazard. Mater.*, 2008, **153**, 709–717.
3. I. Tan, A. Ahmad and B. Hameed, *J. Hazard. Mater.*, 2008, **154**, 337–346.
4. M. Begum, A. Rahman, M. Molla and M. Rahman, *Int. J. Environ. Sci. Technol.*, 2022, 1–14.
5. V. Gupta, N. Bhardwaj and R. Rawal, *Int. J. Environ. Sci. Technol.*, 2022, **19**, 2641–2658.
6. N. R. Khalili, M. Campbell, G. Sandi and J. Golaś, *Carbon*, 2000, **38**, 1905–1915.
7. H. Teng, T.-S. Yeh and L.-Y. Hsu, *Carbon*, 1998, **36**, 1387–1395.
8. H. Teng and H. C. Lin, *AIChE J.*, 1998, **44**, 1170–1177.
9. B.-L. Xing, H. Guo, L.-J. Chen, Z.-F. Chen, C.-X. Zhang, G.-X. Huang, W. Xie and J.-L. Yu, *Fuel Process. Technol.*, 2015, **138**, 734–742.
10. F. Zhou, J. Cheng, J. Liu, Z. Wang, J. Zhou and K. Cen, *Fuel*, 2016, **170**, 39–48.
11. C. Lei, N. Amini, F. Markoulidis, P. Wilson, S. Tennison and C. Lekakou, *J. Mater. Chem. A*, 2013, **1**, 6037–6042.
12. M. Kubota, A. Hata and H. Matsuda, *Carbon*, 2009, **47**, 2805–2811.
13. H. An, B. Feng and S. Su, *Carbon*, 2009, **47**, 2396–2405.
14. A. Pastor, F. Rodrıguez-Reinoso, H. Marsh and M. Martınez, *Carbon*, 1999, **37**, 1275–1283.
15. T. Wang, S. Tan and C. Liang, *Carbon*, 2009, **47**, 1880–1883.
16. A. Kumar, B. Prasad and I. Mishra, *J. Hazard. Mater.*, 2008, **152**, 589–600.
17. J. Maciá-Agulló, B. Moore, D. Cazorla-Amorós and A. Linares-Solano, *Carbon*, 2004, **42**, 1367–1370.

18. E. Raymundo-Pinero, D. Cazorla-Amorós, A. Linares-Solano, J. Find, U. Wild and R. Schlögl, *Carbon*, 2002, **40**, 597–608.
19. C. Srinivasakannan and M. Z. A. Bakar, *Biomass Bioenergy*, 2004, **27**, 89–96.
20. X. Ma, H. Yang, L. Yu, Y. Chen and Y. Li, *Materials*, 2014, **7**, 4431–4441.
21. Q.-S. Liu, T. Zheng, P. Wang and L. Guo, *Ind. Crops Prod.*, 2010, **31**, 233–238.
22. A. Ahmad and B. Hameed, *J. Hazard. Mater.*, 2010, **173**, 487–493.
23. S. Mahanim, I. W. Asma, J. Rafidah, E. Puad and H. Shaharuddin, *J. Trop. For. Sci.*, 2011, 417–424.
24. V. Siipola, T. Tamminen, A. Källi, R. Lahti, H. Romar, K. Rasa, R. Keskinen, J. Hyväluoma, M. Hannula and H. Wikberg, *BioResources*, 2018, **13**, 5976–6002.
25. C. Namasivayam and D. Kavitha, *Dyes Pigm.*, 2002, **54**, 47–58.
26. R. E. Franklin, *Acta Crystallogr.*, 1951, **4**, 253–261.
27. G. M. Jenkins and K. Kawamura, *Polymeric carbons: carbon fibre, glass and char*, Cambridge University Press, 1976.
28. H. Marsh and F. R. Reinoso, *Activated carbon*, Elsevier, 2006.
29. C. S. Allen, F. Ghamouss, O. Boujibar and P. J. Harris, *Proc. R. Soc. A*, 2022, **478**, 20210580.
30. M. Kwiatkowski and E. Broniek, *Materials*, 2021, **14**, 4121.
31. A. Ould-Idriss, M. Stitou, E. Cuerda-Correa, C. Fernández-González, A. Macías-García, M. Alexandre-Franco and V. Gómez-Serrano, *Fuel Process. Technol.*, 2011, **92**, 261–265.
32. B. P. Kumar, K. Shivakamy, L. R. Miranda and M. Velan, *J. Hazard. Mater.*, 2006, **136**, 922–929.
33. G. Stavropoulos and A. Zabaniotou, *Microporous Mesoporous Mater.*, 2005, **82**, 79–85.
34. J. Acharya, J. Sahu, C. Mohanty and B. Meikap, *Chem. Eng. J.*, 2009, **149**, 249–262.
35. H. Deng, G. Li, H. Yang, J. Tang and J. Tang, *Chem. Eng. J.*, 2010, **163**, 373–381.
36. J. Oubagaranadin and Z. V. P. Murthy, *Activated Carbon: Classifications, Properties and Applications*, 2011, pp. 239–266.
37. S. Mopoung, P. Moonsri, W. Palas and S. Khumpai, *Sci. World J.*, 2015, **2015**, 415961.
38. O. Kazak, Y. Ramazan Eker, H. Bingol and A. Tor, *Bioresour. Technol.*, 2017, **241**, 1077–1083.
39. F. Benstoem, D. Mousel and J. Pinnekamp, *Abrasion of granular activated carbon used for elimination of micropollutants in municipal waste water treatment*, 2015.
40. P. Satya Sai and K. Krishnaiah, *Ind. Eng. Chem. Res.*, 2005, **44**, 51–60.
41. C. Hu, S. Sedghi, S. H. Madani, A. Silvestre-Albero, H. Sakamoto, P. Kwong, P. Pendleton, R. J. Smernik, F. Rodríguez-Reinoso and K. Kaneko, *Carbon*, 2014, **78**, 113–120.

42. L. Fan, J. Wang, L. Zhao, N. Hou, T. Gan, X. Yao, P. Li, Y. Zhao and Y. Li, *Electrochim. Acta*, 2018, **284**, 630–638.

43. A. B. García, A. Martínez-Alonso, C. A. L. Y Leon and J. M. Tascón, *Fuel*, 1998, **77**, 613–624.

44. M. Belhachemi, R. V. Rios, F. Addoun, J. Silvestre-Albero, A. Sepulveda-Escribano and F. Rodriguez-Reinoso, *J. Anal. Appl. Pyrolysis*, 2009, **86**, 168–172.

45. A. Yaumi, M. A. Bakar and B. Hameed, *Energy*, 2018, **155**, 46–55.

46. A. Daifullah, S. Yakout and S. Elreefy, *J. Hazard. Mater.*, 2007, **147**, 633–643.

47. C. Pongener, P. C. Bhomick, A. Supong, M. Baruah, U. B. Sinha and D. Sinha, *J. Environ. Chem. Eng.*, 2018, **6**, 2382–2389.

48. J. R. Kastner, J. Miller, D. P. Geller, J. Locklin, L. H. Keith and T. Johnson, *Catal. Today*, 2012, **190**, 122–132.

49. S. Wong, N. Ngadi, I. M. Inuwa and O. Hassan, *J. Cleaner Prod.*, 2018, **175**, 361–375.

50. K. Koehlert, *Chem. Eng.*, 2017, **124**, 32–40.

51. G. Jordan, M. Predotova, M. Ingold, S. Goenster, H. Dietz, R. G. Joergensen and A. Buerkert, *J. Plant Nutr. Soil Sci.*, 2015, **178**, 218–228.

52. Ü. Geçgel, G. Özcan and G. Ç. Gürpınar, *J. Chem.*, 2013, **2013**, 614083.

53. G. L. Seegert and A. S. Brooks, *J. Fish. Res. Board Can.*, 1978, **35**, 88–92.

CHAPTER 2

Industrial Applications of Activated Carbon

MUHAMMAD SAJID(ⓘ 0000-0001-9471-8395)[a,b]

[a] Faculty of Materials and Chemical Engineering, Yibin University, Yibin 644000, Sichuan, China; [b] Department of Chemical Engineering, University of Gujrat, Gujrat 50700, Pakistan
Email: engr.sajid80@gmail.com

2.1 Introduction

Activated carbon (AC) is a versatile material which is generally prepared from sustainable carbon resources and is sometimes referred to as activated charcoal, activated coal, biochar, and active carbon. It can be prepared through thermochemical treatment of carbonaceous feedstock such as biomass, coal, coke, biosolids, and various forestry wastes as shown in Figure 2.1.[1] Among these available carbon resources, lignocellulosic biomass (agricultural wastes and forestry residues) and biosolids are promising due to their abundant availability, low-cost index, and non-food nature.[2,3] Additionally, the industrial use of these waste materials is an engineering solution that helps to maintain sustainable waste management of these waste materials.

The AC prepared from various feedstocks has been used for numerous industrial- and bench-scale applications for a long time.[4] AC infused with carbon moieties is more adaptable, and can produce a network with various organic and inorganic species. This behaviour of the produced AC is utilized in various applications involving adsorption and absorption phenomena.[5]

Activated Carbon: Progress and Applications
Edited by Chandrabhan Verma and Mumtaz A. Quraishi
ⓒ The Royal Society of Chemistry 2023
Published by the Royal Society of Chemistry, www.rsc.org

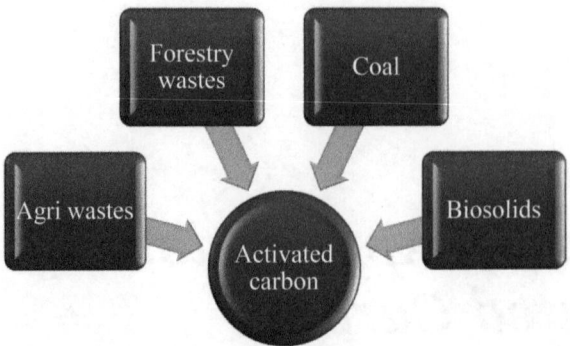

Figure 2.1 Production of AC from various feedstocks.

The activation procedure produces particles having minute pores with high surface area. Thermochemical activation produced a surface area ranging from 950 $m^2 g^{-1}$ to 2000 $m^2 g^{-1}$.[5] Phosphoric acid treatment of olive-derived AC improved the active surface area up to 1000 $m^2 g^{-1}$.[6] This improvement is attributed to the well-developed porous structure of the modified AC. Additionally, the chemical activation improved the surface properties and pore structure which improved not only the adsorption/absorption properties but also the mechanical strength. Thus, the adsorption ability is improved and the developed material is capable of adsorbing even very small molecules in a very dilute solution. This property has enabled the use of AC in water and wastewater treatment applications as well as in catalyst development applications.[7]

The use of AC has several advantages.[8] It is sustainable and renewable except that the AC is prepared from coal. Currently, the production of AC from coal is discouraged due to associated carbon footprints.[9,10] Coconut shells, woody plants and herbaceous plants are currently more promising materials to produce AC. Adsorption is the chief industrial application of produced AC. The process of layering of molecules without any chemical reaction on the contacting surface of any material is defined as adsorption, so adsorption is classified as a surface phenomenon. On the other hand, absorption is molecular penetration into the structure due to pores; hence, it is a bulk phenomenon. If chemical transformation is involved, the process is defined as chemisorption. AC has the potential to observe all the processes simultaneously during any purification scheme; hence, it is highly selective. For the removal of drugs from waste solution or the purification of drug solution, adsorption is one of the most significant methods. The forces concerned in this type of contact can be either physical or chemical in nature or a combination of both. Hence, AC is the best choice. Along with these, AC has numerous industrial applications which will be discussed in detail in this chapter. The current status of AC adaptability, progress, and prospects will also be discussed.

2.2 Applications of AC

AC has a highly active surface due to its well-developed surface area and amended pore structure. Therefore, AC has extended utilization as a medium for purification, dechlorination, deodorization, and decolorization of vapours, gases, and liquids. Additionally, AC is an economical and persuasive adsorbent for various industrial and residential applications. Its application ranges from the domestic water filter to the most selective medium for gold recovery from ore processing. The principal industrial applications include food and beverages, drinking water purification, wastewater treatment, cosmetology, cleaning of gases, off-gas purification, soil enrichment, and recovery of precious metals.[11] Additionally, AC has applications as an adsorbent material in pharmaceuticals, chemicals, military applications, and the agricultural and environmental industries to remove undesired components from process streams as well as from the final products. It is also used to neutralize the toxicity of various compounds in processing lines and effluents. Furthermore, it has applications in the capturing of soundwaves, radiowaves, and microwaves.[12,13] It is also an important component of air filters and masks used in both the medical profession and the chemical industry. Along with these, the application of AC as a base material of catalysts and active catalysts with various impregnants is a huge research and development application which still has a wide market share. Some important industrial applications are summarized in Figure 2.2.

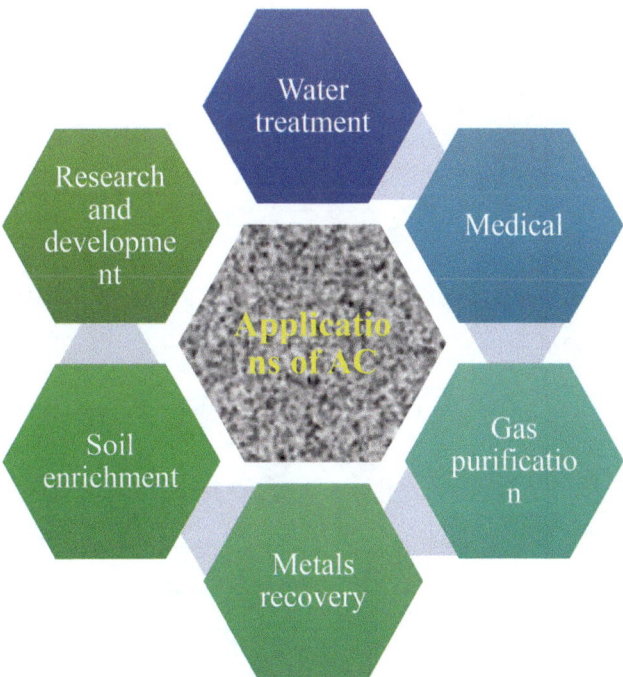

Figure 2.2 Summary of industrial applications of AC.

Special preparation and modification methods are applied to produce AC for specific applications. Particle size and shape as well as the overall physical structure of the prepared AC are important for specific applications (Figure 2.3). Various product configurations are selected for better process efficiency. Typically, AC is prepared in the following physical structures for best application and performance.

Granular AC has the largest size with a relatively reduced active surface area among ACs. This type is particularly efficient in gas and vapour adsorption due to its high diffusion rate.[14,15] Therefore, granular AC is selected for water treatment applications. Powdered AC is composed of finely ground particles made from crushed carbon materials, especially coal. This size is generally used in rapid mixing applications such as clarifiers and carbon block filters. The fusion of powdered AC particles with some binding agents produced extruded AC. The developed AC has a tough structure with specific spherical or cylindrical shaped particles. Therefore, it is particularly selected for low-pressure drop applications such as gas cleaning and purification. Extruded AC has high mechanical strength and reduced dust formation potential.[12] The impregnation of various inorganic particles produces special antimicrobial properties. Therefore, impregnated AC is selectively used in water purification services: for example, Ag-impregnated AC is used in drinking water filters to impede the growth of microbes. Bead AC is usually

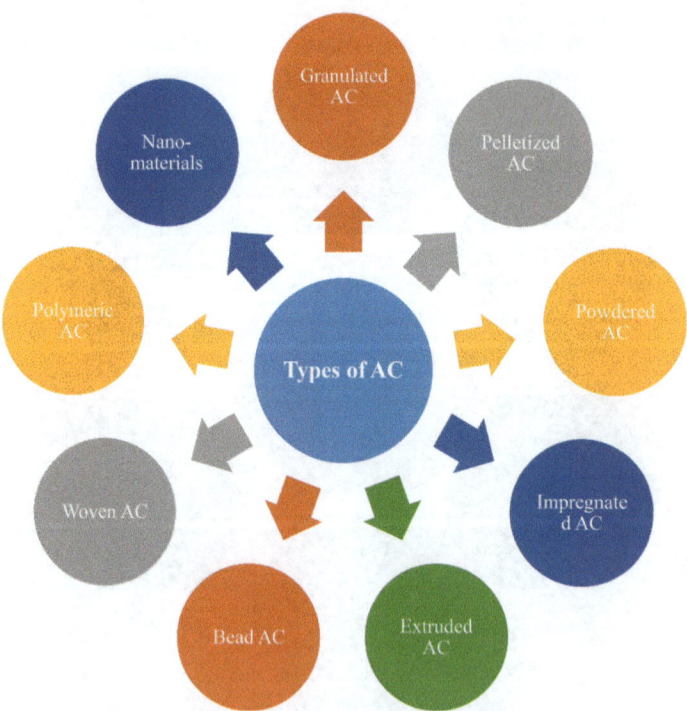

Figure 2.3   Types of AC.

produced from petroleum pitch and is very similar in properties to extruded AC. It also has low dust formation potential and high mechanical strength. Its spherical shape makes it ideal for low-pressure drop applications. Woven AC is a specially designed structure in which AC cloth is formed by weaving AC into technical rayon fibre. The AC cloth thus produced exhibits filtering properties and is mostly used in various odor removal applications. It also has applicability in many defence applications. Similarly, polymer AC is a composite of biocompatible polymer and AC in which polymer is coated over AC particles. This coating produces a permeable smooth shell without hindering the pores and adsorption capacity of the parent AC. Polymer AC materials are specially designed adsorbents for medical applications such as hemoperfusion (medical treatment). Biocomposites are specially selected and designed to produce polymeric AC for various medical applications. AC and extruded AC as opposed to powdered AC (PAC) are expected to dominate the market due to their flexible applications and high strength. While GAC-based purification systems are long-term and cost-effective, PAC-based systems exhibit lower initial costs and can be much more easily operated. The physical adsorption forces associated with AC are not always sufficient to adsorb a given compound.

The selection of size and type of AC is a crucial task that affects the process efficiency significantly. There are no typical selection criteria; however mostly, the selection of AC for a particular service is governed by

    (i) The type of reaction/process
    (ii) The concentration of the target compound
    (iii) The service length
    (iv) The severity and toxicity of the compound

2.2.1 Water Industry

AC is a recognized universal adsorbent and is heavily used in water and wastewater treatment applications (both domestic and industrial).[16] Due to increased industrialization and mineral processing activities, waterways are contaminated especially with various persistent organic pollutants (POPs) and inorganic metals.[17] The presence of these metallic components especially heavy metals is highly dangerous for human health and environmental ecology.[18] Therefore, careful treatment of municipal water supplies, aquarium waters, and swimming pools, and recycling of precious water on the orbiting space station is a crucial task. On the other hand, as far as the drinking water supplies are considered, they further need special care and treatment methodologies.[19] The adsorption process because of its process simplicity and reduced cost is considered to be the most acceptable water treatment method in developing countries. Principally, AC is a compulsory component of every adsorptive treatment methodology and mostly municipal supplies are processed through various AC filters, along with other filters such as gravel and sand filters. However, for drinking water application, filtration and removal of contaminated components through adsorption and absorption are followed by other treatment methods to ensure complete water safety.[20] The microporous size dispersal and adsorbent volume are vital components for the efficient adsorption of some small molecular weight (MW) toxins.[21] The structure and microporous size of AC are generally optimized experimentally in which maximum adsorptive removal takes place. These results are then declared as guidelines for using the optimum size region concerning the molecular dimensions of contaminants. However, the existence of naturally dissolved matter in streams and its influence on the adsorption of the target compound is sometimes inevitable. Researchers are continuously struggling to define the optimal pore region for newly developed synthetic chemicals considering their molecular dynamics. These studies are further optimized considering the effect of various dissolved matters. Although the developed ACs exhibit hydrophobic surface properties with very low water adhesion, still the molecular binding of H_2O molecules may hinder the adsorption of pollutants. Meanwhile, oxidants used also hurt AC adsorption.[22] All these factors must be considered during the design phase of any separation and purification applications.

Treatment of industrial wastewater is usually considered to be an additional expenditure. Therefore, the most reliable economical process is always a choice. Continuous efforts led to the development of AC filters as the most economical filter designs for removal of a wide variety of contaminants from industrial wastewater. AC has the potential to remove organic and inorganic pollutants, POPs, volatile organic components (VOCs), and metals from industrial and municipal wastewater. Investigation of cheaply available adsorbents has employed various waste biomasses as feedstocks to prepare AC. Various materials have been investigated and it was elucidated that AC produced from indigenous sources is very helpful for removing emergent impurities. Among them, coconut shells, herbaceous materials such as Cannabis Sativum Hemp (CSH), and biorefinery wastes have been proven excellent materials which can be converted to AC economically.[23] Biorefinery waste materials have the advantage of very low chemical treatment for surface modification. The filter designs for drinking water and industrial water treatment applications are shown in Figure 2.4 which have the potential to achieve the following objectives in water treatment applications, *i.e.* removal of

  (i) odor-, taste-, and colour-causing compounds
 (ii) dissolved organic and inorganic materials from putrefying plants and other resources
(iii) undesired products due to disinfection such as chlorination
 (iv) cyanotoxins and toxins due to aquatic culture (*e.g.* anatoxin-A, microcystin-LR, and cylindrospermopsin)
  (v) endocrine-disrupting compounds
 (vi) contamination caused by drugs and personal care products
(vii) perfluorooctane sulfonate (PFOS) and perfluorooctanoate (PFOA)

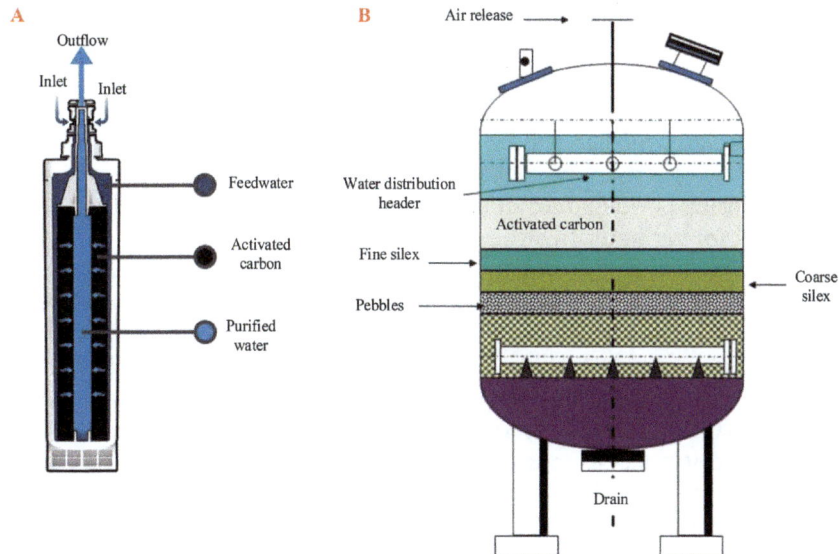

Figure 2.4 AC filters: (A) drinking water filter and (B) industrial filter.

(viii) pesticide-based compounds
(ix) metal, heavy metals, and metallic ions
(x) any other adverse effects

Water treatment based on the aforementioned objectives is continuously progressing. Boehm *et al.*[24] postulated that weakly basic compounds show strong adsorption under both simulated gastric and intestinal conditions, while weakly acidic compounds show healthy adsorption at gastric pH. As the extent of ionization of any compound is increased, its efficiency of adsorption decreases. This kinetic behaviour is exploited to remove pharmaceutical and organic complexes from water during water treatment applications. It was confirmed experimentally that the maximum adsorption capacity can be achieved at pH 7.5.[5] It was learned that the adsorption process reaction kinetics follows first-order kinetics and the factors that affect the adsorption of organics are the amount of AC used, temperature, and pH. AC prepared from coconut shells was used for the removal of various pollutants from drinking water supplies. Results elucidated that an amount of 10 mg L^{-1} is adequate to attain >90% contaminant removal.[25] Ridder *et al.* observed the removal of >50% of pharmaceuticals employing wastewater-preloaded granular AC. As ultrapure water contains no ions to shield the surface charge, charge effects were found to be more distinct. 0–58% and 32–98% removal were observed for negatively and positively charged pharmaceuticals, respectively, at a granulated AC dose of 6.7 mg L^{-1}.[26] Powdered AC has been used extensively for the removal of pharmaceutically activated compounds and endocrine-disrupting compounds. Similarly, cyanobacterial toxins can be removed with AC in drinking water treatment, and a hybrid system of powdered AC and membrane filtration is much more effective for large-scale municipal water supplies.[27] Several adsorption isotherms have been studied to optimize the process design in various studies, which concluded that the AC adsorption process is the most efficient. Because there are no transformation products, >90% of naproxen can be removed using powdered AC from post-sedimentation water emerging from a full-scale drinking water treatment plant.[27]

Similarly, AC has been used to remove metallic ions, heavy metals, and various fungicides and bacteria from water (Table 2.1). The extent of treatment depends on the final use of water, and drinking water needs careful treatment and rigorous testing before final supply to municipalities. AC produced with high pyrolysis temperature has the potential to reduce bacterial mobility from water reservoirs.[28] The used AC can be regenerated; however, the existence of natural organic matter may hinder the replenishment as well as removal efficiency of AC-based systems.[29] Therefore, periodical testing and reverse treatment of filters are necessary to process stability and effectiveness.

2.2.2   Food and Beverage Industry

AC is used in the food industry to remove the colours, tastes, and odorous compounds from solutions and beverages. The first application of AC in the food and beverage industry was witnessed in 1974 when AC was first applied to decolour a newly produced syrup. Later, when AC was declared a safe substance for the food industry by a special committee of scientists in 1981, its use was further expanded. Nowadays, AC has wide applications in food and beverage production enterprises. For example, in the wine industry AC is used to deodorize the product solution,[30,31] as a decolouring agent in sugar refineries,[32,33] for the removal of colouring and odorant pigments from vegetable oils and molasses,[34] and for the debittering of protein hydrolysates.[35] Table 2.2 summarizes the objectives and characteristics of AC in various food industries. It has been elucidated in these research investigations that the selection of the appropriate AC type, dose, and process parameters can remove all the undesired flavours and colours from the food

Table 2.1   Different types of AC and their specialized applications in water treatment.

AC type	Applications
Granulated AC	Dechlorinates water, adsorbs organic material, and produces a chemically reducing environment
Ni-impregnated AC	Inhibits bacterial growth in water
Extruded AC	Removes chemicals and dechlorinates
Powdered AC	Eliminates of traces of synthetic chemicals, odor, and taste-triggering moieties resulting from chemical spills and/or algal buds

Table 2.2 Specialized objectives of AC treatment in various beverage streams.

Beverage	Treatment objective
Wine	Removal of haze, odor, and decolorization
Beer	Modification of taste
Fruit juices	Decolorization and deodorization
Coffee and tea	Decaffeination
Soft and carbonated drinks	Decolorization of sugar syrup
Sweeteners	Purification of sucrose, glucose, and fructose syrups/solutions
Citric acid solution	Purification for fermentation
Monosodium glutamate (MSG)	Decolorization and stabilization
Lactic acid syrup	Decolorization of extracts
Gelatin	Decolorization and purification
Glycerin	Decolorization and odor removal
Edible oils	Removal of contaminants, catalyst traces, and colour modification

stream without impacting the food quality. AC is usually applied as a bed material in filters to remove these compounds from the feed stream similar to water treatment applications. Two-stage AC treatment of the beverage unit is outlined in Figure 2.5.

Yeasts present in wines such as *Dekkera* and *Brettanomyces* produce an unpleasant aroma, which is mainly due to the decarboxylation of ferulic, hydroxycinnamic, and coumaric acids into 4-ethyl guaiacol and 4-ethylphenol by these yeasts.[36] The aroma caused by these products is highly undesirable in wines and produces an unpleasant sensation for customers; therefore, the removal of these off-flavour-causing molecules is strongly desirable. LuísFilipe-Ribeiro *et al.* studied the potential of AC to remove the colouring and odor-causing pigments from red wine.[31] In this experimental study, AC prepared from lignocellulosic biomass was used to remove the 4-ethylphenol and 4-ethyl guaiacol, and a high removal efficiency (>73%) was recorded. It was further elaborated that the application of AC is advantageous to remove these contaminants without disturbing the wine characteristics and colour of red wine. AC characteristics depend on the activation process and mechanism; therefore, they can be altered to achieve the desired objectives. It has been used successfully to decolour and de-odorize the grape pulp to produce fine grades of white wine. Hence, it can be used successfully to remove volatile phenolic compounds from beverages as well as from wines.[36] Results of AC treatment of red wine to remove volatile phenols employing different ACs are provided in Figure 2.6.

Along with these physical processes, AC has been used as a base material to prepare different treatment technologies to remove/detoxify various substances from food and beverage products.

Bisphenol A (BPA) is an important organic molecule. It has been recognized as an endocrine-disrupting compound that can disturb the endocrine system by stashing large quantities of body hormones like estrogen.[37]

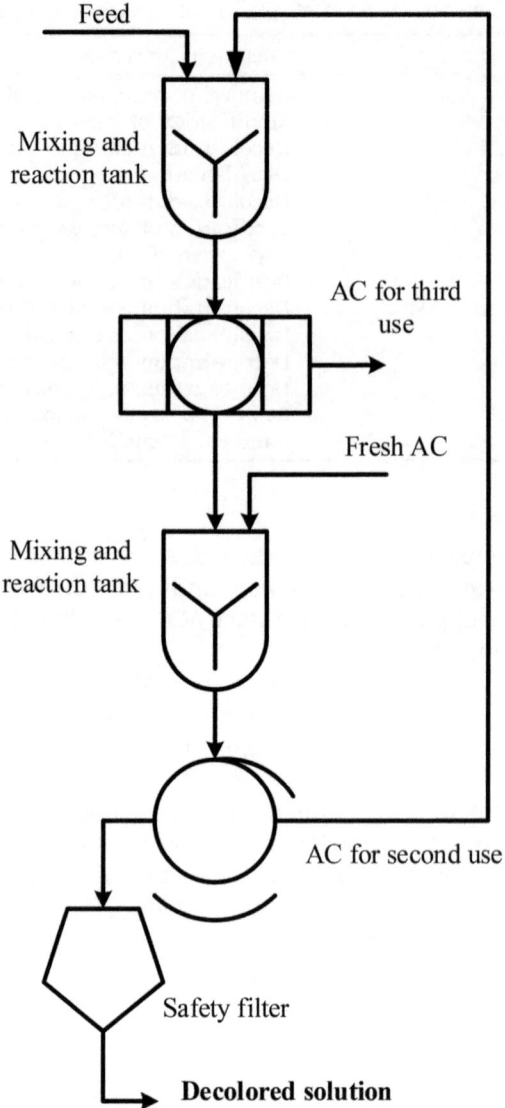

Figure 2.5   Two-stage AC treatment of beverages.

Its limits in food and beverages are strictly monitored, and the European Food Safety Authority has defined 4 mg kg^{-1} as the maximum allowable limit, whereas the US Environmental Protection Agency has defined it as 0.05 mg kg^{-1} body weight per day. If exceeded, it will result in various health disorders, such as behavioural syndromes, tumours, developmental problems, metabolic disorders, reproductive complications, infertility, and respiratory disorders.[38] Because of these problems, efforts have been devoted to removing BPA from foods and beverages up to trace levels. Liang *et al.*

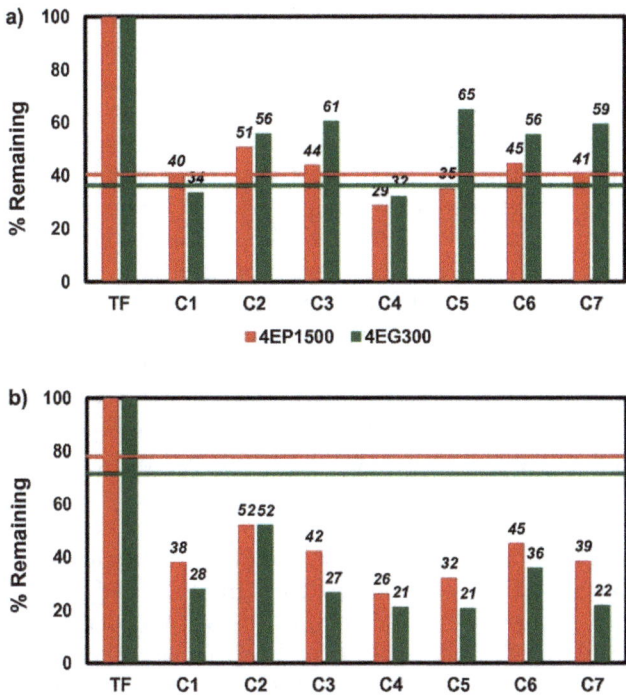

Figure 2.6 AC treatment of wines to remove the 4-ethylphenol (4EP) and 4-ethylguaiacol (4EG) at (a) high concentration and (b) low concentration (TF: fresh sample of red wine and C1–C7: samples treated with ACs; odor detection threshold for 4EP (—) and 4EG (—)). Adapted from ref. 31 with permission from Elsevier, Copyright 2017.

prepared a nanocomposite by mixing cobalt with magnetized AC and used it to produce an electrode for the detoxification of BPA.[39] The result showed a very high electrochemical oxidation potential of the as-prepared electrode toward BPA and stable oxidation efficiency was observed in a very dilute solution (up to 10 nM BPA).

It can be concluded from the earlier discussion and literature survey that AC provides the best solution to remove odoring and colouring components in the food and beverage industry without impacting the aroma, taste, and quality of food. However, as far as the wine industry is considered, it does not provide a perfect solution for the removal of volatile phenols.

2.2.3 Medical Field and Pharmaceutical Industry

AC has had wide medical and pharmaceutical applications since ancient times.[40] AC has been used as an adsorbent in various medical products such as masks, dialysis machine filters, hemoperfusion columns, and wound bandages. Its medical use is defined as internal use if applied as an oral adsorbent and external use when used in various extracorporeal devices.[41] Broadly, the pharmaceutical use of AC is due to its non-specific powerful adsorption potential which is matchless to any other material. AC is applied to remove objectionable and harmful elements – toxins – from the human body. AC is equally beneficial to remove these toxins no matter whether these are externally injected substances that enter the body *via* environmental impact through skin, eyes, breathing, feed, or body-worn devices or produced internally. If we consider the human body as an intimate part of the environment, toxins produced – whether internally or externally – are body pollutants just like ecological pollutants and hence need to be removed. AC finds application in the removal of these toxins and handling of inebriation and harm caused by external (exogenous) actions. The medical and medicinal use of AC has various pathways and classifications. Considering the medical, medicinal, and pharmaceutical applications of AC, the principal applications are as follows.[42] (a) The use of AC as an orally administered moiety is classified as medicinal use, or AC as an extracorporeal circuit like hemoperfusion columns for body fluids refinement is regarded as a medical device.[43] The oral introduction of AC accelerates the removal of most of the low molecular weight substances from the body, sanctioning the truly universal antidote behaviour of AC. (b) An AC-filled hemoperfusion column is used for the purification of body fluids. The complete extracorporeal device is composed of a composite engineering design including the system to control the blood flow, avoiding air bubble formation and additional lines feeding supplementary nutrients and drugs into the blood circulation as shown in Figure 2.7. (c) A gas mask containing AC is an

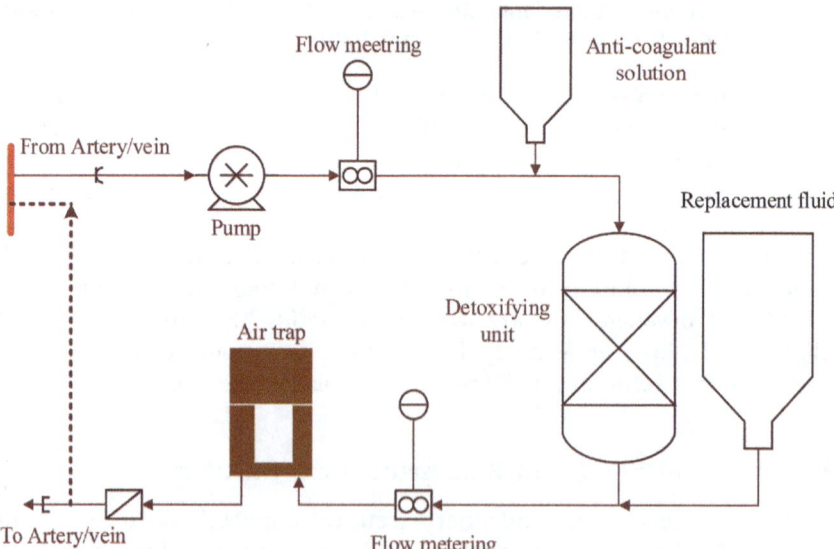

Figure 2.7   Schematic illustration of the blood purification mechanism.[40] Reproduced from ref. 40 with permission from Elsevier, Copyright 2006.

engineered product that also shields humans from exogenous poisoning and viral infections, but it is rarely defined as a medical device. However, with the emergence of the pandemic COVID-19, masks are now defined as medical devices and come under medical regulations.[44] Still, there is no clear definition of masks in many countries. Regardless of their administrative designation, masks are now an integral medical component and are treated together with other medical devices. The use of AC filters for pharmaceuticals and medical gases is daily increasing. The schematic illustration of the blood purification unit is provided in Figure 2.7. A similar mechanism is also adopted for the urine purification (dialysis) system.

AC prepared for use in medical appliances needs special care and treatment following the medical protocols. Simple AC with chemical modification is not suitable and not recommended for clinical use. Specially prepared AC is termed a "medical grade AC", which must obey the first medical principle of "do no harm". Therefore, the prepared AC must be biocompatible and hemocompatible. In actuality, the term biocompatibility means "the ability of a material, device, or a system to perform without a clinically significant host response in a specific application".[40] Additionally, any material to be used as a medical aid[45]

(i) should be free from thrombogenic, toxic, allergic, or inflammatory responses
(ii) should not harm blood cells
(iii) should not change plasma proteins and enzymes
(iv) should not elicit immunological reactions
(v) should be free from carcinogenic effects
(vi) should not damage adjacent tissue

Given these principles, the medical AC should not produce any hazard to the human body or body fluids and it should not remove necessary metabolites and nutrients from the body. However, the management of "do not remove necessary nutrients" is very difficult because it is against the nature of AC. Therefore, the use of AC is usually accompanied by other buffers and nutrient supplements to accommodate nutrient losses.

2.2.4 Purification of Gases

AC has high application potential in filters for gas purification and off-gas treatment. Use of AC is the safest, most efficient gas separation and purification technology which can be used in industrial, municipal, and commercial applications. AC-based purification systems are able to handle sporadic flow/loading designs because AC becomes active immediately after start-up and rapidly responds to changes in odor concentration.[46] Therefore, the demand for AC in gas purification and gas separation applications is continuously increasing due to the strengthening of environmental protection guidelines, as well as to satisfy the mandate of purified gas application. A growing demand for AC is expected in gas phase separations. AC continues to be applied in various industrial gas purification processes (ranging from medical gases to gases used in the food and beverage industry) as well as in various chemical processes. Granular AC and extruded AC as opposed to powdered AC are expected to dominate the market due to their flexible applications and high strength.

The physical structure and pore volume of any adsorbent are critical parameters which strongly influence the efficiency of the gas separation and purification process. The physiochemical features of amorphous C together with various surface modifications make it an excellent adsorbent to remove unwanted components during the separation and purification mechanism. Mostly, AC is used in the purification of gases through the elimination of undesired components through adsorption and chemisorption. As the adsorption of molecules is governed by attractive forces and pore volume, the physical forces due to AC are not strong enough and the molecules in gaseous streams have relatively high energy, and therefore, it is difficult to capture them. This problem is handled by using the internal structure of AC as a carrier for reactive species for chemisorption and/or catalytic reactions. The principal applications of AC in the gas industry are summarized in Table 2.3.

Mostly, industrial applications AC are as follows: AC filters, AC pellets, AC clothes, and AC beds in a column. Sometimes AC is packed in a semi-permeable membrane and placed in a gas circulation system to remove the undesired components from circulating gas or air.[47] Typically, AC filters are used to clean the flue gases in coal-fired power plants, cement plants, and many other air pollution prevention applications. AC impregnated with speciality chemicals has been used to adsorb acid gases to prevent acid rain and other toxic gases such as amines, ammonia, and aldehydes. Additionally, carbon filters are also helpful to reduce the contamination of radioactive elements with off-gases such as radioactive mercury and iodine. HCN,

Table 2.3 Typical applications of AC in gas separation and purification applications.

Field	The objective of AC treatment
Solvent recovery	Conditioning of emitted vapours and recovery of entrapped molecules
Industrial respirators	Adsorptive removal of inorganic/organic vapours
Carbon dioxide (CO_2)	Removal and purification of CO_2 evolved during processes such as fermentation and metallurgy
Off-gas treatment	Recovery of metal ions and dioxins from effluent gases of waste incineration
Cigarette filters	Elimination of tar and control of nicotine
Air conditioning	Decontamination of air imperiling heating, ventilation, and air conditioning systems in offices, hospitals, and airports
Semiconductors	Air purification produces ultrahigh-quality air
Refrigeration systems	Elimination of food odors

H$_2$S, siloxanes, VOCs, arsine, and phosphines can also be recovered using an AC incorporated gas filtration system.[47] AC treatment can eliminate the traces of unwanted components from gas streams, which in turn remove the foaming and corrosion potential of the gas streams. AC treatment removes the colour and impurities of natural gas, which makes the handling safe during its compression to make liquified natural gas (LNG).

AC-fixed beds are the most common industrial application of gas separation and purification. In this case, granulated and impregnated AC is used in a bed of fixed height. The AC volume and bed height are designed through the balancing of the breakthrough curve and flow balancing against the bed. The flow rate is optimized in view of the concentration of the pollutants needed to be removed. Initially, the flowing stream is directed toward the fixed bed of AC, which is contaminant free at the start. Gradually as the flow proceeds, the concentration of contaminants increases, forming a layer on the interior and exterior of the AC bed due to the migration of contaminants from the gas stream toward the AC bed. The mass transfer zone moves through the AC bed as the adsorption of the contaminant proceeds with time and flow. When the exit stream concentration is equal to the AC-bed concentration, the breakthrough curve will be obtained. The breakthrough time can be calculated through mass balance. The obtained curve is similar to an S-shaped curve as shown in Figure 2.8. When the concentration of the exit flow stream is nearly equal to the concentration of the inlet gas stream, the bed should be changed. However, considering the safety factor and to implement the economic balance, the breakthrough concentration (C_b) is mostly selected with a safety factor. The AC-bed life is determined by the nature and concentration of contaminants, the volume of the gas stream, regulatory requirements, and the frequency of use. Generally, AC beds are equipped with a long shelf-life and mostly provided as parallel arrangements. Apart from these, AC is also applied in gas separation/purification

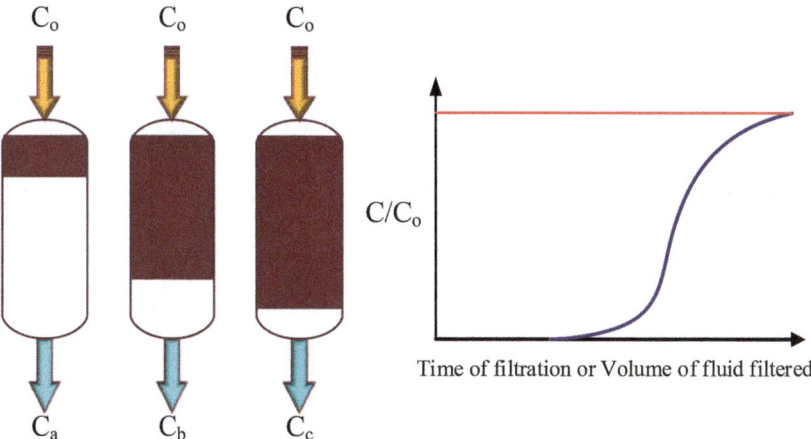

Figure 2.8   Characteristics of fixed bed filtration.

as AC fibres. These fibres are more flexible and more efficient but less selective; due to their high production costs, they are used only in special applications.

2.2.5 Mineral and Metal Recovery

AC is also applied to recover various precious metals from their lean solution and process streams in which these metals are mixed due to catalyst erosion and/or metallurgical processing. AC has been used for a long time to recover gold from gold mines.[48,49] Due to the variety of contaminants in minerals, the process sustainability of AC depends on how the developed AC holds its surface properties for a long time and how many cycles can be performed without the AC losing its adsorption potential. Therefore, the selection of feedstock to produce AC for the mineral industry is of paramount importance. Because of economic concerns, it is usually the principal objective to recover these metals in a safe way.[50] Therefore, the applied AC must have high strength and longevity for a continuous recovery and recycling process. Having a very low attrition rate throughout the absorption circuit and process, the AC results in an improved metal/mineral recovery rate. It has been observed that coconut shells are the best available feedstock to produce AC for mineral applications. They are durable and can withstand rigorous carbonization and chemical activation processes; therefore, they have been used to manufacture AC for mineral extraction, especially for gold extraction.[51] Such AC can be used in carbon in pulp, carbon in column, and carbon in leach applications successfully without affecting its surface properties and adsorption potential.[52] The typical properties that an AC must have for selective mineral applications are as follows:

   (i) Accelerated kinetics of the adsorption process
  (ii) Typically, high loading capacity for useful minerals, *e.g.* gold (Au)
 (iii) Tremendously low mineral solubility losses
  (iv) Highly prescribed for "preg-robbing" ores
   (v) Excellent attrition resistance
  (vi) Insignificant carbon losses in the processing environment
 (vii) Uniform grain activation/reactivation
(viii) Lower deterioration rate in the regeneration process
  (ix) Highly selective spongy outline avoiding blinding and nailing of
       screens
   (x) Nominal head losses
  (xi) Augmented sorption reactions

Along with these direct industrial applications, AC has been used as supporting media in various applications. AC is the major base material for the production of various catalysts as well as has been used as a carrier for various active radicals for catalysis. Hence, it can be declared that AC has the highest application in research and development. Furthermore, AC and biochar are also used in soil amendment which is an important aspect of modern agriculture. Hence, it can be concluded that AC is the most promising industrial adsorbent and absorbent in a vast range of fields.

References

1. L. Liang, F. Xi, W. Tan, X. Meng, B. Hu and X. Wang, *Biochar*, 2021, **3**, 255–281.
2. M. Sajid, U. Farooq, G. Bary, M. M. Azim and X. Zhao, *Green Chem.*, 2021, **23**, 9198–9238.
3. M. Sajid, A. Raheem, N. Ullah, M. Asim, M. S. Ur Rehman and N. Ali, *Renewable Sustainable Energy Rev.*, 2022, **168**, 112815.
4. B. Kayranli, *Alexandria Eng. J.*, 2022, **61**, 443–457.
5. M. Sajid, S. Bari, M. Saif Ur Rehman, M. Ashfaq, Y. Guoliang and G. Mustafa, *Alexandria Eng. J.*, 2022, **61**, 7203–7212.
6. F. J. García-Mateos, R. Ruiz-Rosas, M. D. Marqués, L. M. Cotoruelo, J. Rodríguez-Mirasol and T. Cordero, *Chem. Eng. J.*, 2015, **279**, 18–30.
7. S. Mulk, M. Sajid, L. Wang, F. Liu and G. Pan, *J. Environ. Chem. Eng.*, 2022, **10**, 106613.
8. H. Liang, C. Zhu, S. Ji, P. Kannan and F. Chen, *Biochar*, 2022, **4**, 1–13.
9. M. Sajid, G. Bary, M. Asim, R. Ahmad, M. I. Ahamad, H. Alotaibi, A. Rehman, I. Khan, Y. Guoliang, M. Irfan Ahamad, H. Alotaibi, A. Rehman, I. Khan and Y. Guoliang, *Alexandria Eng. J.*, 2022, **61**, 3069–3092.
10. M. Sajid, A. Chowdhury, G. Bary, Y. Guoliang, R. Ahmad, I. Khan, W. Ahmed, M. F. S. Khan, A. M. Alqahtani and M. N. Alam, *J. Math.*, 2022, **2022**, 6989612.
11. M. Sajid, M. I. Ahmad, S. S. Shafqat, M. K. Pasha and M. Asim, *Univ. Wah J. Sci. Technol.*, 2020, **4**, 1–7.
12. K.-D. Henning and H. von Kienle, in *Industrial Carbon and Graphite Materials Volume 1: Raw Materials, Production and Applications*, ed. H. Jaeger and W. Frohs, Wiley-VCH, 1st edn, 2021, pp. 491–531.
13. O. T. Diejomaoh Abafe, M. M. Azim and M. Sajid, in *Ionic Liquid-Based Technologies for Environmental Sustainability*, Elsevier B.V., 2022, pp. 225–233.
14. R. Ahmad, J. Zhang, A. Farooqi, M. Nauman Aslam and M. N. Aslam, *Numer. Math.*, 2019, **12**, 1231–1245.
15. G. Bary, P. Ru and W. N. Zhang, *J. Phys. G: Nucl. Part. Phys.*, 2019, **46**, 115107.
16. J. Przepiórski, *Interface Sci. Technol.*, 2006, **7**, 421–474.
17. S. A. Carmalin and C. L. Eder, *Ecotoxicol. Environ. Saf.*, 2018, **150**, 1–17.
18. M. I. Ahamad, J. Song, H. Sun, X. Wang, M. S. Mehmood, M. Sajid, P. Su, A. J. Khan and A. J. K. Ping Si, *Int. J. Environ. Res. Public Health*, 2020, **17**, 1070.
19. G. Bary, L. Ghani, M. I. Jamil, M. Arslan, W. Ahmed, A. Ahmad, M. Sajid, R. Ahmad and D. Huang, *Sci. Rep.*, 2021, **11**, 19683.

20. M. R. Dilshad, A. Islam, B. Haider, M. Sajid, A. Ijaz and R. Ullah, *Chem. Pap.*, 2021, **75**, 3131–3153.

21. A. Rehman, J. Song, F. Haq, M. Irfan, M. Sajid and Z. Zahid, *Appl. Geogr.*, 2021, **135**, 102550.

22. V. H. Nguyen, S. M. Smith, K. Wantala and P. Kajitvichyanukul, *Arabian J. Chem.*, 2020, **13**, 8309–8337.

23. M. S. U. Rehman, I. Kim, N. Rashid, M. A. Umer, M. Sajid and J.-I. Han, *Clean: Soil, Air, Water*, 2016, **44**, 55–62.

24. J. J. Boehm and R. C Oppenheim, *Aust. J. Pharm. Sci.*, 1977, **6**, 107–111.

25. D. J. De Ridder, A. R. D. Verliefde, S. G. J. Heijman, J. Q. J. C. Verberk, L. C. Rietveld, L. T. J. Van Der Aa, G. L. Amy and J. C. Van Dijk, *Water Sci. Technol.*, 2011, **63**, 416–423.

26. A. Rossner, S. A. Snyder and D. R. U. Knappe, *Water Res.*, 2009, **43**, 3787–3796.

27. L. F. Delgado, P. Charles, K. Glucina and C. Morlay, *Sci. Total Environ.*, 2012, **435–436**, 509–525.

28. S. M. Abit, C. H. Bolster, P. Cai and S. L. Walker, *Environ. Sci. Technol.*, 2012, **46**, 8097–8105.

29. S. K. Mohanty and A. B. Boehm, *Water Res.*, 2015, **85**, 208–215.

30. M. Sajid, M. R. Dilshad, M. S. Rehman, D. Liu and X. Zhao, *Molecules*, 2021, **26**, 2208.

31. L. Filipe-Ribeiro, J. Milheiro, C. C. Matos, F. Cosme and F. M. Nunes, *Food Chem.*, 2017, **229**, 242–251.

32. H. Coklar and M. Akbulut, *Food Sci. Technol.*, 2020, **40**, 179–189.

33. M. Sajid, Y. Bai, D. Liu and X. Zhao, *Waste Biomass Valorization*, 2021, **12**, 3271–3286.

34. M. Bernal, M. O. Ruiz, R. M. Geanta, J. M. Benito and I. Escudero, *Chem. Eng. J.*, 2016, **283**, 313–322.

35. A. T. Idowu and S. Benjakul, *J. Food Biochem.*, 2019, **43**, e12978.

36. J. Milheiro, L. Filipe-Ribeiro, F. Cosme and F. M. Nunes, *J. Chromatogr. B: Anal. Technol. Biomed. Life Sci.*, 2017, **1041–1042**, 183–190.

37. L. A. D. Gugoasa, *J. Electrochem. Soc.*, 2019, **167**, 37506.

38. Å. Bergman, J. Heindel, S. Jobling, K. Kidd, R. T. Zoeller, M. A. Philbert, D. R. Bell and S. Safe, *Toxicol. Lett.*, 2012, **211**, S3.

39. M. Liang, X. Hou, Y. Xian, Y. Wu, J. Hu, R. Chen, L. Wang, Y. Huang and X. Zhang, *Food Chem.*, 2022, **376**, 131948.

40. S. V. Mikhalovsky and V. G. Nikolaev, *Interface Sci. Technol.*, 2006, 7, 529–561.

41. S. S. Awad, P. B. Rich, S. Kolla, J. G. Younger, C. A. Reickert, V. P. Downing and R. H. Bartlett, *ASAIO J.*, 1997, **43**, M745–M749.

42. V. G. Nikolaev and V. A. Samsonov, *Artif. Cells, Nanomed., Biotechnol.*, 2014, **42**, 1–5.

43. O. Petuhov, T. Lupascu, D. Behunová, I. Povar, T. Mitina and M. Rusu, *C*, 2019, **5**, 31.

44. A. Chowdhury, M. Sajid, N. Jahan, T. Isaac, P. Maitra, G. Yin, X. Wu, Y. Gao and S. Wang, *Biomed. Pharmacother.*, 2021, **142**, 111956.

45. J. Schwartz, A. Padmanabhan, N. Aqui, R. A. Balogun, L. Connelly-Smith, M. Delaney, N. M. Dunbar, V. Witt, Y. Wu and B. H. Shaz, *J. Clin. Apheresis*, 2016, **31**, 140–338.
46. T. Saliev, *J. Carbon Res.*, 2019, **5**, 29.
47. A. L. Cukierman, *ISRN Chem. Eng.*, 2013, **2013**, 1–31.
48. M. K. Al Mesfer, M. Danish, M. I. Khan, I. H. Ali, M. Hasan and A. El Jery, *Processes*, 2020, **8**, 1–16.
49. J. B. Zadra, A. L. Engel and H. J. Heinen, *U.S. Bureau of Mines*, *R.I.* 4843, 1952.
50. R. L. Paul, A. O. Filmer and M. J. Nicol, The recovery of gold from concentrated aurocyanide solutions, *3rd International Symposium on Hydrometallurgy*, Extractive and Process Metallurgy Program Committee of the Metallurgical Society of AIME and the Mineral Processing Division of SME-AIME held at the 112th AIME Annual Meeting, Atlanta, Georgia, 6–10 March 1983.
51. A. Mehmet, W. A. M. te Riele and D. W. Boydell, *JOM*, 2012, **38**, 23–28.
52. CAC, Activated carbon for metal (gold) recovery|Carbon Activated Corporation, https://activatedcarbon.com/applications/metal-recovery (accessed 14 May 2022).

CHAPTER 3

Medical Applications of Activated Carbon

PAYAL B. JOSHI,[a,d] MURTHY CHAVALI,[*b]
GAGAN KANT TRIPATI[c] AND SURABHI TONDWALKAR[d]

[a] Shefali Research Laboratories, 203/454, Sai Section, Ambernath
(East)-421501, Mumbai, Maharashtra, India; [b] Office of the Dean
(Research) & Division of Chemistry, Department of Science, Faculty of
Science & Technology, Alliance University, Chandapura-Anekal Main Road,
Bengaluru 562106, Karnataka, India; [c] Centre for Nanotechnology,
Rajiv Gandhi Proudyogiki Vishwavidyalaya, 462003 Bhopal, MP, India;
[d] Department of Chemistry, Mithibai College of Arts, Chauhan Institute of
Science and Amrutben Jivanlal College of Commerce & Economics,
Mumbai 400 056, Maharashtra, India
*Email: ChavaliM@gmail.com

3.1 Introduction

Very high surface area (950 to 2000 $m^2\,g^{-1}$), excellent porosity, and non-specific adsorption capacity are the key features that make activated carbon (AC) a unique and versatile physical adsorbent. AC preparation involves carbonization of feedstock by gradual heating under anaerobic conditions below 600 °C. Recently, a detailed literature review was documented on the preparation of AC.[1] Despite the huge commercial success of AC in environmental, purification, and medicinal applications, there are few reports on the detailed structural pattern of this material. An interesting study by Harris *et al.* investigated the atomic structure of AC using the aberration-corrected TEM technique. It was revealed that AC has pentagonal rings

Activated Carbon: Progress and Applications
Edited by Chandrabhan Verma and Mumtaz A. Quraishi
© The Royal Society of Chemistry 2023
Published by the Royal Society of Chemistry, www.rsc.org

(fullerene-like structure), thereby explaining its higher porosity and enhanced adsorption capacity.[2]

Since ancient times, AC has been used in medicine. However, the interest in medical applications of AC has been intriguingly slow. The neglect of using AC as medicine is due to its physical characteristics, which make it difficult to administer to a patient. AC was infrequently used orally to treat drug overdose and poisoning. Numerous *in vivo* studies with humans and animals have demonstrated the efficacy of AC in its binding capacity with drugs and poisons. Before delving into the mechanism of action of AC as a medicine, we need to understand the terminologies related to AC. The reported literature survey reveals that 'activated charcoal' and 'active charcoal' are the terms used interchangeably. In this chapter, the authors will use the phrase 'activated carbon', which shall be abbreviated as AC for brevity. Several editorials and commentary papers discuss the advent of AC as medicine and its importance in clinical practice.[3-5]

3.1.1 Beginnings of AC in Medicine

Early physicians believed in the therapeutic value of AC. The earliest description of AC in medicine was its oral use for treating various diseases. Egyptian Papyrus (1550 BC), Hippocrates (400 BC), and Pliny (50 AD) used wood charcoal to treat diseases such as anthrax, epilepsy, vertigo, and chlorosis.[6] It was recommended by early physicians to administer a dosage of 1/16 oz AC for treating fever. Besides its internal applications, many physicians studied the external applications of AC to treat wounds and ulcers, which are even practised in modern medicine. D. H. Kehl (1793) suggested the use of AC dispensed in water as a mouthwash. Claudius Galen, a prolific Roman physician, studied the use of charcoal obtained from vegetable and animal origins to treat various diseases. In 1813, Michel Bertrand was the first physician to showcase the miracles of oral AC. He publicly consumed 5 g of arsenic trioxide mixed with AC and sugar. He demonstrated certain discomfort but there were no untoward symptoms of toxicity. Following Bertrand's footsteps, in 1831, Pierre Touéry consumed strychnine at 10 times its lethal dose mixed with AC and showcased no untoward toxicity in front of the French Academy of Medicine.[7] Yet, the medical panel were not impressed with the AC and its oral medication prowess. In the mid-1960s, there was renewed interest amongst toxicologists to demonstrate the oral use and medical efficacy of ACs.

3.1.2 Route of Administration

There is no internationally accepted guideline for the administration of AC to patients. However, there are pioneering papers of the American Academy of Clinical Toxicology (AACT) and the European Association of Poisons Centres and Clinical Toxicologists (EAPCCT) on single- and multi-dose ACs for the treatment of acute poisoning that serve as guidelines for AC-drug usage.[8] It is known that the rate of uptake of a drug or poison by charcoal will be inversely proportional to the particle size squared. Numerous studies are reported that describe the drug:AC concentration ratio for efficacy especially in poison treatment. Tsuchiya and Levy (1972) performed *in vitro* and *in vivo* adsorption studies to compare the efficacy of powdered charcoal and tablet formulations on test drugs, *viz.*, aspirin, salicylamide, and phenylpropanolamine.[9] Remmert *et al.* (1990) reported *in vivo* studies to compare the efficacy of tablets, capsules, and suspensions of ACs on acetaminophen. They showed that suspensions and tablets of ACs were more effective than capsules.[10]

AC can be employed as a medical adsorbent externally or internally. AC is popular as an oral medicine to treat poisoning. However, using AC as an oral drug presents several challenges such as gritty taste, unappealing physical characteristics, biocompatibility, and cytotoxicity, which are dealt with later in this chapter. Speaking of external usage, AC is usually employed as a wound bandage and hemoperfusion device. A hemoperfusion device is a typical extracorporeal device used in blood purification where anticoagulated whole blood is passed through a column containing AC. This process allows the removal of toxins and cytokines from poisoned and septic patients, respectively.[11]

3.2   Medical Grade AC

AC is used internally as an oral adsorbent. Clinical physicians recommended immediate administration of AC in case of poison treatment in larger doses of 50–100 g followed by smaller doses of 25–50 g every few hours to the patient. There are certain requirements for AC to be termed 'medical grade'. The prerequisite of AC to be used as a medicine is its haemocompatibility if used in direct contact with skin, mouth, or blood. As a biomaterial, AC must satisfy the following criteria as laid down by medical experts:[12]

(a) should not damage blood cells,
(b) should not result in adverse immunological reactions such as inflammation and allergies,
(c) should not lead to carcinogenic effects,
(d) should not release impurities and toxic substances in the bloodstream, and
(e) should not adsorb and remove important metabolites and nutrients from the body.

Of the aforementioned requirements, it is challenging to prevent AC from eliminating important metabolites and nutrients from the body. This is attributed to its non-specific adsorptive capacity. Also, if other drugs are administered along with AC, it tends to neutralize the drug action. Hence, it is advisable to avoid taking other drugs and antidotes along with AC.[13]

3.2.1   Haemocompatibility of AC

The direct contact of AC with blood results in the reduction of calcium, potassium, and phosphate concentration along with blood cell damage.

Unlike the gastrointestinal (GI) tract, blood is highly sensitive toward AC when administered for blood purification. Blood serum contains about 60–80 mg mL^{-1} of proteins that act as biological surfactants.[14] When AC encounters blood surfactants, they release microparticles easily due to their poor adherence to the AC surface. These microparticles release into the bloodstream and block the blood vessels, causing microembolism.

To overcome the problem of incompatibility of AC with blood, surface modification has become an essential pursuit. AC surfaces coated with biocompatible materials such as albumin and cellulose layers were reported.[15] However, coated AC with a thickness of about 5 μm slowed down the external diffusion to the adsorbent surface. Further, it restricted larger molecules to come into contact with the AC surface completely.[16] Uncoated phenol-formaldehyde resin-based pyrolyzed carbon has demonstrated good haemocompatibility.[17] Polystyrene and poly(vinyl pyridine) are used to prepare AC that can withstand attenuation during transport and storage, thereby inhibiting the release of microparticles in the bloodstream.[18] Radomski *et al.* demonstrated that the poor biocompatibility of AC is related to its rough surface.[19] Hence, AC with a smooth surface has been studied using encapsulation technology. Encapsulation of AC avoids debris from leaving AC, yet the mesopores of AC get blocked, thereby reducing its adsorptive properties.[20] Hence, in hemoperfusion, there is a constant tussle between adsorptive capacity and biocompatibility, making AC a difficult choice of material in oral drugs. Recent developments are directed toward carbon-based graphene materials for hemoperfusion, which has solved compatibility problems to some extent.[21]

3.2.2 Palatability of Oral AC

One of the major problems in administering oral AC is its poor palatability. Further, this problem is especially a cause of concern amongst paediatric patients.[22] It leaves a gritty sensation in the mouth, decolorizes gums and teeth, and sticks to the throat. As AC is used in large amounts for poison treatment, encapsulation is not a plausible solution to render palatability. Various flavouring agents have been added to enhance the appeal of AC as an oral drug. A suspension of sorbitol with ACs for poison treatment was reported.[23] Other flavouring agents to enhance the palatability of AC are bentonite,[24] saccharin sodium,[25] gelatin,[26] ice cream,[27] and milk chocolate.[28] Dagnone *et al.* (2002) compared the palatability of common flavouring agents using a visual analogue scale with a facial hedonic scale. Children rated higher palatability for AC with cola drinks rather than orange or chocolate flavouring agents.[29] Though palatability is a serious issue, a comparison of different groups of poisoned patients was done where the volunteers rejected ACs at a higher rate than in actual overdose cases. Spiller *et al.* (2001) reported that patients who have potentially toxic overdoses exhibit more likelihood of taking ACs than those who have milder symptoms of overdose irrespective of their palatability.[30] Wendy *et al.* (2010) performed a prospective, open-label, three-way, crossover trial to demonstrate the effect of charcoal cookie formulation on the absorption of cimetidine and its comparison with the standard aqueous charcoal product. They concluded that the AC cookie formulation was effective in reducing cimetidine absorption associated with enhanced palatability.[31]

3.3 Wound Healing Using AC

ACs have been employed in clinical medicine, as absorbent dressings for superficial and chronic wounds and in post-operative trauma restoration (Figure 3.1). AC-based wound dressings are made of fibres incorporated with carbonized cellulose products. These bandages help to create a physiological environment that promotes quick wound healing while also preventing infection transmission.

AC develops strong van der Waals forces that drain secretions, bacteria, and odours from inflamed, granulating, and chronic wounds.[32] ACs speed up the haemocoagulation process, increase exudate build-up on the surface, establish a barrier against outside microbe penetration, and prevent wound dehydration when used in conjunction with the dressing material. A dry wound dressing called 'xerodressing' was applied on venous leg ulcers; it showed good efficacy with reduced ulcer size.[33]

The surface of ACs can be impregnated with biologically active chemicals to increase biological activity. Silver-incorporating AC has demonstrated an antibacterial effect in numerous studies. Activated charcoal dressings such as Actisorb® (Johnson & Johnson), InvaSorb (Hasti Medic), and Tecasorb (FWDS Research Company) are some of the examples commonly employed in wound dressing that uses Ag-incorporating AC. Lin *et al.* (2014) demonstrated the use of a silver-based AC dressing with good antibacterial effects and biocompatibility on excision wounds.[34] Apart from wound healing, malodour is a common problem seen in patients with diabetic foot ulcers.[35] The blood supply to the chronic wound results in necrosis. Necrotic tissues

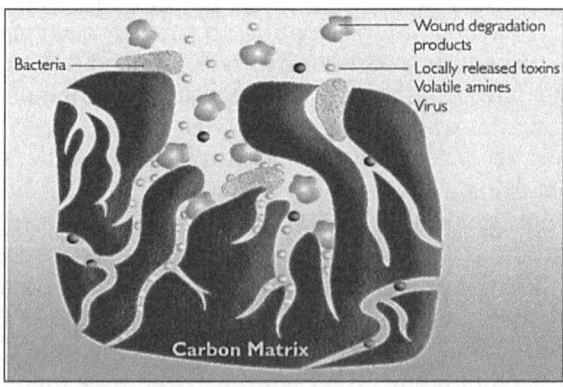

Figure 3.1 Wound healing process using AC.

become infected with anaerobic bacteria that release volatile fatty acids which are malodorous. Chakravarthi *et al.* demonstrated in their double-blind experiment that activated charcoal along with sodium bicarbonate adsorbs and neutralizes volatile molecules, thereby decreasing odour.[36] Minsart *et al.* used acrylate-end-capped, urethane-based precursors (AUPs) and methacrylated alginate (AlgMOD), which were blended with AC. These blends were then processed in hydrogel sheets and applied to the wound. It was demonstrated that hydrogel-based odour absorbing ACs dressing reduced wound exudate and malodour.[37] Sebastian *et al.* demonstrated that a sterile polyacrylate wound pad with AC cloth treatment was superior in diabetic wound healing to applying a non-adhesive hydrocellular foam dressing with silver.[38]

3.4   Intoxication Treatment Using AC

Intentional or accidental swallowing or drinking of a toxic substance is called poisoning. The most common cause of hospitalization is acute poisoning that needs emergency medical intervention.

3.4.1   AC as an Antidote

The use of AC in gastric contamination was the first successful attempt toward bringing a dawn of AC in medicine. In 1929, William P. Hort was the first reported physician to use AC in a poisoned patient. When a 40-year-old male patient accidentally ingested mercuric chloride, he suffered from intense pain in his upper GI tract, followed by stomachache, diarrhoea, and hypotension. Initially, he was administered magnesium sulphate which did not provide aid to the patient. At last, when one teaspoon of pulverized AC was administered every hour and the patient showed signs of recovery shortly after the first dose. Activated charcoal was continued for a few more days and the patient showed full recovery.[39] Sir Alfred B. Garrod carried out the first controlled study of purified animal charcoal for treating intoxication due to strychnine with emphasis on the molar ratio of AC and strychnine.[40] Benjamin Rand reported minimal toxicity of various poisons such as strychnine, hydrocyanic acid, digitalis, and morphia when co-ingested with AC.[41] The single-dose AC (SDAC) risk:benefit ratio during gastric decontamination follows a typical pattern called Bailey's GI decontamination triangle as shown in Figure 3.2. This triangle is a good reference to decide whether the patient should undergo GI decontamination.[42]

Recently, numerous studies have reported clinical studies showing the efficacy of AC for poison treatment. Spiller *et al.* (2006) reported a multicentre study on acetaminophen overdose using AC with *N*-acetylcysteine showing greater efficacy amongst patients.[43] Isbister *et al.* (2007) demonstrated that single-dose activated charcoal administered post-citalopram overdose has lower chances of developing abnormal QT prolongation.[44] Over the years, use of ACs in gut decontamination in poisoned patients

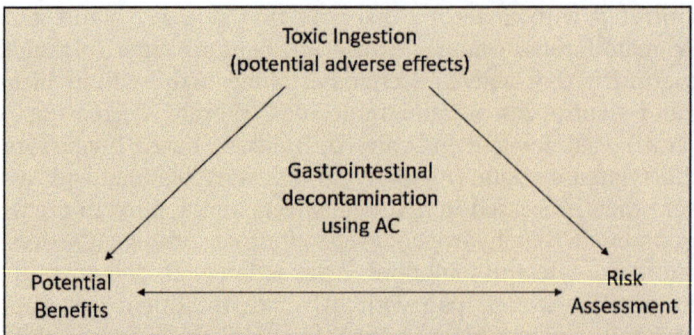

Figure 3.2   GI decontamination triangle with three corners depicting risks, benefits, and adverse effects of using activated charcoal.

has declined. The oral use of ACs leads to complications such as emesis, constipation, aspiration, and bowel obstruction. Thus, AC is recommended only in case of life-threatening intoxication cases or as a multi-dosage formulation. Randhawa *et al.* (2021) reported using multi-dose AC to treat carbamazepine acute poisoning.[45] Hatanaka *et al.* (2022) reported multi-dose AC treatment for lamotrigine poisoning.[46]

The major challenge in administering AC is its cytotoxicity, which can potentially impede its applications in medicine. Barnes *et al.* (2004) reported a colony formation assay to determine the cytotoxicity of ACs. It was observed that ACs removed some nutrients from cell growth media and were suitable for use in the treatment of sepsis and renal failure.[47a] The same team worked on investigating the cytotoxic potential of novel AC adsorbents developed for extracorporeal therapies. They reported no apparent cytotoxic response toward the V79 cell line under modified conditions.[47b]

Some poisons exhibit poor affinity with AC such as alcohols, nicotinic acid, iron, and hydrocarbons.[48] This reduces the efficacy of AC. Recently, magnetic AC has gained attention. Danalıoğlu *et al.* (2017) reported the use of a magnetic AC/chitosan composite material for the removal of toxic antibiotics, namely, ciprofloxacin, erythromycin, and amoxicillin.[49] Oliveira *et al.* (2018) synthesized hydrophilic magnetic nanoparticles *via* a greener route and further coated them with carbon-based shells for drug delivery applications. The nanocomposites were also functionalized with nitric acid that showed enhanced drug delivery with a pH dependence.[50]

3.4.2   ACs as Drug Delivery Systems

Earliest attempts to use ACs as drug delivery agents were use of a trypsin/AC complex as a prolonged-release medicine and as a subcutaneous injection to treat carcinoma in dogs.[51] Roivas and Neuvonen reported activated charcoal as a suitable matrix for the sustained release of nicotinic acid.[52] Numerous efforts are made to utilize ACs as drug delivery agents with reduced cytotoxicity. Recently, Miriyali *et al.* (2017) investigated the application of AC as a drug carrier for amorphous drug delivery using paracetamol and ibuprofen as model drugs. They studied the cytotoxicity of AC on human colon carcinoma (Caco-2) cells, which was found to be low at the concentrations tested from 10 to 800 $\mu g\,mL^{-1}$.[53]

Román *et al.* (2018) compared the interaction of pure and Al-doped AC with 5-fluorouracil in anticancer drugs. They found that doping ACs increased their adsorption capacity by enhancing their interactions with polar atoms of the adsorbate, thus improving their adsorption capability.[54] Yadavalli *et al.* (2019) demonstrated the use of AC to topically deliver acyclovir in the treatment of herpes simplex virus (HSV) infection. They also reported that highly porous AC (HPAC) acted as an inhibitor of viral entry and the adverse effects due to viral replication through prophylactic, neutralization, and treatment models.[55]

3.4.3 ACs in COVID-19

Respiratory masks have become a common means to inhibit droplet transmission among humans in the COVID-19 pandemic. Table 3.1 lists various pathogens, types of ACs, and their effects on pathogen reduction.

Using respiratory masks with ACs provided efficient protection from droplet transmission of COVID-19 infection. The hydrophobicity of the virus, surface charge on ACs, and electrophoretic repulsive forces between ACs and viral surfaces dictate the efficiency of viral load removal. As SARS-CoV-2 has a spike S-protein with a hydrophobic character, ACs are highly efficient adsorbents of the spiked virus particles.[56] Recently, highly efficient CNT-field-effect transistor-based biosensors were demonstrated to selectively detect SARS-CoV-2 infection at a limit of detection of 10 fM.[57] CNTs are far superior to carbon-based materials than AC. Their surfaces are easily functionalized through noncovalent or covalent interactions and thus act as drug delivery agents of different molecules. CNTs have been successfully applied as drug carriers in cancer therapy,[58] rheumatoid arthritis,[59] brain disorders,[60] and vaccine delivery agents.[61]

Table 3.1 Various pathogens, types of ACs, and their effects on pathogen reduction.

Type of AC	Pathogen	Outcome/effects
Granular ACs	Bacteriophage MS2	Inefficient removal of viruses
AC fibre composite	Bacteriophage MS2	Efficient removal of viruses
Super-powdered ACs (S-PAC)	Bacteriophage Qβ, MS2	Decreased electrophoretic repulsion between pathogen surfaces and ACs enhances viral load removal
Multi-walled carbon nanotubes (MWCNTs) loaded with NPs	Bacteriophage MS2	Cu_2O-sensitized MWCNTs showed higher viral removal activity compared to titanium or iron NPs

3.5   Conclusions

AC is undoubtedly a successful material to treat intoxication and drug overdose and has shown promise as a drug delivery system. Some of its unique applications are in protective clothing, masks, and biosensor devices. AC in medicine has come a long way with humble beginnings as the antidote to being used as drug delivery agents. ACs and their applications in medicine will have some trade-offs in terms of cytotoxicity and biocompatibility. With alarmingly high numbers of reports on advanced activated materials such as graphene and CNTs, it shall be necessary to classify them in appropriate categories against ACs rather than replace them entirely. CNTs, graphene, and fullerenes are efficient antimicrobial agents far superior to ACs. Consequently, extensive research on toxicity and biocompatibility is essential to understand the mechanism of action of these AC-based materials.

Abbreviations

AACT	American Academy of Clinical Toxicology
AC	Activated carbon
AlgMOD	Methacrylated alginate
AUPs	Acrylate-end-capped, urethane-based precursors
Caco-2	Human colon carcinoma
CNTs	Carbon nanotubes
EAPCCT	European Association of Poisons Centres and Clinical Toxicologists
GI	Gastrointestinal
HPAC	Highly porous AC
HSV	Herpes simplex virus
MWCNTs	Multi-walled carbon nanotubes
SDAC	Single-dose AC
S-PAC	Super-powdered AC

References

1. Z. Heidarinejad, M. H. Dehghani and M. Heidari, *et al.*, Methods for preparation and activation of activated carbon: a review, *Environ. Chem. Lett.*, 2020, **18**, 393–415.
2. P. J. F. Harris, Z. Liu and K. Suenaga, Imaging the atomic structure of activated carbon, *J. Phys.: Condens. Matter*, 2008, **20**, 362201.
3. J. Greensher, H. C. Mofenson and T. R. Caraccio, Ascendency of the black bottle (activated charcoal), *Pediatrics*, 1987, **80**(6), 949–951.
4. D. A. Spyker, Activated charcoal reborn. Progress in poison management, *Arch. Intern. Med.*, 1985, **145**(1), 43–44.
5. R. Spector and G. D. Park, New roles for activated charcoal, *West. J. Med.*, 1986, **145**(4), 511–512.
6. T. Gupta, Historical Production and Use of Carbon Materials: The Activated Carbon, in *Carbon*, Springer, Cham, 2018, DOI: 10.1007/978-3-319-66405-7_2.

7. S. G. Marketos and G. Androutsos, Charcoal: from antiquity to the charcoal artificial kidney, *J. Nephrol.*, 2004, **17**, 453–456.

8. American Academy of Clinical Toxicology, European Association of Poisons Centres and Clinical Toxicologists, Position statement and practice guidelines on the use of multi-dose activated charcoal in the treatment of acute poisoning, *J. Toxicol., Clin. Toxicol.*, 1999, **37**, 731–751.

9. T. Tsuchiya and G. Levy, Drug adsorption efficacy of commercial activated charcoal tablets *in vitro* and man, *J. Pharm. Sci.*, 1972, **67**, 624.

10. H. P. Remmert, M. Oiling, W. Slob, W. F. van der Giesen, A. van Dijk and A. G. Rauws, Comparative antidotal efficacy of activated charcoal tablets, capsules, and suspension in healthy volunteers, *Eur. J. Clin. Pharmacol.*, 1990, **39**, 501.

11. J. F. Winchester, N. B. Harbord, E. Charen and M. Ghannoum, Use of dialysis and hemoperfusion in the treatment of poisoning, in *Handbook of Dialysis*, ed. J. T. Daugirdas, P. G. Blake and T. S. Ing, Lippincott, Williams, & Wilkins, Philadelphia, 5th edn, 2015, p. 368.

12. H. J. Gurland, A. M. Davison, V. Bonomini, D. Falkenhagen, S. Hansen, T. Kishimoto, M. J. Lysaght, J. Moran and A. Valek, Definitions and Terminology in biocompatibility, *Nephrol., Dial., Transplant.*, 1994, **9**(S2), 4–10.

13. K. Eckert, P. Eyer and T. Zilker, Activated charcoal—first aid treatment in oral poisoning, *Dtsch Arztebl Int.*, 1999, **96**, A-2826–A-2830.

14. M. Leeman, J. Choi, S. Hansson, M. U. Storm and L. Nilsson, Proteins and antibodies in serum, plasma, and whole blood—size characterization using asymmetrical flow field-flow fractionation (AF4, *Anal. Bioanal. Chem.*, 2018, **410**(20), 4867–4873.

15. S. V. Mikhalovsky, Emerging technologies in extracorporeal treatment: focus on adsorption, *Perfusion*, 2003, **18**(1S), 47–54.

16. E. Denti and J. M. Walker, Activated carbon: properties, selection, and evaluation, in *Sorbents and their clinical applications*, ed. C. Giordano, Academic Press, New York, 1980, pp. 101–116.

17. S. R. Sandeman, C. A. Howell, G. J. Phillips, A. W. Lloyd, J. G. Davies and S. V. Mikhalovsky, *et al.*, Assessing the in vitro biocompatibility of a novel carbon device for the treatment of sepsis, *Biomaterials*, 2005, **26**, 7124–7131.

18. W. Qin, L. Xiao-Yi, Z. Rui, L. Chao-jun, L. Xiao-jun, Q. Wen-ming, Z. Liang and L. Li-chengy, Preparation of polystyrene-based activated carbon spheres and their adsorption of dibenzothiophene, *New Carbon Mater.*, 2009, **24**(1), 55–60.

19. A. Radomski, P. Jurasz, D. Alonso-Escolano, M. Drews, M. Morandi, T. Malinski and M. W. Radomski, Nanoparticle-induced platelet aggregation and vascular thrombosis, *Br. J. Pharmacol.*, 2005, **146**, 882–893.

20. J. Miao, F. Zhang, M. Takieddin, S. Mousa and R. J. Linhardt, Adsorption of doxorubicin on poly(methyl methacrylate)-chitosan-heparin-coated activated carbon beads, *Langmuir*, 2012, **28**(9), 4396–4403.

21. Z. Li, X. Huang, K. Wu, Y. Jiao and C. Zhou, Fabrication of regular macro-mesoporous reduced graphene aerogel beads with ultra-high

mechanical property for efficient bilirubin adsorption, *Mater. Sci. Eng., C*, 2020, **106**, 110282.

22. L. West, Innovative approaches to the administration of activated charcoal in pediatric toxic ingestions, *Pediatr. Nurs.*, 1997, **23**(6), 616–619.
23. P. Eyer and M. Sprenger, Oral administration of activated charcoal-sorbitol suspension as first aid in the prevention of poison resorption, *Klin. Wochenschr.*, 1991, **69**(19), 887–894.
24. R. P. Navarro, K. R. Navarro and E. P. Krenzelok, Relative efficacy and palatability of three activated charcoal mixtures, *Vet. Hum. Toxicol.*, 1980, **22**(1), 6–9.
25. D. O. Cooney, Saccharin sodium as a potential sweetener for antidotal charcoal, *Am. J. Hosp. Pharm.*, 1977, **34**(12), 1342–1344.
26. S. Noro, F. Ishii and K. Saegusa, The Influence of Cross-Linking Time on the Adsorption Characteristics of Microcapsules Containing Activated Charcoal Prepared by Gelatin-Acacia Coacervation, *Chem. Pharm. Bull.*, 1985, **33**(11), 4649–4656.
27. G. Levy, D. M. Soda and T. A. Lampman, Inhibition by ice cream of the antidotal efficacy of activated charcoal, *Am. J. Hosp. Pharm.*, 1975, **32**(3), 289–291.
28. T. FEisen, P. AGrbcich, P. G. Lacouture, M. Shannon and A. Woolf, The adsorption of salicylates by a milk chocolate-charcoal mixture, *Ann. Emerg. Med.*, 1991, **20**(2), 143–146.
29. D. Dagnone, D. Matsui and M. J. Rieder, Assessment of the palatability of vehicles for activated charcoal in pediatric volunteers, *Pediatr. Emerg. Care*, 2002, **18**(1), 19–21.
30. H. A. Spiller and G. C. Rodgers Jr, Evaluation of administration of activated charcoal in the home, *Pediatrics*, 2001, **108**(6), E100.
31. W. Klein-Schwartz, S. Doyon and T. Dowling, Drug Adsorption Efficacy and Palatability of a Novel Charcoal Cookie Formulation, *Pharmacotherapy*, 2010, **30**(9), 888–894.
32. J. Dissemond, M. Augustin and S. A. Eming, *et al.*, Modern wound care – practical aspects of non-interventional topical treatment of patients with chronic wounds, *J. Dtsch. Dermatol. Ges.*, 2014, **12**(7), 541–554.
33. U. Wunderlich and C. E. Orfanos, Treatment of venous ulcera cruris with dry wound dressings. Phase overlapping use of silver-impregnated activated charcoal xerodressing, *Hautartz*, 1991, **42**, 446.
34. Y. H. Lin, W. S. Hsu and W. Y. Chung, *et al.*, Evaluation of various silver-containing dressing on infected excision wound healing study, *J. Mater. Sci.: Mater. Med.*, 2014, **25**, 1375–1386.
35. G. Lee, S. C. Anand, S. Rajendran and I. Walker, Overview of current practice and future trends in the evaluation of dressings for malodorous wounds, *J. Wound Care*, 2006, **15**(8), 344–346.
36. A. Chakravarthi, C. R. Srinivas and A. C. Mathew, Activated charcoal and baking soda to reduce odor associated with extensive blistering disorders, *Indian J. Dermatol. Venereol. Leprol.*, 2008, **74**(2), 122–124.

37. M. Minsart, A. Mignon, A. Arslan, I. U. Allan, S. V. Vlierberghe and P. Dubruel, *Macromol. Mater. Eng.*, 2021, **306**(1), 2000529.
38. S. Probst, C. Saini, C. Rosset and M. B. Skinner, Superabsorbent charcoal dressing versus silver foam dressing in wound area reduction: a randomised controlled trial, *J. Wound Care*, 2022, **31**(2), 140–146.
39. W. P. Hort, Case of poisoning with corrosive sublimate, in which the administration of charcoal afforded great relief, *Am. J. Med. Sci.*, 1834, **6**, 540–541.
40. A. B. Garrod, Purified animal charcoal: an antidote to all vegetable and some mineral poisons, *Trans. Med. Soc. London*, 1846, **1**, 195–204.
41. B. H. Rand, On animal charcoal as an antidote, *Med. Examiner*, 1848, **4**, 528–533.
42. B. Bailey, Gastrointestinal decontamination triangle, *Clin. Toxicol.*, 2005, **43**(1), 59–60.
43. H. A. Spiller, M. L. Winter, W. Klein-Schwartz and S. A. Bangh, Efficacy of activated charcoal when administered more than four hours after acetaminophen overdose, *J. Emerg. Med.*, 2006, **30**(1), 1–5.
44. G. K. Isbister, L. E. Friberg, B. Stokes, N. A. Buckley, C. Lee, N. Gunja, S. G. Brown, E. Macdonald, A. Graudins, A. Holdgate and S. B. Duffull, Activated charcoal decreases the risk of QT prolongation after citalopram overdose, *Ann. Emerg. Med.*, 2007, **50**, 593–600.
45. M. S. Randhawa, P. Sharma, S. K. Angurana and A. Bansal, Acute carbamazepine toxicity in a child: A case report, *J. Pediatr. Crit. Care*, 2021, **8**, 299–301.
46. K. Hatanaka, Y. Kamijo, T. Kitamoto, T. Hanazawa, T. Yoshizawa, H. Ochiai and Y. Haga, Effectiveness of multiple-dose activated charcoal in lamotrigine poisoning: a case series, *Clin. Toxicol.*, 2022, **60**(3), 379–381.
47. (a) L. M. Barnes, M. C. Murphy, M. Melillo, G. J. Phillips, J. G. Davies, A. W. Lloyd *et al.*, The cytotoxic assessment of carbon adsorbents. In Transactions – 7th World Biomaterials Congress, (Transactions – 7th World Biomaterials Congress, 2004; (b) L.-M. Barnes, G. J. Phillips, J. G. Davies, A. W. Lloyd, E. Cheek, S. R. Tennison, A. P. Rawlinson, O. P. Kozynchenko and S. V. Mikhalovsky, The cytotoxicity of highly porous medical carbon adsorbents, *Carbon*, 2009, **47**(8), 1887–1895.
48. (a) L. Roivas and P. J. Neuvonen, Reversible Adsorption of Nicotinic Acid onto Charcoal, *J. Pharm. Sci.*, 1992, **81**, 917–919; (b) J. F. Wiley, II and K. C. Osterhoudt, Poisonings, in *Pediatric Emergency Medicine Secrets*, ed. S. M. Selbst, Saunders, Philadelphia, 2015, pp. 312–332.
49. S. T. Danalıoğlu, Ş. SenaBayazit, Ö. Kuyumcu and M. A. Salam, Efficient removal of antibiotics by a novel magnetic adsorbent: Magnetic activated carbon/chitosan (MACC) nanocomposite, *J. Mol. Liq.*, 2017, **240**, 589–596.
50. J. Oliveira, R. Rodrigues, L. Barros, I. Ferreira, L. Marchesi and M. Koneracka, *et al.*, Carbon-Based Magnetic Nanocarrier for Controlled Drug Release: A Green Synthesis Approach, *C*, 2018, **5**, 1.
51. E. Falk and A. Sticker, Carbenzyme, *Munch. Med. Wochenschr.*, 1910, **57**, 4.

52. L. Roivas and P. J. Neuvonen, Reversible adsorption of nicotinic acid onto charcoal in vitro, *J. Pharm. Sci.*, 1992, **81**, 917.
53. N. Miriyala, D. Ouyang, Y. Perrie, D. Lowry and D. J. Kirby, Activated carbon as a carrier for amorphous drug delivery: Effect of drug characteristics and carrier wettability, *Eur. J. Pharm. Biopharm.*, 2017, **115**, 197–205.
54. G. Román, E. Noseda Grau, A. Díaz Compañy, G. Brizuela, A. Juan and S. Simonetti, A first-principles study of pristine and Al-doped activated carbon interacting with 5-Fluorouracil anticancer drug, *Eur. Phys. J. E: Soft Matter Biol. Phys.*, 2018, **41**, 107.
55. T. Yadavalli, J. Ames, A. Agelidis, R. Suryawanshi, D. Jaishankar, J. Hopkins, N. Thakkar, L. Koujah and D. Shukla, Drug-encapsulated carbon (DECON): A novel platform for enhanced drug delivery, *Sci. Adv.*, 2019, **5**(8), 1–12.
56. I. M. Ibrahim, D. H. Abdelmalek, M. E. Elshahat and A. A. Elfiky, COVID-19 Spike-host Cell Receptor GRP78 Binding Site Prediction, *J. Infect.*, 2020, **80**(5), 554–562.
57. M. Thanihaichelvan, S. N. Surendran, T. Kumanan, U. Sutharsini, P. Ravirajan, R. Valluvan and T. Tharsika, Selective and electronic detection of COVID-19 (Coronavirus) using carbon nanotube field-effect transistor-based biosensor: A proof-of-concept study, *Mater. Today: Proc.*, 2022, **49**(7), 2546–2549.
58. K. H. Son, J. H. Hong and J. W. Lee, Carbon nanotubes as cancer therapeutic carriers and mediators, *Int. J. Nanomed.*, 2016, **11**, 5163–5185.
59. C. Kofoed Andersen, S. Khatri, J. Hansen, S. Slott, R. Pavan Parvathaneni, A. C. Mendes, I. S. Chronakis, S.-C. Hung, N. Rajasekaran and Z. Ma, Carbon Nanotubes—Potent Carriers for Targeted Drug Delivery in Rheumatoid Arthritis, *Pharmaceutics*, 2021, **13**, 453.
60. M. Çetin, E. Aytekin, B. Yavuz and S. Bozdağ-Pehlivan, Nanoscience in targeted brain drug delivery, in *Nanotechnology Methods for Neurological Diseases and Brain Tumors*, ed. Y. Gürsoy-Özdemir, S. Bozdağ-Pehlivan and E. Sekerdag, Academic Press, Cambridge, MA, USA, 2017, ch. 7, pp. 117–147.
61. K. Kostarelos, A. Bianco and M. Prato, Promises, facts, and challenges for carbon nanotubes in imaging and therapeutics, *Nat. Nanotechnol.*, 2009, **4**, 627–633.

CHAPTER 4

Analytical Applications of Activated Carbon

S. SHAFI,[a] S. ZAFAR[a] AND T. RASHEED*[b,c]

[a] Institute of Chemistry, The Islamia University of Bahawalpur, Bahawalnagar Campus 62300, Pakistan; [b] Interdisciplinary Research Center for Advanced Materials, King Fahd University of Petroleum and Minerals (KFUPM), Dhahran 31261, Saudi Arabia; [c] School of Chemistry and Chemical Engineering, Shanghai Jiao Tong University, Shanghai 200240, China
*Email: tahir.rasheed@kfupm.edu.sa

4.1 Introduction

Activated carbon (AC) has been around for so long in human history that pinpointing its exact origin is impossible. Wood char, coal char, and highly developed porous structures were used before, what are now known as AC (which has an elevated adsorption ability). AC was simply a slightly devolatilized carbonaceous substance that was employed. The term "adsorbent" refers to a material that absorbs water. The first known example dates back to 3750 BCE. Wood char was used by both Egyptians and Sumerians to lower the amount of water in their bodies. Copper, zinc, and tin ores are used to make bronze, as well as are sources of no-smoke copper, zinc, and tin fuel.[1] The first evidence of carbon's therapeutic use was discovered in a papyrus record from 1550 BCE in Thebes (Greece). Hippocrates (about 400 BCE) later advocated that water be cleaned using wood char before ingestion to remove poor taste and odor and to treat ailments such as epilepsy, chlorosis, and anthrax.[2]

Activated Carbon: Progress and Applications
Edited by Chandrabhan Verma and Mumtaz A. Quraishi
© The Royal Society of Chemistry 2023
Published by the Royal Society of Chemistry, www.rsc.org

Dr D. M. Kehl employed charred wood to minimize the odors coming from gangrene in 1793, which was the first known application of charcoal as a gas phase adsorbent. In England, AC was first utilized in the industrial sector in 1794; it was first used as a bleaching agent in the sugar industry. Biochar was first used for gas adsorption in the mid-19th century. In 1854, the Mayor of London ordered that all wastewater ducts be fitted with natural wood charcoal purifiers to eliminate foul odors. In 1872, the chemical industry first used gas hoods with AC to protect workers from mercury vapor inhalation.[2]

During the twentieth century, modern society prospered by rapidly expanding the production and use of activated charcoal, especially in the second half of the century, due to increasingly stringent regulations and strict standards for water reserves, clean air implementation, air quality assurance, grid supply, and economical reuse of waste chemicals. Further, apart from the replacement of petroleum-based products, the search for new markets for agro-industrial products necessitated the use of hemicellulose materials and many other biomass resources for the manufacture of biochar. Over 300 000 tons of biochar produced from wood and coconut shells are exported each year to other countries.[2] However, this is only a minor part of the global demand for AC, which totaled 12 804 000 tons in 2015.[2]

Scientific production of carbon materials, on the other hand, has been quite active in the past 20 years. Additionally, carbon nanotubes and graphene have seen exponential increases in contributions. During the studied period, however, the number of research papers published on carbon fibers and AC increased significantly, with 18 600 and 17 516 research papers produced, respectively. Additionally, 1668 manuscripts were published in 2016 that dealt with AC. In this chapter the most common applications of lignocellulosic precursors and biomass wastes for the production of AC, as well as the methods for analyzing their physical, chemical, and microstructural properties, are discussed.

4.2 Heavy Metal Removal Methods

Textile wastewater with heavy metals can be treated physically, chemically, and biologically.[3] Given the peculiar characteristics of yard goods wastewater and how operators may help their facilities, we have selected a treatment method that is both inexpensive and efficient.[4] An increasing number of approaches, technologies, and methods have been developed to recover the effluents of textile wastewater.[5] A few examples are transmembrane purification, coagulation/flocculation, and electrodialysis.[6] There was the possibility of cathodic recovery, electrodeposition, ionic operations, or Soxhlet extraction.[7,8] The process of removing pollutants from the environment is termed casing machinery, microbioremediation, phytobiological restoration, and phytoremediation.[9] Unfortunately, these technologies pose disposal challenges. Compared to the other processes, adsorption has the broadest range of applications. The adsorption process is much more versatile than the other processes listed. It produces a lot of sludge, creates dangerous working conditions, and has high disposal costs in addition to its effectiveness.[4,10,11]

A cost-effective method for extracting metallic ions from sludges with only a few dissolved solids is sorption. The adsorption process produces little sludge and is cost-effective.[4] As little as 1 $mg\,L^{-1}$ of heavy metal can be removed from liquids by adsorption.[12] Several companies offer botanical residues in substantial quantities for use as adsorbents. Pretreatment of desiccants impacts productivity, economy, and flexibility.[13] In sewage treatment, adsorption is probably the most common method, where effluents are combined with porous structures such as charcoal and clay. Churning the filter bed makes sewage flow through gritty debris. Studies have been conducted on the utilization of biomass wastes to treat water and sewage. It may also be hard to get rid of the large quantities of feedstock that are generated. Heavy metals can be removed from water using these wastes.[14]

4.3 Treatment of Industrial Sewage Using Adsorbents

Due to its effectiveness in removing persistent pollutants that are impossible to remove with a variety of conventional methods, adsorption has become increasingly popular as a method of dealing with hydrodynamic waste that is less expensive.[15] The word "adsorption" was coined by Kayser in 1881.[16] Whenever the molar mass of a substance is greater than that of the dominant species, it adsorbs on top of the dominant species.[17] Kayser published his primary quantitative research method in 1773 on the assimilation of gases by clay and charcoal.[16] In 1785, Lowitz conducted a follow-up study in order to establish the degree of decolourization caused by carbon.[18] Adsorption is the process whereby components, such as gases, water vapor, and fluids, voluntarily come together at a surface without undergoing any reaction (Figure 4.1).[19] Physical and chemical treatment of sewage is an imperative part of the process, as it is a method of handling the increasing

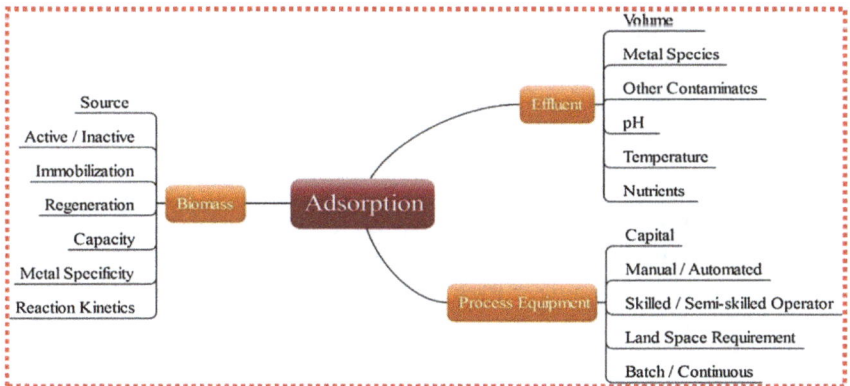

Figure 4.1 The adsorption process of heavy metals. Reproduced from ref. 19 with permission from Elsevier, Copyright 2021.

wastewater treatment demands and reusing water efficiently.[20,21] It has been suggested that physicochemical approaches could be used to remove almost all natural and inorganic pollutants from industrial effluents. Both physical and chemical methods are effective and cost-effective only when textile effluents contain high concentrations of solutes.[3] For heavy metal removal, adsorption has been widely utilized since the 1990s as a physicochemical procedure. It also uses green energy and is flexible in its design to ensure the highest quality of wastewater treatment. Regenerating the desiccant is also possible.[10,22,23] Using adsorbents, the water processing costs can be reduced from \$200 to \$5 per cubic metre. On the other hand, other methods might cost from \$10 to \$450 per m^3 of treated water.[24] A significant downside of the adsorption method is the increasing cost of adsorbent materials, which increases wastewater treatment costs.[25] Furthermore, the regeneration rate of adsorbent materials is low, which limits their use in the treatment system. Adsorption of heavy metals should be based on low-cost, high-efficiency, and environmentally friendly adsorbents.[17,26] In Malaysia, coal is usually used as an adsorbent for surface- and wastewater treatment.[27] The adsorption procedure involves waste water, adsorbents, and metal ions for removal of heavy metals. For the removal of heavy metals and dyes, adsorption has been extensively investigated. The distinctiveness and structure of adsorbents play important roles in diverse applications, according to a comprehensive bibliometric review of published literature on the adsorption process.[28]

The adsorption rate, large surface area, particle density, tunable pores, and diameter of adsorbents can all affect the elimination of metal ions.[29] Biochar and carbon black can be utilized to soak up heavy metals effectively.[30] Sorption capacity is determined by a diversity of factors, together with the interaction time or retention time, the stability of a solution (adsorbate) in a fluid (wastewater), the affinity of the solvent to the adsorbent (carbon), and several activated carbon atoms in the carbon.[31,32] According to Dehariya and Lal (2018), the pH, temperature, interaction period, metal ion concentration at the beginning, adsorbate concentration, and agitation rate can all contribute to the optimal treatment of metal-contaminated wastewater. In the case of heavy metal adsorption from aqueous solutions, pH plays a significant role in influencing the chemical diversity and surface charge of the adsorbent.[33] Based on the adsorption hypothesis, the adsorption efficiency decreases as particles trapped on the surface desorb from the façade side at increased temperature. Temperature gradients influence ion adsorption, indicating some ion-exchange activity.[34,35] It has been reported that a temperature of 40 °C is optimal for adsorption (range: 10 to 45 °C). This temperature increases the reactivity of the exothermic and spontaneous adsorption processes by increasing the solubility of compounds in treatment methods for sewage. The maintenance of kinetic energy is easier when the temperature is lower. The process in concern, according to Dixit *et al.* (2014), is physical adsorption. Adsorption occurs as a physical or chemical process at various temperatures and in a variety of media.[36] In an exergonic adsorptive system, reducing the temperature will increase metal cation elimination, according to H. Gupta (2016), as due to possible padding impacts, the implosion strength of bubbles decreases at elevated temperatures, and thus adsorption decreases.[37]

During the process, desiccants most likely affect adsorption capability, so the low temperature directly affects adsorption.[38,39] According to Tiadi *et al.*,[40] concentrating adsorbents increases the accessibility of additional catalyst surfaces for sorption, allowing heavier metals from aqueous solutions to penetrate the surface more easily. By starting with large quantities of metal ions, the adsorption capacity can be increased. This is because metal ions are the key factor in overcoming the transition barrier between the solid and liquid phases.[41] Furthermore, stirring led a crucial role in the adsorption procedure. Due to this factor, the solute distribution in the solution phase and the configuration of the outside boundary film are affected.[42] Commonly speaking, agitation regulates ion elimination, while swirling increases the rate of absorption. As a result of high speeds and intense stirring, the adsorbent disperses readily in solution, and mass transfer is reduced.[43,44] It has been reported that adsorption efficiency is influenced by the ratio of adsorbent to adsorbate. Adsorption is affected by the intermediates that are used to manufacture the adsorbent, the intrinsic properties of the sorbent material, the type of adsorbate, and the environmental solution chemistry.[37]

4.4 Removal of Organic Dyes

Arecaceae by-products were employed to prepare colour adsorbents for use in the colour treatment of dye sludge in this work, which resolved the problems of waste management for preventing water pollution. Recovering and recycling garbage not only provides social and economic benefits but also helps to preserve nature. Kinetic and isothermal models and impacting factors are shown in Table 1.

Nethaji *et al.* synthesized AC extracted from the *Borassus aethiopum* floral part and used it as a malachite green (MG) dye adsorbent in 2010. The sorption capacity was measured using three distinct crystallite sizes of 100, 600, and 1000 m, a 10 g L^{-1} concentration, pH 6.78, and a temperature of 300 K. In comparison to the Temkin and Freundlich models, the Langmuir isotherm model was found to be the best fit for the adsorption data. Biosorption concepts, the Elovich model, the pseudo-first-order method, and the pseudo-second-order method were utilized to examine the adsorption behaviour and calculate the adsorption rate constant.[50] Kini *et al.* employed palm tree male flowers (PTMF) as an agent for removal of methylene blue (MB) from an aqueous medium in 2014. For biosorption tests, medium PTMF granules with a size of 150 μm were utilized. The soaking efficiency of MB dye increased from 40.75 to 91.65% when the adsorbent concentration was increased from 0.05 to 0.30 g at equilibrium. Furthermore, the pseudo-second-order kinetics model was found to be adequate for the MB adsorption rate, with a high correlation coefficient. The Langmuir isotherm was used to explain the adsorption isotherm of MB onto PTMF, which had an adsorption capacity of 157.3 mg g^{-1} at 323 K.[51] Jawad *et al.* employed

Table 4.1 Optimum conditions and treatment efficiency of AC produced from Arecaceae species.

Adsorbent	Pollutant	C_0 (mg L^{-1})	Dose (g L^{-1})	Temp. (K)	Time (min)	Adsorption capacity (mg g^{-1})	Adsorption efficiency (%)	Adsorption model	Ref.
Palmyra palm fruit seeds	Cadmium	20	—	303	300	—	92.5	PFO and Langmuir	45
Areca nut seeds	Fluoride	20	20	—	180	4.8	75	Langmuir	46
Date palm leaflets	Ciprofloxacin	200	0.5	298	880	6100	83	Langmuir and PSO	47
Coconut fronds	Carbofuran insecticide	250	0.2	303	240	—	80	Langmuir and PSO	48
Borassus flabellifer shells	Naphthalene	200	—	313	720	—	76	Langmuir and PSO	49

fallen coconut leaves, a readily available form of agricultural waste, as a soaking agent for cationic dye (MB) from a solution phase in 2015. The leaf bits and pieces were ground into a powder with particle sizes ranging from 150 to 212 microns. The best parameters – a stirring duration, initial MB concentration, adsorbent dose, and initial solution pH of 5.00 min, 19.01 mg L^{-1}, 1.26 g L^{-1}, and 8.65, respectively, resulted in an MB adsorption efficiency of up to 86.38%. A comparison of the correlation coefficients for the pseudo-quadratic model and the pseudo-first-order model under optimal conditions of analysis showed a superior correlation coefficient for the pseudo-quadratic model.[52] Coconut shells (CSs) continue to be one of the most accepted materials for the construction of AC. In 2016, to prepare AC from CSs, Islam *et al.* used potassium hydroxide as the activator. 100% of the methyl orange dye was adsorbed for 12 minutes, and the mechanism followed the pseudo-second-order adsorption model ($R^2 > 0.995$).[53]

In a study published in 2019, Youssef *et al.* conducted research using by-products from date palm rock bottom used in tartlet factories in Shubra Al Khaimah in the preparation of AC. A 24 hour equilibration time was confirmed as optimal for MB removal, in conjunction with pH 7, and increasing temperature with equilibration time. In addition, the correlation coefficient reached 0.9943 for the MB adsorption of AC based on the pseudo-second-order kinetics model. Based on the Langmuir isotherm model, three adsorbents (CP212, CP214, and CP124) were probed for MB adsorption, and the adsorption capacities were determined to be 19.2, 20, and 80 mg g^{-1}, respectively.[54] Saudi Arabia had 33 million date palm trees in 2021, which will remain an irreplaceable symbol for locals for years to come. Alhogbi *et al.* utilized palm tree fiber (PTF) to prepare AC (Figure 4.2). The AC was used to remove an anionic dye (congo red [CR]) as well as a cationic dye (Rhodamine B [RhB]) from polluted water. Pseudo-second-order reactions and Langmuir models can both be fitted to adsorption kinetics and isotherm studies. PTFAC removed RhB (99.86%) and CR (98.24%) efficiently and it

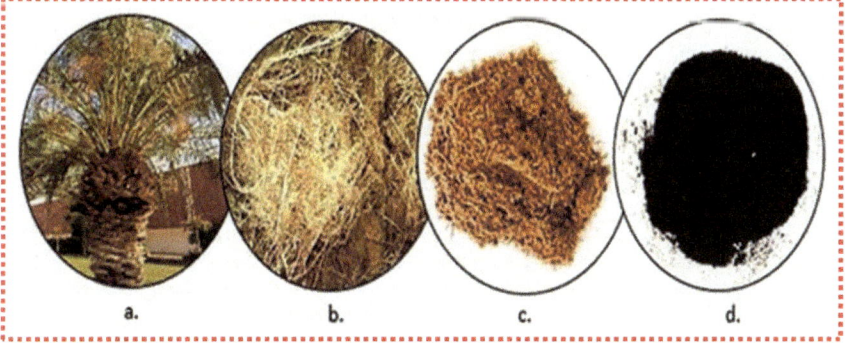

Figure 4.2 (a) Date palm tree, (b) palm fiber, (c) palm fiber powders, and (d) AC powder. Reproduced from ref. 55, https://doi.org/10.3390/pr9030416, under the terms of the CC BY 4.0 license https://creativecommons.org/licenses/by/4.0/.

could be reused up to five times.[55] Muniyandi *et al.* investigated the adsorption ability of MB dye from textile wastewater using palmyra shell biochar (palm). An optimal experimental process was achieved with 50 minutes time, 55 mg initial concentration, a pH of 10, and a dosage of 5 g of adsorbent. The cationic MB dye adsorbed to the surface of the adsorbent almost completely due to dielectric repulsion between the adsorbent's surface and the adsorbent's pores.[56] The adsorption ability was not affected by temperature, concentration, time, or content. The adsorption efficiency for the cationic dye was about 80% at pH > 6. As for the anionic dye, pH < 6 gives an adsorption efficiency of about 90%. Historically, the pH of the medium has been a dominant factor in determining the adsorption process. Despite the porous structure and the huge surface area of the adsorbent, the efficiency of the process did not change significantly.[56]

4.5 Detoxifying the Environment of Toxic Organics

The Rebecca Manesco Paixao team used the AC derived from Babassu coconuts in 2017 to eliminate nitrate from water. At 45 °C, the adsorption capacity was observed to be 10.13 mg g^{-1}. A pH of 2, a material weight of 0.2 g, and a period of 90 min were observed to be the most effective adsorption conditions. A nitrate concentration of 100 mg L^{-1} was also found to be the most effective.[57] Mohammed Hassan in his research used the material based on AC-based membranes for removing *Escherichia coli* microbes from water (Figure 4.3). The membrane filters were effective at removing *E. coli* bacteria (96–99%).[58]

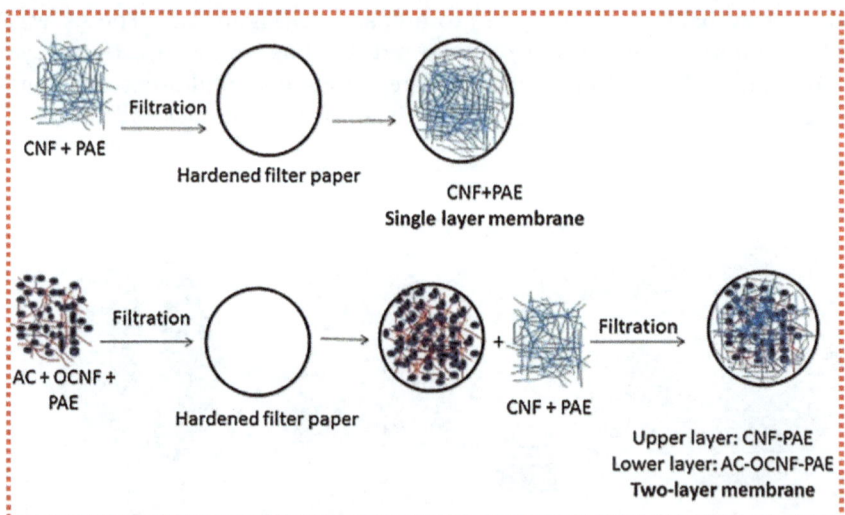

Figure 4.3   Illustration of single-layer and double-layer membranes from cellulose nanofibers, oxidized cellulose nanofibers, and AC, and polyamide-amine-epichlorohydrin (PAE). Reproduced from ref. 58, https://doi.org/10.3390/polym9080335, under the terms of the CC BY 4.0 license https://creativecommons.org/licenses/by/4.0/.

The Abdul Hai group used AC extracted from date seeds to remove NaCl from saline solution in 2019. An adsorption capacity of 22.5 mg g^{-1} was recorded when 5–10 minutes time, 250 mg L^{-1} concentration, and 2 g L^{-1} dosage were taken into account. Adsorption occurred on a continuous column with a flow rate of 10 mL min^{-1}.[59] The Adeline Lim team used AC from oil palm trunks to remove tannins from aqueous solutions in 2020. According to the kinetic and isothermal models of adsorption, the material was evaluated for its manufacturability. It was found that the AC sample fitted the Freundlich and pseudo-first-order models well. There was a maximum adsorption capacity of 1047.47 mg g^{-1} at pH 2 and 1087.28 mg g^{-1} at pH 4. The material displayed good performance in acidic environments.[60] Using date AC, Sahmarani Rayane in 2021 treated water contaminated with organochlorine pesticides with continuous column adsorption. When 50 grams of adsorbent, natural pH, a flow rate of 2.5 mL min^{-1}, and an inlet concentration of 57 mg L^{-1} were utilized for the adsorption procedure, an optimal adsorption capacity between 70 and 100% was achieved.[61]

4.6 Removal of Antibiotics

The use of antibiotics has not been limited to treating and preventing human diseases. Antibiotics have been used for extravagance and avert diseases in animals and plants and to promote livestock growth.[62,63] Large amounts of antibiotic residues are released into ecosystems from all of the aforementioned activities. This is similar to the problem of heavy metal pollution. The same holds for antibiotics, which are naturally occurring compounds found in many ecosystems. Antibiotics increase bioavailability, thereby polluting ecosystems when they become readily available to humans. Hence, alternatives have been sought. To meet interchangeability, alternatives must be available at a low cost, for example, soot, ebony, coffee dregs, rice bran, sugarcane, peach pits, fish, manure squander, and rubber tire waste.[64–66] Xinbo Zhang *et al.* (2020) made bio-adsorbents from coffee ground waste to remove antibiotic sulfonamides (SAs). The two adsorbent materials hydrochar (HC) and biochar (HC) differed significantly in their adsorption performances (Figure 4.4). The utmost adsorption capacity of HC was 130.1 l g^{-1} for sulfamethoxazole (SMX) and 121.5 l g^{-1} for sulfadiazine (SDZ) compared to BC, which was 85.7 l g^{-1} for SMX and 82.2 l g^{-1} for SDZ at 298 K.[67]

According to another study, the maximum tetracycline (TET) removal capacity was 20.40 mg g^{-1} in a pH 7 environment at 500 °C. TET adsorption is governed by hydrogen bonding, electrostatic interactions, and π–π electron donor–acceptor (EDA) interactions.[68] By using organic waste as an adsorbent, biochar has been used to get rid of a large array of antibiotics from water. Accordingly, there was over 90% removal efficiency for clarithromycin (CLA), 70% for TET, and 15% for ampicillin (AMP). Biochar from coffee grounds did not bind to the three germicides SMX, ofloxacin (OFL), and trimethoprim (TMP).[69] The latest study conducted by Pei-Ling Yen *et al.*

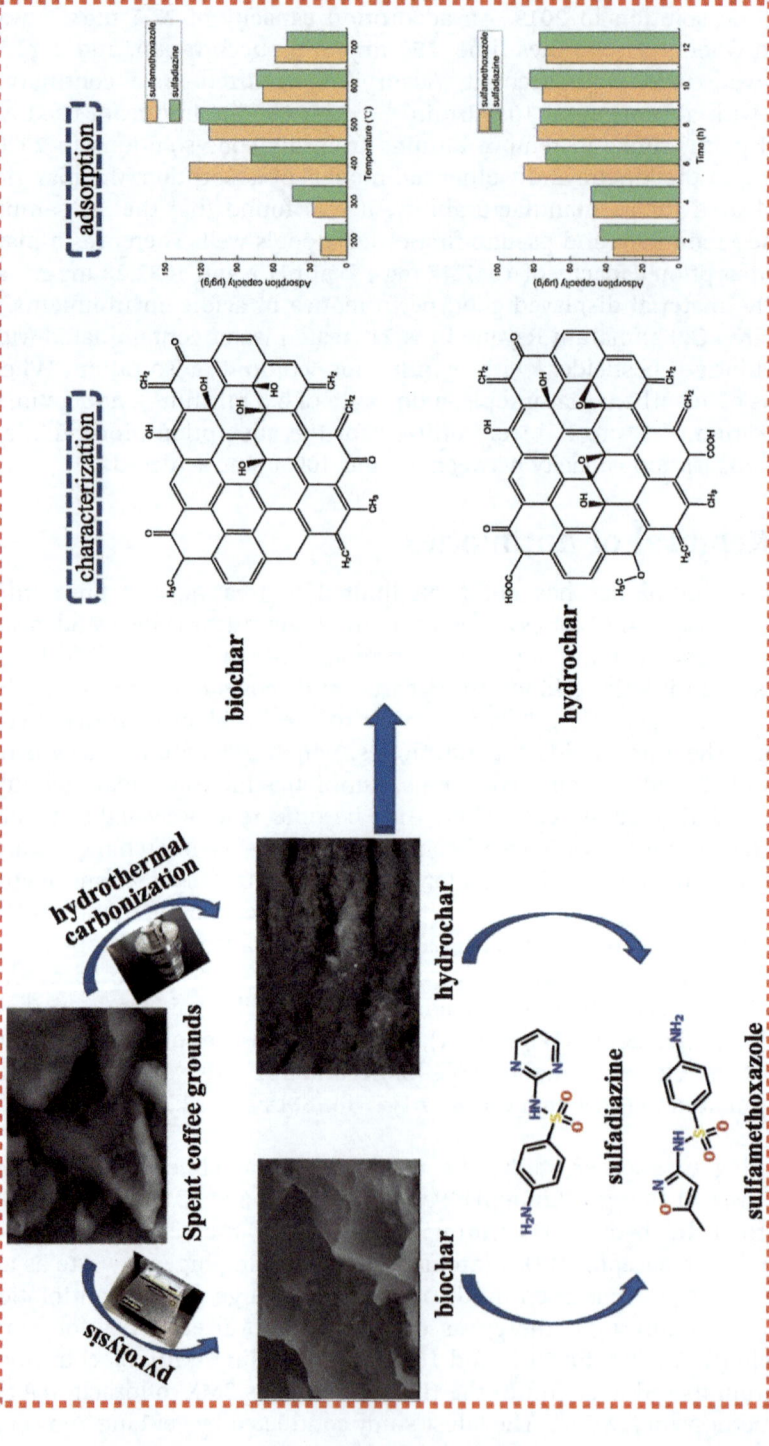

Figure 4.4 Illustration of the sulfonamide antibiotic adsorption capacities of spent coffee ground-based biochar and hydrochar. Reproduced from ref. 67 with permission from Elsevier, Copyright 2020.

(2021) showed that coffee grounds can be used effectively to remove norfloxacin (NOR) from water. The pH, adsorbent dose, and NOR concentration factors for the best adsorption efficiency were found to be 6.26, 1.32 gL^{-1}, and 24.69 mgL^{-1}, respectively, and the NOR removal efficiency reached 99.85%. The Langmuir isotherm model was used to explain the adsorption capacity of the coffee grounds, with a correlation coefficient of 0.974, and the utmost adsorption capacity of 69.8 mgg^{-1}.[70] Antibacterial-adsorbed coffee brans have not been premeditated in-depth, and hardly some papers have been in print in recent times. It would be useful to conduct studies on the decontamination of antibiotic contamination based on the use of biosorbents from coffee grounds, which may provide a solution to this problem.[71]

4.7 Gas Adsorption/Storage

One of the most significant properties of charcoal is its ability to adsorb gases. Renewable and widely available biomass can be used to make AC at a low price. This adsorbent is extremely selective for CO_2, can be readily regenerated, and has excellent moisture stability, unlike zeolites and MOFs.[72] Carbon adsorbents obtained from a range of lignocellulosic resources can be used for CO_2, CH_4, and H_2 storage. Several possible strategies have been identified to regulate and decrease atmospheric CO_2 levels, such as carbon dioxide sequestration and capturing from point sources (power plants, fuel extraction, dispensation, transportation, combusting biofuels, retail use, and industrial operations). CO_2 can be separated and recovered through a variety of processes, including solvent absorption, the Arctic approaches, ultrafiltration, solid bioreactors, and temperature fluctuation adsorption.[73] To date, chemical absorption with amine- and ammonia-based sorbent materials have acknowledged an immense pact of consideration due to its high productivity improvements, and is therefore extensively used by manufacturers.[74] CO_2, NO_2, CH_4, and halon discharges can be lowered on a massive scale by developing efficient systems for detaching and collecting CO_2. The success of this strategy depends, however, on creating a recyclable and durable desiccant with a high CO_2 adsorption efficiency. For many years, it has been known that the surface morphology of charcoal affects its adsorbent dosage. It has been suggested that adding Lewis bases to porous carbon surfaces may improve the carbon's ability to collect CO_2 since CO_2 is acidic (weak Lewis acid).[73,74] In contrast, methane, though on a different scale, has attracted much attention as a possible form of clean energy for the future because of its abundant availability, overall safety (as compared to most fuels), economical nature, and low carbon footprint. There are several compound classes which have been identified as potential candidates for portable storage in future transportation solutions. These compounds should have high storage capacity, fast storage and release kinetics, and long-term cycle efficiency under moderate thermodynamic conditions. According to the US Department of Energy, the methane adsorption capacity is 180 v (STP)/v (standard temperature and pressure equivalent volume of methane per volume of adsorbent: 35 bar and ambient air temperature).[75] The reason for this is that there are many types of charcoal, hydrates, and metal–organic framework (MOF) compounds with exceptionally porous structures and specialized mesopore thicknesses that are suitable for meeting such conditions. Several MOFs have been reported to have a large surface area (3000 $m^2 g^{-1}$) which can be polymerized, as well as walls with pores that can be polymerized.[76]

Despite this, there was still some degree of destabilization caused by ionic or hydrophobic interactions.[77] Since they are a novel class of substanceThs. Recent results demonstrate that storage of CH_4 using ACs generated from wood pulp, coconut fibers, or resins can be similar to that of liquefied petroleum at 250 bar, with an exceptional uptake of 200 v/v at 27 °C and the pressures ranging from 34 to 48 bar.[75] Additionally, hydrogen is a promising biofuel. In recent decades, fuel cells have been industrialized largely because of hydrogen vehicles, in terms of both manufacture and storage. The US DOE established a benchmark of 5.5 wt% H_2 and 40 g H_2 per L for gravimetric and volumetric system targets for 2017.[78] Collection boxes with sufficient volumetric and gravimetric densities are required for portable hydrogen storage systems to run fuel cells utilizing H_2. As a result, the issue is not just developing efficient fuel cells or a lucrative technique for producing H_2, but also storing it. Every vehicle must be equipped with a safe, inexpensive, and small onboard H_2 accumulator. Compression, cryogenic and cryo-compressed storage, metal hydrides, and large surface area sorbents (for example, charcoal) are now used for H_2 storage and transfer. Many metals, alloys, and intermeshed metallic compounds may also reversibly adsorb substantial quantities of H_2. But, none of these has been proven to be useful for portable reservoirs.[79] For the time being, there is no reliable method for predicting the adsorption of H_2 into carbonaceous materials.[80]

References

1. F. Derbyshire, M. Jagtoyen and M. Thwaites, *Porosity in Carbon*, Edward Arnold, London, 1995, p. 227.
2. M. Inagaki and J. Tascón, *Interface Science and Technology*, Elsevier, 2006, vol. 7, pp. 49–105.
3. M. A. Kamaruddin, M. S. Yusoff, H. A. Aziz and C. O. Akinbile, *Int. J. Sci. Res. Knowl.*, 2013, **1**, 60–73.
4. M. E. Goher, A. M. Hassan, I. A. Abdel-Moniem, A. H. Fahmy, M. H. Abdo and S. M. El-sayed, *Egypt. J. Aquat. Res.*, 2015, **41**, 155–164.
5. S. Gunatilake, *Methods*, 2015, **1**, 14.
6. M. Zhao, Y. Xu, C. Zhang, H. Rong and G. Zeng, *Appl. Microbiol. Biotechnol.*, 2016, **100**, 6509–6518.
7. A. A. Al-Gheethi, I. Norli, J. Lalung, A. Megat Azlan and Z. Nur Farehah, *Clean Technol. Environ. Policy*, 2014, **16**, 137–148.
8. A. S. Naje, S. Chelliapan, Z. Zakaria, M. A. Ajeel and P. A. Alaba, *Rev. Chem. Eng.*, 2017, **33**, 263–292.

surface area. One gram of active carbon has a surface area of 3000 m^2, or 32 000 sq. ft (calculated by gas adsorption analysis). The specific surface area of charcoal, or non-AC, ranges from 2.0 to 5.0 m^2 g^{-1}. By attaining a high surface area by activation and/or by altering its surface characteristics and functionalization through chemical treatment, AC's prospective uses may be adequately suited. Coconut husks and carbon-based wastes from paper mills are two typical sources of AC. Activated coal and activated coke are the names for the ACs made from coal and coke, respectively. Numerous industries, including those related to the environment, fuel storage, analytical chemistry, gas adsorption, chemical purification, agriculture, gas storage, skincare, the purification of alcoholic beverages, mercury scrubbing, food additives, *etc.*, could benefit from the use of AC (Figure 5.1). The current chapter provides a thorough explanation of the uses of AC in the environment.

By physically or chemically activating carbonaceous sources such as coconut husks, coal, lignite, bamboo, willow peat, wood, coir, and petroleum pitch, AC may be manufactured. Physical activation can be accomplished by oxidizing the materials in the presence of oxygen at temperatures greater than 250 °C, typically 600–1200 °C, or by a process known as carbonization, which involves the pyrolysis of carbon-rich materials at high temperatures, typically 600–900 °C, under inert conditions, such as in a nitrogen or argon atmosphere. The most common way to achieve physical activation using the oxidation approach is to heat carbon-rich materials for an hour at 450 °C in a muffle furnace. By impregnating carbon-rich materials with appropriate

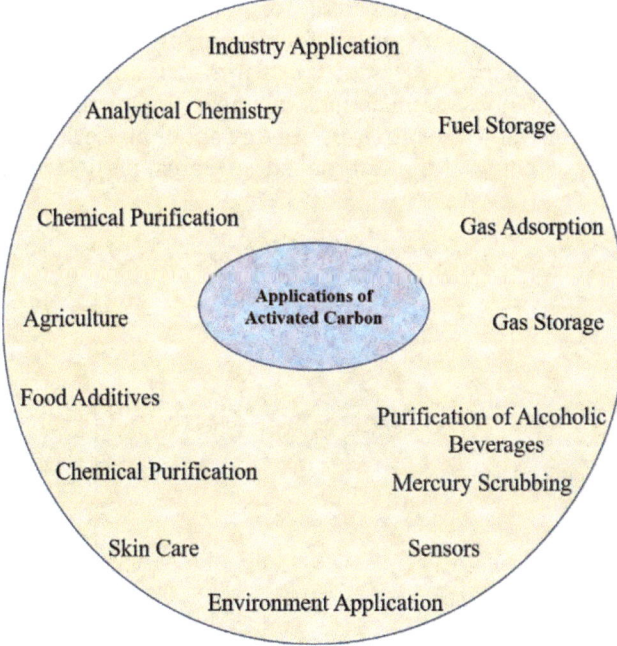

Figure 5.1 Numerous applications of AC.

chemical species, primarily strong acids, strong bases, and their salts (5% NaOH, 5% KOH, 5% CaCl$_2$, 25% H$_3$PO$_4$, and 25% ZnCl$_2$), one might accomplish chemical activation. After that, the newly created carbon is heated at a high temperature, typically between 250 and 600 °C, to activate it. The carbon is activated by heating, which makes it more porous and creates more openings. Because of their high consistency, low operating temperature, better and more uniform activation, and quicker response activation time, chemical activations are preferable over physical ones.

Based on their properties, surface features, nature, and other factors, ACs are very complex materials that are challenging to categorize. Based on their use, synthesis, and size, various general categories have been suggested. They may be broadly grouped into the following categories: granular AC (GAC), powdered AC (PAC), bead AC (BAC), extruded AC (EAC), polymer-coated carbon (PCC), impregnated AC (ImC), and woven carbon (WoC) (Figure 5.2). Nearly all of the carbon particles in PAC, a finer kind of AC, may pass through a sieve with a chosen mesh size. PAC is formed of crushed or ground carbon particles. According to the ASTM classification, PAC refers to substances or particles that can fit through an 80-mesh sieve and have diameters of 0.177 mm or less. Other process units such as fast mix basins, clarifiers, raw water inputs, and gravity filters are often combined with PAC. Compared to PAC, GAC is larger and has a smaller external surface area. Rapid gas diffusion is made possible by GAC, which makes it effective as vapour and gas adsorbents. It might also be used to separate chemical species from flowing systems and quick mix basins, as well as for air water treatment and filtration. Different sizes of GAC are indicated; some are helpful for liquid phase applications, while others are useful for vapour phase applications. As an example, GAC with particle sizes of 8×30 mm, 20×40 mm, or 8×20 mm is helpful for liquid phase applications, whereas GAC with particle sizes of 4×10 mm, 4×8 mm, or 4×10 mm is useful for vapour phase applications. Due to the optimal balancing of the external surface area, particle size, and head

Figure 5.2 Synthesis and classification of AC.

loss properties, the most common and practical GACs for use in aqueous phases are those with particle sizes of 8×30 mm and 12×40 mm.

EAC is the result of combining PAC with the appropriate binder. EAC is an AC block (ACB) with a diameter ranging from 0.8 to 130 mm. Due to its low dust concentration, high mechanical strength, and minimal pressure drop, EAC is mostly used to purify gas. EAC is marketed as a CTO (*i.e.*, chlorine, taste, and odour) filter in addition to other ways. Similar to EAC, BAC has high mechanical strength, low pressure drop, and low dust content, but its particle size is smaller than that of EAC. BAC is mostly formed from petroleum pitch and it acquires sizes between 0.35 and 0.80 mm. Water purification and filtration processes often employ BAC. Silver and iodine are two inorganic impregnates found in the matrix of ImC. Numerous cations, including Al^{3+}, Mg^{2+}, Zn^{2+}, $Fe^{3+/2+}$, Li^+, and Ca^{2+}, have been included to create various ImC for a variety of purposes, particularly for the reduction of air pollution. The good antibacterial and antimicrobial properties of the silver-loaded ImC make them excellent choices for the purification and filtration of residential water. H_2S and thiol-based molecules are both often eliminated *via* the ImC. The potential for H_2S adsorption on the ImC is as high as 50%. To create PCC, porous carbon-rich materials are coated with a biocompatible polymer to create a smooth, permeable covering devoid of obstruction-causing pores. Hemoperfusion benefits from PCC. The AC made from rayon fibre is referred to as WoC. Compared to activated charcoal, WoC, sometimes referred to as carbon cloth, has a considerably greater adsorption capacity. WoC has a surface area that ranges from 500 to 150 $m^2 g^{-1}$ and pore volumes that range from 0.3 to 0.8 $cm^2 g^{-1}$.

5.2 Environmental Applications of AC

One of the major problems facing the world today is the environmental pollution brought on by industrial waste from the textile industry. Chemical contaminants such as pigments, dyes, sizing agents, heavy metal salts, softening agents, complexing agents, and stiffening agents are all present in textile effluents. The wastewater from the textile industry has a very dangerous effect on the environment, which hurts human life. Utilizing textile waste directly to dispose of it safely and turning it into smart adsorbing materials like ACs are two creative recycling methods. ACF is often utilized as sophisticated adsorbent materials and smart filters to adsorb and eliminate numerous textile industry pollutants because of its enormous surface area and extensive pore size distribution. Fibre wastes, including acrylic, cotton, and viscose fibres, may be used to make ACFs. Near the end of the 18th century, charcoal's adsorptive properties were first noted. It was found that charcoal may remove the colour from several drinks. The first industrial application of charcoal was made possible by this discovery in a sugar refinery in England in 1794. AC is employed in a variety of industrial processes nowadays, like gas and air cleaning using conventional reusable material recovery methods. New applications have developed as a result of increased environmental consciousness and the implementation of tight emissions regulations, most notably in the field of air pollution removal. Additionally, AC is used more and more for the treatment of water, including service water, drinking water, groundwater, and wastewater. Its main function here is to adsorb dissolved organic impurities and other organic contaminants that impact odour, taste, and colour. AC is used to alleviate environmental issues affecting air and water, and new uses are always being discovered in quick succession. The reactivation and subsequent reuse of used AC and the recycling of AC that cannot be renewed or reactivated are of special importance in the area of environmental protection. The treatment, filtration, and decolorization of liquids is yet another vast area of use for AC and is crucial in the food, pharmaceutical, beverage, and other sectors.[1]

Aniline, caffeine, suspended particulates, and other contaminants may all be removed with AC. Isocyanate is created by using aniline, a major component in the petrochemical sector. Additionally, it is utilized in the manufacturing of antioxidants and accelerators and the pharmaceutical, herbicide, and rubber industries.[2] The alkaloid caffeine, which may be found in coffee, tea, chocolate, soft drinks, and certain types of sweets, is utilized as a stimulant for the heart, brain, and respiratory system. Its average daily intake is roughly around 200 mg per person, and if consumed, about 1 to 10% is eliminated *via* urine when processed within the body.[3] ACs from apricot stones that were chemically activated with $ZnCl_2$ and H_3PO_4 were used for removing aniline, and the former was found to perform better than the latter, with surface areas of 1111 and 1382 $m^2 g^{-1}$, respectively.[2] Coconut shell AC, durian shell AC, orange peel AC, and coffee residue AC could adsorb benzene (212.77 $mg g^{-1}$), and remove toluene (874 $mg g^{-1}$), *p*-nitrotoluene (20 $mg g^{-1}$), and formaldehyde (245 $mg g^{-1}$), respectively.[4] With a suitable surface area of 1031 $m^2 g^{-1}$, the phosphoric acid-activated pineapple leaf AC absorbed caffeine (155.5 mg per gram).[3]

5.2.1 Water Treatment

Due to the scarcity of freshwater, AC has been utilized more and more to remove impurities and pollutants from water. The use of liquid phase or aqueous solutions is said to be the case for over 80% of all ACs worldwide.[5] ACs are known to be among the most efficient and significant adsorbents for inorganic and organic contaminants. The primary reasons for using ACs are their adaptability in terms of adsorption capacity, surface area, and porous structure.[6] Electric charges are generated owing to the interaction of functional groups of AC with the ions present in the aqueous medium. Depending on the pH of the solution and the adsorbents' surface properties, the dissociation or intensity of attraction will change.[4] Many studies have lately recognized the application of nanocomposites to eliminate contaminants from water by incorporating nanoparticles on ACs to improve adsorption efficiencies and absorption capacities. The removal of dye and metallic contaminants is the principal application for ACs, which were loaded with tungsten oxide nanoparticles,[7] zero-valent iron nanoparticles,[8] iron oxide nanoparticles,[9] and Ag nanoparticles.[10] The pH of the solution is among the most crucial factors affecting the charge density as well as the cationic or anionic nature of the surface. As a result, solutions with high pH values will attract cations, whereas those with low pH values will do the opposite. As a result, adjusting the pH of the solution may affect the absorption of the charged inorganic groups. Acidic carbons will be better at holding onto cations, whereas basic carbons are efficient at removing anions.

5.2.1.1 Elimination of Inorganic Pollutants

(a) Elimination of Metallic Contaminants

Metallic pollutants, like metalloids and heavy metals, pose a threat to water resources because they are non-biodegradable and prone to biomagnification in food chains. Their presence is harmful and may build up in living things or the human body, leading to long-term health problems and illness.[11] Additionally, they feature intricate and microscopic structures that are problematic since it is hard to remove them. The primary sources of these metals are industrial and agricultural processes including mining, fertilizer use, electroplating, and vehicle production.[12] Numerous types of research have shown the utilization of ACs for the quick and inexpensive elimination of metallic contaminants.[11] As a result, ACs are often used to remove metals from water, including lead(II), cadmium, iron, copper, and chromium(VI).[13] Due to the tiny size of metallic pollutants and their repeated charging in solution, which lead to significant interactions with AC during the electrostatic adsorption process, the elimination of metallic contaminants from water is a straightforward process.[14] The porosity and surface area, the interaction between the structure and the metal ion, the pH and point of zero charge of the solution, the size of the ion, and the functioning of surface oxygen are all aspects affecting the adsorption mechanism on AC. Dispersive forces, electrostatic forces, and non-electrostatic forces interact with one another when the adsorbent and adsorbate are in contact with each other, and these forces are crucial to the adsorption process. According to a review of the literature, *Acacia mangium* AC has a chromium removal capacity of 37.16 $mg\,g^{-1}$, compared to *Acacia nilotica* bark AC's removal rate of 93.1%.[15,16] One of the most significant elements used in industrial production is lead, which is used to make pipes as well as additives and accumulators for paints and fuels.[17] Instances of lead doses up to 2 $mg\,l^{-1}$ discharged into drinking water from domestic pipes due to the use of lead in pipework have been reported. According to reports, lead tends to build up in living things and causes poisoning symptoms in people.[18] As a result, in this instance, it is necessary to assess this dangerous contaminant that has to be eliminated from the water. Except for tamarind wood AC, which has a pH of 10, the lead adsorption solution has an average pH of 5. Tamarind wood AC formed by $ZnCl_2$ activation has the largest surface area, which is estimated to be 1322 $m^2 g^{-1}$,[4] and tropical almond shell AC produced by a combination of CO_2 and steam activation had a maximum adsorption capacity of 114.8 $mg g^{-1}$.[19] Lead reportedly disappears (100%) when exposed to watermelon peel AC, and it does so with a 99.8% elimination efficiency when exposed to oil palm shell AC.[20]

Highly toxic water is also affected by cadmium.[21] It is only allowed to have a concentration of 5–10 ppm in water since it is toxic to aquatic ecosystems and living things even at low concentrations.[22] Cadmium contamination is mostly caused by the manufacture of batteries, improper battery disposal, cadmium plating, and the use of fertilizers high in cadmium. These industries' waste effluents harm the environment and water resources.[11] With a high removal efficiency of 99.5%, oil palm shell AC was employed to remove cadmium.[20] ACs obtained from date pits when chemically activated at pH 6 and temperatures between 25 and 30 °C were capable of eliminating Cd, with the adsorption efficiencies ranging from 118.1 $mg g^{-1}$ to 127.0 $mg g^{-1}$.[23] Mercury is found in aqueous waste from industry, as much as lead and cadmium.[24]

Walnut shell AC with physically activated Fe[4,25] and *Ceiba pentandra* hull AC used for chromium removal were shown to have extremely poor adsorption capacities of less than 30 $mg g^{-1}$, while chemically activated chestnut oak shell AC and date pit AC[23,26] had greater adsorption capacities of 85.47 $mg g^{-1}$ and 120.5 $mg g^{-1}$, respectively. This demonstrates that chemical activation is preferred to enhance the adsorption process for the elimination of chromium. Copper and iron can only be eliminated using chemically activated watermelon peel AC, with the corresponding efficiencies of 91% and 99%, respectively, whereas physically activated date pit AC can hardly eliminate copper, iron, and aluminium.[20,23] Another research study that uses Fe_3O_4/CSAC (where CSAC is cigarette soot AC) as an adsorbent has shown that it is superior to CSAC alone for the elimination of As(v) and As(III).[9]

(b) Elimination of Non-metallic Contaminants

Anions may be helpful and are necessary for human bodies, but too much of certain anions, like molybdate, phosphate, and fluoride, can be dangerous and harmful to living things. Fluorosis has affected millions of individuals owing to excessive fluoride levels in drinking water, particularly in poor nations.[27] For example, Viswanathan *et al.* assessed the water's fluoride ions and determined that they were excessive (3.24 ppm), and they stated in their paper how a high fluoride concentration in the water is positively correlated with fluorosis in children. Similarly, molybdate may be carcinogenic to humans and is easily absorbed by grazing cattle and incorporated into water and food chains.[28] Therefore, to achieve a permitted level of concentration in the water, the concentration of anions should be carefully monitored and evaluated regularly.

The utilization of agricultural waste coir pith with $ZnCl_2$ activation for the elimination of anion contaminants from water such as phosphate, molybdate, and thiocyanate was explored in three distinct research investigations by Namasivayam and Sangeetha.[28-30] For the removal of thiocyanate, phosphate, and molybdate, the maximum adsorption capacities are in the range of 16–18 $mg\,g^{-1}$, 5.1 $mg\,g^{-1}$, and 4–8 $mg\,g^{-1}$ at an optimal pH of 3-4, respectively. The authors concluded that coir pith, being a cheap adsorbent, may be a useful precursor used to make ACs for the elimination of anion contaminants and that the adsorption capacities are impacted by ion exchange and chemisorption on the surfaces of the ACs.

5.2.2 Removal of Organic Pollutants from Water

5.2.2.1 Colour/Dye Removal

In a variety of sectors, including paint, textiles, cosmetics, and paper goods, dyes are widely used and applied. According to environmentalists, the textile sector is the main contributor to water contamination.[31] For every manufacturing, 2–20% of the 3600 different types of dyes used in the industry are lost in the effluent outflow. As a result, dyes are among the biggest and worst contaminants in wastewater. These effluents include hazardous and carcinogenic dyes that may contaminate rivers and other water resources. Their presence in the water might hinder aquatic life's ability to produce oxygen, which kills the marine environment. Additionally, consuming dyes may be harmful since doing so results in changes to the human body.[32]

The elimination of colours from aqueous solutions is a useful application for AC. The capacity for removal relies on the pore size, surface area, and functional groups of the AC, the solubility, polarity, and molecular size of the dyes; pH; and other ions in the solution.[33] The ability of ACs derived from *A. mangium* and *A. nilotica* to eliminate numerous dyes was studied. These ACs had a methyl orange dye removal efficiency of 90.5%[34] from cloth and a methylene blue adsorption capacity of 250 $mg\,g^{-1}$.[35] The most frequent dye to be eliminated from wastewater is methylene blue. According to reports, the majority of adsorption capacities exceeded 100 $mg\,g^{-1}$. KOH-activated distiller's grain AC, with a surface area of 1430 $m^2\,g^{-1}$, achieved the highest adsorption (934.579 $mg\,g^{-1}$) of methylene blue from an aqueous medium at pH 5.8 and 55 °C.[36]

Date pits have been utilized extensively in the manufacture of AC, and dyes including remazol yellow, methylene blue, maxilon blue, and methyl orange and have been removed using both physical and chemical activations. AC made from bamboo cane powder showed a tremendous adsorption capacity of 2600 $mg\,g^{-1}$ for the eradication of lanasyn orange from water.[32] The adsorption of methyl orange recorded on date pit AC generated by activation with $ZnCl_2$ was 434.8 $mg\,g^{-1}$, whereas methylene blue was best removed from aqueous solution using KOH-activated date pit AC, which adsorbed 316.1 $mg\,g^{-1}$ of dye.[23]

To remove the dyes rhodamine B and methylene blue from an aqueous medium, Tuan *et al.*[10] and Anfar *et al.*[7] studied the use of green nanocomposites made from almond and coconut shells activated with WO_3 and Ag nanoparticles, respectively. Rhodamine B elimination with WO_3/AC was high (adsorption capacity: 1666.67 $mg\,g^{-1}$) in comparison to methylene blue removal with Ag/AC (240 $mg\,g^{-1}$).

5.2.2.2   Elimination of Phenolic Compounds

Most phenolic compound-containing effluents are produced by the chemical and petrochemical industries. These contaminants pose a major threat to aquatic ecosystems and living things since they are very poisonous and carcinogenic. Therefore, phenolic compound concentrations in water bodies have been rigorously controlled by environmental and health protection authorities, and the removal of these contaminants is given priority. The maximum permissible levels of phenols in drinking water established by the World Health Organization (WHO) and Environmental Protection Agency (EPA) are 0.001 $mg\,L^{-1}$ and 0.1 $mg\,L^{-1}$, respectively. Surface chemistry is more important for the adsorption of the phenolic compound by AC as compared to porosity due to the significant increase in surface functionality in comparison to the distribution of pore size. The amount of phenol adsorbed decreases for both low and high pH values, depending on how the pH of the solution impacts it. Less adsorption occurs because the phenol's high pH and anionic nature cause the molecules to repel one another off the carbon surface. Acidic solutions with lower pH levels produce protons, which interact with carbonyl groups to reduce adsorption.[37]

Argan nutshell AC was used by Zbair *et al.*[38] to study the adsorption of Bisphenol A. According to results, the maximum adsorption capacity at 293 K was 1250 $mg\,g^{-1}$. In terms of adsorption, the nutshell AC performed better than commercialized ACs. Durian peel is a further precursor employed in this adsorption. Although the adsorption procedure took a day, the reported adsorption efficiency with durian peel AC was extremely low (4.2 $mg\,g^{-1}$), with a removal effectiveness of 69%.[20] For the adsorption of phenols, lignocellulosic biomass, like kenaf, rapeseed, avocado kernels, baobab wood, date pits, and *Acacia* seeds, has been utilized to produce AC. The baobab wood AC activated with H_3PO_4 was noted to be the greatest adsorbent among these ACs. At an ideal pH of three and 50 °C, the maximum adsorption was accomplished in 24 hours. According to Bansal *et al.*,[39] basic surface functional groups promote adsorption, whereas acidic surface functional groups on ACs may inhibit the adsorption efficacies of AC. Another work that used chemical activation of date pit AC with both H_3PO_4 and

KOH to remove phenolic chemicals (*o*-nitrophenol and *p*-cresol) provides support for this finding. AC treated with base improved the adsorption of phenolic compounds, while acidified AC was only capable of removing 142.9 mg g^{-1} from an aqueous medium.[23]

5.2.2.3 Elimination of Pesticides

Although pesticides are necessary for contemporary agriculture, their frequent use may harm the water's quality. The most widely used pesticides in farming and gardening are carbofuran, benzon, and 2,4-dichlorophenoxyacetic acid. They are regarded as risks to the well-being of aquatic life and humans because of their mutagenicity and carcinogenicity, and their highest levels in water are established at specific amounts. For carbofuran, bentazon, and 2,4-dichlorophenoxyacetic acid, the acceptable quantities in tap water, drinking water, and tap water are 0.09, 0.05, and 0.1 mg L^{-1}, respectively. These findings indicate that one of the best separation methods for removing pesticides from water resources is adsorption. The amount of organic matter and the continual rate of flow throughout the adsorption process, together with the carbon dosages, all affect the pesticides' ability to be absorbed by AC.

The adsorption capacities of physicochemically activated date seeds ACs were studied by Salman *et al.*[40] The values for the adsorption of carbofuran and bentazon, respectively, were 137.04 mg g^{-1} and 86.26 mg g^{-1} for date seeds that had undergone physicochemical treatment with KOH and CO_2. They attributed the discrepancy in adsorption capacity to the higher molecular size of bentazon compared to carbofuran, which facilitates facile adhesion and diffusion onto the surface of the AC. Similar outcomes from similar experiments were reported in different research studies with findings for bentazon and carbofuran. They also included a 2,4-dichlorophenoxyacetic acid adsorption experiment. This research demonstrated that AC could eliminate 175.4 mg g^{-1} of pesticides from an aqueous medium.[41] To create ACs for the elimination of bromopropylate, physically activated maize cob, olive kernel, soya, and rapeseed stalks were used. For each of the precursors employed in this investigation, the removal efficiency ranged from 90 to 100%.[42]

5.2.2.4 Pharmaceutical Compound Removal

Some of the most pervasive environmental toxins today are pharmaceutical substances which are utilized extensively in aquaculture, agriculture, and human life. Pharmaceutical use has grown over time, and this has resulted in continuous discharge into aquatic environments. Pharmaceutical substances are very stable and hydrophilic in water bodies, which makes them potentially persistent in water and environmentally sensitive. Even while drugs are normally present in very little amounts in water, they may nonetheless hurt the ecosystem. The removal of pharmaceutical molecules (pharmaceuticals) using AC relies on the bridging method, operating conditions, the kind of adsorbent as well as precursor, the activation method, the characteristics of the adsorbate, *etc*. Additionally, the adsorption of AC is greatly influenced by operational factors such as adsorbent dosage, adsorbate characteristics, operating temperature, pH of the solution, ionic capacity, and organic assembly. Most scientists have shown that the Langmuir and Freundlich isotherm models have been controlled by the adsorption of the medicinal ingredient by AC.

To develop ACs, mugwort leaves, mung bean husk, olive stones, and cork powder are activated chemically and physically. Steam-activated cork powder was shown to have the best results among these, removing 378.1 $mg\,g^{-1}$ of ibuprofen from water. When compared to the cork powder treated with K_2CO_3, the steam-activated powder was better. Olive stones that had been CO_2 activated also performed somewhat better than the other ACs, with an adsorption capacity of 282.6 $mg\,g^{-1}$ as opposed to the 200 $mg\,g^{-1}$ that the other adsorbents managed to have. For the elimination of naproxen, waste olive AC and waste apricot AC were also be compared. Apricot trash treated with $ZnCl_2$ and olive waste treated with H_3PO_4 managed to eliminate 106.4 $mg\,g^{-1}$ and 39.5 $mg\,g^{-1}$ of the drug in 1 hour and 26 hours of adsorption, respectively,[32] and SMX, CBZ, and PAR all showed excellent extraction efficiencies of 75, 85, and 84%, respectively.[43] Tetracycline, clofibric acid, and paracetamol decreased by 97.03, 83, and 84%, respectively, when waste textile (cotton) AC activated with H_3PO_4 was used.[44]

5.2.2.5 Drinking Water Purification

AC is often used in drinking water purification together with other treatment techniques including flocculation, oxidation, filtration, *etc.* When employed in this manner, AC may perform several tasks.

5.2.2.6 Adsorption of Solute Organic Matter

The primary method for removing organic solutes is adsorption. If the water to be treated includes biodegradable materials, settling microorganisms on AC may improve the filtration process. Water may include a variety of organic components from various sources. There are times when even ground and surface water contain humic particles in amounts high enough to alter the flavour. Even after bank filtration, surface water, especially river water, still includes various organic molecules, of which only a small fraction are often recognizable. Organic chemicals and chlorinated hydrocarbons from insecticides, herbicides, and pesticides employed in rigorous agriculture have been shown to contaminate groundwater to an increasing amount. This specific kind of man-made water pollution, in addition to altering the flavour and odour of the water, may also have long-term harmful or cancerous effects on people when ingested over an extended period. Utilizing AC can effectively remove even the tiniest residues of such dangerous compounds from water. For use in water purification applications, unique, highly active, abrasion-resistant, and readily regeneratable ACs have been produced.

5.2.2.7   Dechlorination/Deozonation

In water filtration systems, oxidizing chemicals including chlorine, hypochlorite, hydrogen peroxide, chlorine dioxide, and ozone are widely employed for the decontamination and removal of organic materials. The majority of the time, the water must subsequently be purified of extra oxidants. This may be done successfully by catalyzing the oxidants' breakdown on the surfaces of GAC. The dechlorination capability increases with the size of the AC particle.

5.2.2.8   Groundwater Cleanup

For groundwater remediation filtration, AC is often used to remove undesirable organic pollutants from the soil and groundwater. AC filter vessels are used in conjunction with sand filters to clean up polluted sites *via* remediation. VOCs and other odorous substances that may be emitted during the cleanup process may be eliminated using carbon filters. AC remediation and groundwater treatment will continue to be a significant and expanding market for AC filtration and associated services as "brown-field" lands are sought after for new residential and commercial projects. Unfortunately, experience has shown that organic solvents, volatile chlorinated hydrocarbons, and organic compounds employed for agricultural uses often contaminate groundwater. Tri- and tetrachloroethylene, 1,1,1-trichloroethane, and dichloromethane are the chlorinated hydrocarbons that are most often discovered in groundwater. Some of these compounds are so persistent that decades after being introduced into groundwater and distributed over a large region, they may still be found there. AC is usually often used in the rehabilitation of groundwater polluted with such dissolved organic compounds, for example, as part of the treatment of groundwater to produce drinking water. When there are significant amounts of contaminants present, a stripper unit may be added before the AC filter.

5.2.2.9   Service Water Treatment

The issues with service water treatment are essentially the same as those with drinking water treatment. Service water (cooling water, aquarium water, boiler feed water, condensate water, and pool water) is often considered to be utilized water that has to be treated before being used again.

(a) Boiler Feed Water Treatment
   To prevent corrosion, the water used to fill high-pressure boilers must be oxygen-free. The addition of hydrazine hydrate transforms the oxygen in the feed water into nitrogen and water (N_2 and H_2O). Filtration with GAC quickens this process in cold water.
(b) Condensate and Contact Water Treatment
   Lubricants may sometimes be present in a hot condensate. Before reusing the condensate, especially in high-pressure boilers, these oils

must be removed as completely as possible. Before treating the condensate with AC in cases of significant oil concentrations, it is recommended to de-oil the condensate by employing a mechanical oil separator. GAC may then be used for fine cleaning. Treatment of condensate water and that of contact water are closely linked processes. In systems that use solvent recovery, the steam regeneration process results in a condensate that is highly solvent-laden and typically has both organic and aqueous components. Solvents saturate the aqueous phase (contact water), which is regularly cleaned by passing it through layers of GAC.

(c) Treatment of Swimming Pool Water

Recirculation systems purify swimming pool water physically and chemically. For the treatment of swimming and bathing pool water, a variety of combinations of techniques (specified in DIN 19643-1 to 5) are available. Nitrogen–chlorine complexes and organic and halogenated organic compounds (AOX, THM) may all be effectively removed by filtration using AC. The remaining amounts of ozone and chlorine are also removed.

(d) Aquarium Water Treatment

Toxins from animal waste and decaying food pollute freshwater and saltwater aquariums' water supplies. Filtration with AC is a method that may be used to get rid of these contaminants and medicine leftovers.

5.2.2.10   Waste Water and Sewage Treatment

The relevance of AC adsorption is continuously increasing among the many chemical, physical, and biological processes employed in wastewater treatment. In situations where dangerous compounds or contaminants that are difficult to biodegrade must be removed, the use of AC is strongly advised. In the paper, textile, and petrochemical sectors, AC is virtually usually used in the treatment of wastewater. Rarely is it feasible to distinguish the individual components of combinations found in wastewater in real-world settings.

5.2.2.11   Landfill Seepage Treatment

Water seeping from landfills is treated using several different procedures. A substantial role is currently played by AC adsorption in addition to processes including biological treatment, chemical oxidation, reverse osmosis, flocculation/precipitation, and vaporization. Rarely is PAC utilized in contemporary facilities. GAC is often used in fixed beds, followed by reactivation. After other pollution-reduction procedures, adsorption is employed, which boosts the efficiency of the AC. The difficult to biodegrade COD and AOX molecules are removed by adsorption. The influent concentration is the main factor determining achievable pollution loads.

5.2.2.12 Industrial and Municipal Wastewater Treatment

In both municipal and industrial settings, AC plays a crucial role in maintaining the quality of the water supply. It is used in wastewater treatment facilities in municipal settings to remove a variety of soluble organics, as well as certain inorganic substances and heavy metals, following the first treatment. AC is used in industrial settings to filter out a wide range of unwanted substances from different effluent streams. AC acts as a catalyst in both situations to keep dangerous organic or inorganic substances out of drinking or surface water.

5.3 Oil Spill Cleanup

On a global scale, oil is one of the most significant sources of energy and raw materials for synthetic chemicals and polymers. Oil spills have the potential to have a large negative influence on the environment whether it is investigated, transported, stored, or utilized as a derivative. Environmental impact and human health result from oil spills in bodies of water. Due to the coating properties of these materials, petroleum oil pollution has an impact on the economy, tourism, and recreational activities.[45,46] The strong stink from oil spills may be smelled from kilometres away, and the excessive development of green algae changes the sea's hue and the scenery. According to estimates, 3.2 million tons of oil are released into the ocean each year.[47,48] A broad range of weathering processes, such as dissolution, evaporation, photochemical oxidation, dispersion, microbiological degradation, and agglomeration, are applied to crude oil that has leaked into the marine environment. Adsorption is a widely used method for cleaning up petroleum-contaminated water.[49] Chemicals that cause pollution cling to the surface of solids *via* the physical process of adsorption. Most adsorbents typically exist in monolayers and have porous structures with many small holes. AC is one of several materials for water remediation that have been used in the adsorption of oil. On an industrial scale, AC is often utilized as an adsorbent, primarily in the purification and separation of gases and liquids, as well as a catalyst and catalyst support.[50] Additionally, more recent applications are constantly being developed, notably those related to technological advancement and environmental preservation. A huge network of molecular-sized holes inside the carbon particles defines AC. Non-planar carbon layers make up AC (with some linearly or single-bonded carbons). This constitutes the full carbon specimen and describes how porosity was formed and how it was distributed across the entire carbon specimen.

Tabbakh *et al.* used a chemical activation process using 50% phosphoric acid followed by carbonization in an N_2 environment at 500 °C to produce ACs from wheat straw, and uncooked and cooked maize cobs. The ability of the ACs to remove organic waste, such as oil, is used to gauge their efficacy. Both the environment and people are harmed by oil spills on bodies of water. AC made from fried maize cobs had the greatest oil adsorption capability.

Several active carbons have been generated under certain process conditions that have surface areas that approach BET Nitrogen adsorption/desorption isotherms and Fourier transform infrared (FT-IR) spectroscopy were used to characterize ACs.[51]

Ukpong *et al.* examined the batch adsorption of crude oil from water using potassium hydroxide (KOH) made from coconut coir-AC (CCAC) under a variety of conditions, including adsorbent dose, contact duration, initial oil concentration, temperature, and agitation speed.[52] As clearly shown by FTIR and SEM examination, the morphological change considerably enhanced the hydrophobicity of the adsorbent, resulting in a CCAC with a much superior adsorption capacity for removing crude oil, with a maximum adsorption capacity of 4859.5 $mg\,g^{-1}$ at 304 K. The experimental findings demonstrated that when the adsorbent dose, contact duration, and initial oil concentration were increased, a greater proportion of crude oil was removed. The Langmuir, Freundlich, Temkin, Toth, Sips, and Redlich–Peterson isotherm equations were used to analyze the experimental isotherm data. The Freundlich model had the best match, with a correlation coefficient (R^2) value of 0.999. The Boyd model revealed that the adsorption was controlled by an internal transport mechanism and film diffusion was the main mode of adsorption. The kinetic data were properly fitted into various kinetic models, with the pseudo-second-order model showing the best fit and having an R^2 of 0.999. The adsorption of crude oil from the water was chemisorption and endothermic in nature ($\Delta H^\circ = 134$ $kJ\,mol^{-1}\,K^{-1}$), and the positive value of entropy ($\Delta S^\circ = 0.517$ $kJ\,mol^{-1}\,K^{-1}$) indicated an increase in disorder and unpredictability at the adsorbent–adsorbate interface. At higher temperatures, the feasibility and spontaneity of the adsorption were shown to increase by the reduction in Gibbs energy (ΔG°) with temperature. The produced adsorbent showed strong potential for use as a low-cost, recyclable, and environmentally friendly adsorbent in oil spill cleanup.

By making good use of waste biomass from water hyacinth to create an affordable nano-magnetic adsorbent material efficient for simple oil spill separation through an external magnetic field, Shokry *et al.* displayed environmental awareness and protection.[53] Compared to the shoots, the roots of the water hyacinth had a greater affinity for adsorbing oil spills (2.2 $g\,g^{-1}$). Water hyacinth roots were effectively used to extract nano-AC following zinc chloride treatment, alkaline activation, and carbonization. The nano-magnetic AC hybrid substance (NMAC) was formed by introducing nano-magnetite into the activated carbonized nanomaterials. The crystalline character of both the water hyacinth-AC that was recovered and its magnetic hybrid substance was shown by X-ray diffraction. SEM images suggested that both the generated AC and the hybrid magnetite materials were on the nanoscale. The vibrating sample magnetometer was used to assess the magnetic characteristics of the manufactured NMAC. A maximum selectivity of 30.2 g oil per g for oil adsorption was measured in the magnetic nano-hybrid material. In the presence of 1 $g\,L^{-1}$ of magnetic nano-hybrid, the ideal oil spill of 80% was obtained after 60 minutes. A strong external magnetic field made it simple to separate the oil spill absorbent magnetic nano-hybrid

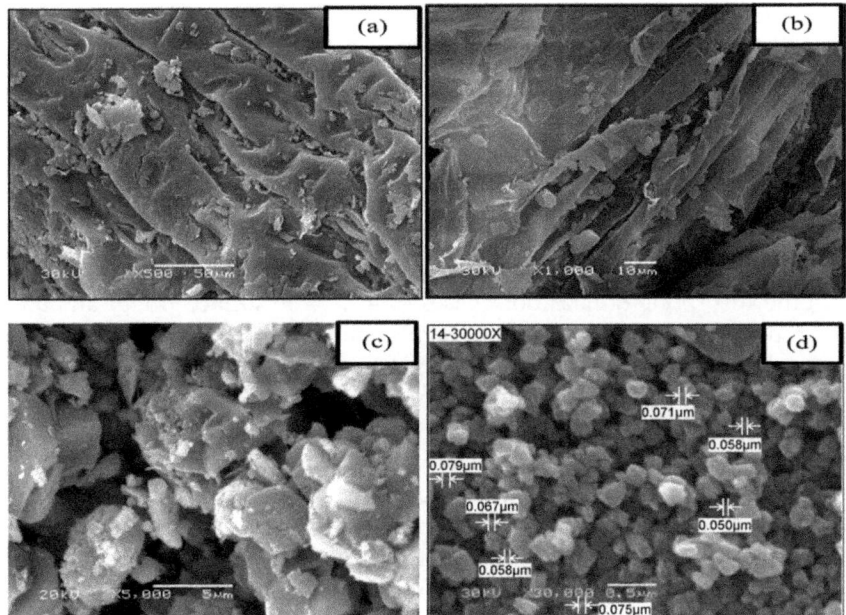

Figure 5.3 SEM micrographs of surface imaging (a) shoots segment of water hyacinth, (b) roots segment of water hyacinth, (c) prepared activated carbon, (d) magnetite immobilized activated carbon.[53] Reproduced from ref. 53, https://doi.org/10.1038/s41598-020-67231-y, under the terms of the CC BY 4.0 license https://creativecommons.org/licenses/by/4.0/.

substance from the treatment medium. The morphological structures of the AC, natural raw water hyacinth, and NMAC product were contrasted using SEM to ascertain the alterations that occurred as a consequence of the activation and magnetization processes. According to the SEM analysis (Figure 5.3), the NMAC's surface morphology differed from those of its parents, natural water hyacinth raw carbon and AC, which showed extensively layered accumulations with porous structures. After the carbonization and magnetization processes, the fibres were gathered and cut into bundles with many fibres, and certain circuitous accumulations of carbon fibres were visible. The pore regions of the AC had smaller holes among the natural fibres, whereas the pores got bigger within the bundles of fibres because of the variety brought about by the carbonization process of water hyacinth. Additionally, it was shown that the prepared AC developed huge holes on its external surface. The evaporation of $ZnCl_2$ during the carbonization process may be the cause of these voids.

5.4 Control of Gas Pollutants

With the rise in energy consumption, the burning of fossil fuels results in the release of greenhouse gases (GHG) and other pollutants into the environment. As a result, AC has also been found to be effective in the adsorption of gases in addition to liquids.[54–56] Due to its many benefits, including easy regeneration, affordability, high gas adsorption capacity in a normal atmosphere, resistance to moisture, adequate pore size distribution, high surface area, high mechanical stability, and minimal energy requirement, AC is one of the promising solid adsorbents that can be used to adsorb CH_4, CO_2, NO_2, H_2S, and H_2 gases.[57] The efficiency of adsorption of polluting gases on AC has been improved as a consequence of dipole–dipole interactions, H–H interactions, and covalent bonds between the functional groups and gas.[58] The adsorption capacity, surface area, micropore structure, reproducibility, and processability of the solution all play a role in the adsorption of hazardous gases by AC.[59] The surface groups and textural properties of the adsorbents are highly complex, leading to the extremely complex adsorption of CO_2,[60] and AC typically has an adsorption capacity of roughly 5 wt% (298 K and 0.1 bar). Through electrostatic interactions, the functional group K^+ ions were crucial for the adsorption of CO_2.[61] The ability to adsorb CO_2 greatly improved as the heating temperature was increased to 873 K. However, at lower pressures (below 0.3 bar), the pore size had a significant impact on the adsorption behaviour, with the micropores having the greatest impact on CO_2 adsorption (1 bar).[62] The arrangement of pores, and the kind and chemical organization of the AC surface are the driving forces behind the use of AC in the adsorption of NO_2. The threshold for hydrogen sulphide (H_2S), a colourless, poisonous, odorous, and detrimental gas for both human and environmental health, is 0.0047 parts per million.[63] The most popular and successful method for removing H_2S involves the use of AC treated with oxidative chemicals (KI and $KMnO_4$) or caustic chemicals (KOH and NaOH).[64] Due to their toxicity, SO_2 and NO_x removal has attracted a lot of attention recently. ACs have a better possibility of effectively removing these hazardous gases due to their improved sorption capabilities. It was discovered that impregnating AC with metal oxides works well to concurrently remove SO_2 and NO_x.[65]

In terms of CO_2 gas adsorption, almond shell AC had a lower adsorption ability than sunflower seed AC (2.7 mmol g^{-1} and 4.6 mmol g^{-1}, with a surface area of 862 m^2 g^{-1} and 1790 m^2 g^{-1}, respectively).[4] *Phoenix dactylifera* seed AC activated with N_2 eliminated 141.14 mg g^{-1} of CO_2[57] and AC from coconut shells removed up to 88.8% of the H_2S.[20]

AC is the safest and most effective method for removing volatile organic compounds (VOCs), airborne chemicals, gaseous pollutants, and other emissions and aromas in industrial, municipal, and commercial applications. AC systems have lower footprints than other odour control technologies and are a cost-effective way to comply with environmental laws. Carbon systems are appropriate for intermittent flow/loading applications since carbon is immediately effective upon startup and reacts instantly to variations in odour concentration.

5.4.1 Odour Control and H₂S Removal

AC treatment of effluent airflow has been identified as one of the most effective methods for removing H_2S.[66] There is widespread use of AC for odour reduction in municipal wastewater treatment facilities using agitation or accumulation of sludge.

5.4.2 VOC Removal

VOCs pose a concern to human health and are among the most dangerous indoor air pollutants. When injected into flue gas, PAC minimizes hazardous emissions to help comply with stricter standards. Vapour phase-AC treatment eliminates pollutants in industrial, municipal, and commercial vapour streams.[67] Volatile organic molecules are absorbed by the medium when the vapour stream passes through vessels containing AC. Due to the presence of BTEX in indoor air, exposure to BTEX concentrations may have a significant impact on human health. Ventilation or filtration is the sole method for removing BTEX from the environment. For the BTEX pollution mitigation technique, AC is the key carbon-rich material. The biomass of *Digitalis purpurea* L. was chosen as a natural lignocellulosic material suitable for carbonization. *Digitalis purpurea* L. was pyrolyzed to produce ACs (DPAC1–60). Chemical activation of the biomass was performed between 500 and 900 °C using potassium carbonate (K_2CO_3), zinc chloride ($ZnCl_2$), phosphoric acid (H_3PO_4), and sulfuric acid (H_2SO_4). The investigation demonstrated that DPAC58 produced from the *D. purpurea* L. biomass was appropriate for BTEX removal from indoor air. For applications involving the elimination of indoor air pollutants, carbon-based materials were suggested as future research directions and possibilities.[68]

5.4.3 Mercury Removal

For mercury removal from flue gas, AC is an excellent sorbent. Numerous years of study and development and more than 50 full-scale demonstrations have shown that activated carbon injection (ACI) can significantly decrease mercury emissions from the majority of designs, even when native mercury removal is poor. ACI is the commercial choice for mercury-specific air pollution reduction, but success at individual locations needs an awareness of variables that might affect performance. The technology for impregnated PAC is intended to extract mercury in its gaseous state from the flue gas stream and transform it into particles that can be collected downstream.[69] This product's unique pore structure makes it perfect for the fast adsorption elimination of mercury from combustion gases over a wide range of temperatures.[70]

5.5 Treatment of Soil Pollution

Cleaning polluted soil is an important application of AC. Solvents contained in the soil are extracted by sucking up ground air and then adsorbed onto AC. Depending on the concentration of the solvent in the air that has been sucked, either traditional adsorbers or solvent recovery devices are used.

AC is utilized during soil bioremediation for the removal of volatile organics (petroleum hydrocarbons, chlorinated solvents, polychlorinated biphenyls (PCB), and pentachlorophenol) and the adsorption of organic pollutants from pumped subsurface fluids. Soil contamination is a global problem due to its negative effects on the environment and human health. With the continuous growth of industry and agriculture, the concentration of hydrophobic organic pollutants in the soil has increased, resulting in severe soil contamination. Urgent action is required to remove hydrophobic organic pollutants from the soil to restore environmental safety and provide enough food and water for communities. Adsorption has been shown to be an efficient and cost-effective approach to eliminating organic pollutants. AC may play a crucial role in preventing air pollution and odour emissions during the removal of any contaminated soil. According to the research by Dewangan *et al.*, AC may be used to remove up to 100% of hydrophobic organic pollutants from soil.[71] Burachevskaya *et al.* revealed in a separate investigation that the addition of carbon sorbents to Haplic Chernozem soil polluted with zinc and copper was a successful remediation strategy.[72]

5.6 Conclusion

ACs have been used for many applications. The information provided in this chapter on the numerous uses of carbon-based materials in environmental technology demonstrates the enormous relevance and potential of their utilization. AC is utilized to treat environmental problems that impact the air and water, and new applications are always being found. Environmental protection places a specific emphasis on the reactivation and subsequent reuse of old AC as well as the recycling of AC that cannot be regenerated or reactivated. Another significant use for AC is the treatment, filtration, and decolorization of liquids, which is essential in the food, pharmaceutical, beverage, and other industries.

References

1. M. S. Reza, C. S. Yun, S. Afroze, N. Radenahmad, M. S. A. Bakar, R. Saidur, J. Taweekun and A. K. Azad, *Arab. J. Basic Appl. Sci.*, 2020, **27**, 208–238.
2. Z. Kecira, O. Benturki, A. Benturki, M. Daoud and P. Girods, *Environ. Prog. Sustain.*, 2020, **39**, e13463.
3. K. K. Beltrame, A. L. Cazetta, P. S. de Souza, L. Spessato, T. L. Silva and V. C. Almeida, *Ecotoxicol. Environ. Saf.*, 2018, **147**, 64–71.
4. P. González-García, *Renewable Sustainable Energy Rev.*, 2018, **82**, 1393–1414.
5. J. Rivera-Utrilla, M. Sánchez-Polo, M. Á. Ferro-García, G. Prados-Joya and R. Ocampo-Pérez, *Chemosphere*, 2013, **93**, 1268–1287.
6. R. Briones, L. Serrano, R. B. Younes, I. Mondragon and J. Labidi, *Ind. Crops Prod.*, 2011, **34**, 1035–1040.

7. Z. Anfar, M. Zbair, H. A. Ahsaine, M. Ezahri and N. E. Alem, *Fullerenes, Nanotubes, Carbon Nanostruct.*, 2018, **26**, 389–397.
8. R. Khosravi, G. Moussavi, M. T. Ghaneian, M. H. Ehrampoush, B. Barikbin, A. A. Ebrahimi and G. Sharifzadeh, *J. Mol. Liq.*, 2018, **256**, 163–174.
9. U. K. Sahu, S. Sahu, S. S. Mahapatra and R. K. Patel, *J. Mol. Liq.*, 2017, **243**, 395–405.
10. T. Q. Tuan, N. Van Son, H. T. K. Dung, N. H. Luong, B. T. Thuy, N. T. Van Anh, N. D. Hoa and N. H. Hai, *J. Hazard. Mater.*, 2011, **192**, 1321–1329.
11. A. E. Burakov, E. V. Galunin, I. V. Burakova, A. E. Kucherova, S. Agarwal, A. G. Tkachev and V. K. Gupta, *Ecotoxicol. Environ. Saf.*, 2018, **148**, 702–712.
12. A. Azimi, A. Azari, M. Rezakazemi and M. Ansarpour, *ChemBioEng Rev.*, 2017, **4**, 37–59.
13. A. S. Yusuff, *Arab J. Basic Appl. Sci.*, 2019, **26**, 89–102.
14. X. Yang, Y. Wan, Y. Zheng, F. He, Z. Yu, J. Huang, H. Wang, Y. S. Ok, Y. Jiang and B. Gao, *Chem. Eng. J.*, 2019, **366**, 608–621.
15. M. Danish, R. Hashim, M. M. Ibrahim, M. Rafatullah and O. Sulaiman, *J. Anal. Appl. Pyrolysis*, 2012, **97**, 19–28.
16. N. Rani, A. Gupta and A. Yadav, *Environ. Technol.*, 2006, **27**, 597–602.
17. S. Afroze, A. Karim, Q. Cheok, S. Eriksson and A. K. Azad, *Front. Energy*, 2019, **13**, 770–797.
18. Z. Z. Chowdhury, S. B. Abd Hamid, R. Das, M. R. Hasan, S. M. Zain, K. Khalid and M. N. Uddin, *BioResources*, 2013, **8**, 6523–6555.
19. L. Largitte, T. Brudey, T. Tant, P. C. Dumesnil and P. Lodewyckx, *Microporous Mesoporous Mater.*, 2016, **219**, 265–275.
20. H. S. Min, M. Abbas, R. Kanthasamy, H. Abdul Aziz and C. Tay, *Activated carbon: prepared from various precursors*, Ideal International E-publication Pvt. Ltd, 2017, pp. 1–20.
21. W. M. H. Wan Ibrahim, M. H. Mohamad Amini, N. S. Sulaiman and W. R. A. Kadir, *Arab J. Basic Appl. Sci.*, 2019, **26**, 30–40.
22. J. Wang, F. Wu, M. Wang, N. Qiu, Y. Liang, S. Fang and X. Jiang, *Afr. J. Biotechnol.*, 2010, **9**, 2762–2767.
23. M. J. Ahmed, *Process Saf. Environ. Prot.*, 2016, **102**, 168–182.
24. M. Attari, S. S. Bukhari, H. Kazemian and S. Rohani, *J. Environ. Chem. Eng.*, 2017, **5**, 391–399.
25. K. Derdour, C. Bouchelta, A. K. Naser-Eddine, M. S. Medjram and P. Magri, *World J. Eng.*, 2018, **15**(1), 3–13.
26. L. Niazi, A. Lashanizadegan and H. Sharififard, *J. Cleaner Prod.*, 2018, **185**, 554–561.
27. G. Alagumuthu and M. Rajan, *Chem. Eng. J.*, 2010, **158**, 451–457.
28. C. Namasivayam and D. Sangeetha, *Bioresour. Technol.*, 2006, **97**, 1194–1200.
29. C. Namasivayam and D. Sangeetha, *Chemosphere*, 2005, **60**, 1616–1623.
30. C. Namasivayam and D. Sangeetha, *J. Colloid Interface Sci.*, 2004, **280**, 359–365.

31. A. A. Oladipo and A. O. Ifebajo, *J. Environ. Manage.*, 2018, **209**, 9–16.
32. S. Wong, N. Ngadi, I. M. Inuwa and O. Hassan, *J. Cleaner Prod.*, 2018, **175**, 361–375.
33. V. Katheresan, J. Kansedo and S. Y. Lau, *J. Environ. Chem. Eng.*, 2018, **6**, 4676–4697.
34. M. Danish, R. Hashim, M. Ibrahim and O. Sulaiman, *Wood Sci. Technol.*, 2014, **48**, 1085–1105.
35. B. Dass and P. Jha, *Int. J. ChemTech Res.*, 2015, **8**, 269–279.
36. H. Wang, R. Xie, J. Zhang and J. Zhao, *Adv. Powder Technol.*, 2018, **29**, 27–35.
37. M. El Gamal, H. A. Mousa, M. H. El-Naas, R. Zacharia and S. Judd, *Sep. Purif. Technol.*, 2018, **197**, 345–359.
38. M. Zbair, K. Ainassaari, A. Drif, S. Ojala, M. Bottlinger, M. Pirilä, R. L. Keiski, M. Bensitel and R. Brahmi, *Environ. Sci. Pollut. Res.*, 2018, **25**, 1869–1882.
39. R. C. Bansal, D. Aggarwal, M. Goyal and B. C. Kaistha, *Indian J. Chem. Technol.*, 2002, **9**, 290–296.
40. J. Salman, V. Njoku and B. Hameed, *Chem. Eng. J.*, 2011, **173**, 361–368.
41. J. Salman and F. Hussein, *J. Environ. Anal. Chem.*, 2014, **2**, 2.
42. O. A. Ioannidou, A. A. Zabaniotou, G. G. Stavropoulos, M. A. Islam and T. A. Albanis, *Chemosphere*, 2010, **80**, 1328–1336.
43. G. Jaria, C. P. Silva, J. A. Oliveira, S. M. Santos, M. V. Gil, M. Otero, V. Calisto and V. I. Esteves, *J. Hazard. Mater.*, 2019, **370**, 212–218.
44. N. Boudrahem, S. Delpeux-Ouldriane, L. Khenniche, F. Boudrahem, F. Aissani-Benissad and M. Gineys, *Process Saf. Environ. Prot.*, 2017, **111**, 544–559.
45. S. Ben Hammouda, Z. Chen, C. An, K. Lee and A. Zaker, *Cleaner Chem. Eng.*, 2022, 100028.
46. O. N. Ngofa, E. V. Liakos, A. N. Papadopoulos and G. Z. Kyzas, *Bio-interface Res. Appl. Chem.*, 2022, **12**, 2701–2714.
47. J. Wang, H. Duan, M. Wang, Q. Shentu, C. Xu, Y. Yang, W. Lv and Y. Yao, *Environ. Res.*, 2022, **207**, 112212.
48. B. T. Atunwa, A. O. Dada, A. A. Inyinbor and U. Pal, *Mater. Today: Proc.*, 2022, **65**, 3538–3546.
49. D. A. de Freitas, J. A. Barbosa, G. Labuto, R. C. F. Nocelli and E. N. V. M. Carrilho, *Environ. Sci. Pollut. Res.*, 2022, 1–11.
50. H. K. Yağmur and İ. Kaya, *J. Mol. Struct.*, 2021, **1232**, 130071.
51. H. Tabbakh and R. Barhoum, *Mater. Sci.: Indian J.*, 2018, **16**, 1–9.
52. A. Ukpong, G. Habor and I. Oboh, *Open Chem. Eng. J.*, 2020, **8**, 36–47.
53. H. Shokry, M. Elkady and E. Salama, *Sci. Rep.*, 2020, **10**, 10265.
54. S. Afroze, A. B. H. Bakar, M. S. Reza, M. A. Salam and A. K. Azad, Polyvinylidene fluoride (PVDF) piezoelectric energy harvesting from rotary retracting mechanism: imitating forearm motion, 7th Brunei International Conference on Engineering and Technology 2018 (BICET 2018), 2018, p. 94.

55. A. Ahmed, M. S. A. Bakar, A. K. Azad, R. S. Sukri and T. M. I. Mahlia, *Renewable Sustainable Energy Rev.*, 2018, **82**, 3060–3076.
56. M. A. Hossain, S. Shams, M. Amin, M. S. Reza and T. U. Chowdhury, *Buildings*, 2019, **9**, 79.
57. A. E. Ogungbenro, D. V. Quang, K. Al-Ali and M. R. Abu-Zahra, *Energy Procedia*, 2017, **114**, 2313–2321.
58. A. E. Creamer and B. Gao, *Environ. Sci. Technol.*, 2016, **50**, 7276–7289.
59. P. Sharma, H. Kaur, M. Sharma and V. Sahore, *Environ. Monit. Assess.*, 2011, **183**, 151–195.
60. B. C. Bai, S. Cho, H.-R. Yu, K. B. Yi, K.-D. Kim and Y.-S. Lee, *J. Ind. Eng. Chem.*, 2013, **19**, 776–783.
61. Y. Zhao, X. Liu, K. X. Yao, L. Zhao and Y. Han, *Chem. Mater.*, 2012, **24**, 4725–4734.
62. S.-Y. Lee and S.-J. Park, *J. Ind. Eng. Chem.*, 2015, **23**, 1–11.
63. Y. Elsayed, M. Seredych, A. Dallas and T. J. Bandosz, *Chem. Eng. J.*, 2009, **155**, 594–602.
64. N. M. Nor, L. C. Lau, K. T. Lee and A. R. Mohamed, *J. Environ. Chem. Eng.*, 2013, **1**, 658–666.
65. N. Le-Minh, E. C. Sivret, A. Shammay and R. M. Stuetz, *Crit. Rev. Environ. Sci. Technol.*, 2018, **48**, 341–375.
66. T. Tuerhong and Z. Kuerban, *J. Environ. Chem. Eng.*, 2022, **10**, 107177.
67. Z. Li, Y. Li and J. Zhu, *Materials*, 2021, **14**, 3284.
68. K. Isinkaralar, *Biomass Convers. Biorefin.*, 2022, **12**, 4171–4181.
69. W. Zhao, X. Geng, J. Lu, Y. Duan, S. Liu, P. Hu, Y. Xu, Y. Huang, J. Tao and X. Gu, *Fuel*, 2021, **285**, 119131.
70. R. Rodriguez, D. Contrino and D. W. Mazyck, *Ind. Eng. Chem. Res.*, 2020, **59**, 17740–17747.
71. S. Dewangan, A. K. Bhatia, A. K. Singh and S. A. Carabineiro, *C*, 2021, **7**, 83.
72. M. Burachevskaya, S. Mandzhieva, T. Bauer, T. Minkina, V. Rajput, V. Chaplygin, A. Fedorenko, N. Chernikova, I. Zamulina, S. Kolesnikov, S. Sushkova and L. Perelomov, *Plants*, 2021, **10**, 841.

CHAPTER 6

Environmental Applications of Activated Carbon

B. GOPAL KRISHNA,*[a] SANJAY TIWARI,[a] DHRITI SUNDAR GHOSH[b] AND M. JAGANNADHA RAO[c]

[a] Photonics Research Laboratory, School of Studies in Electronics & Photonics, Pt. Ravishankar Shukla University, Raipur, India; [b] Department of Physics, Indian Institute of Technology, Bhilai, India; [c] Department of Geology, Andhra University, Visakhapatnam, Andhra Pradesh, India
*Email: krishna_burra85@yahoo.com

6.1 Introduction

The exponential population growth and social civilization expansion, changes in lifestyles and resource consumption, and continuous development of industries and technologies have all been accompanied by sharp globalization and metropolitan growth during the last several decades.[1] Until now, the harmful effects of massive industrial effluent seepage have been largely influenced by a variety of emissions from metal finishing, coal mining, mineral extraction, oil refining, food processing, pharmaceuticals, cosmetics, coffee pulping, dye manufacturing, vehicle exhausts, pulp tanneries, and electroplating distilleries, primarily in the form of carbon dioxide, hydrogen, methane, hydrogen sulphite, ammonia, volatile organic compounds (VOCs), or heavy metals.[2] Uncontrolled tipping or dumping is a convenient way to dispose of industrial effluents; it is an operation in which waste is spread or released to fill in low economic value open dumps on selected pieces of land (inundated swampland, abandoned sand mines, and quarries) or directly into the atmosphere, without regard for the surrounding

Activated Carbon: Progress and Applications
Edited by Chandrabhan Verma and Mumtaz A. Quraishi
© The Royal Society of Chemistry 2023
Published by the Royal Society of Chemistry, www.rsc.org

environment, nor taking any precautions to compact, cover, and prevent the infiltration of contaminants into the environment.[3] Environmental standards and regulations governing the monitoring of pollution from industrial effluent streams by regulatory bodies have recently become more demanding and restrictive, ultimately affecting waste processing plant design, planning, and operation. This has sparked a surge in research into developing a leading selective, dependable, and long-lasting solution for the judicious treatment of extremely polluted entities. The potential of activated carbon (AC), which is primarily characterized by its high reactivity for the complete elimination of recalcitrant pollutants and environmental remediation, has piqued interest. Its synergetic adsorption strength, controllable pore structure, excellent chemical stability, and superior ability to remove a wide range of organic and inorganic pollutants dissolved in aqueous media, even from gaseous environments, have all piqued interest.[4,5]

AC is a term used to describe highly carbonaceous materials with high porosity and sorption capability, such as wood, coconut shells, coal, and cones. AC is one of the most often used adsorbents in various sectors for the elimination of a variety of contaminants from water and air. Since AC has been synthesized from agricultural and waste products, it has been shown to be a great alternative to the non-renewable and expensive sources previously employed. Carbonization and activation are the two primary processes employed in the production of AC. Precursors are heated to high temperatures between 400 °C and 850 °C in the first phase to remove all volatile components. All non-carbon components of the precursor, such as hydrogen, oxygen, and nitrogen, are removed in the form of gases and tars at high temperatures. This method produces char with a high carbon content but low porosity and surface area. The second phase, on the other hand, entails activating the produced char. Opening previously inaccessible pores, new pore formation by selective activation, and expansion of existing pores are all examples of pore size augmentation during the activation process. To achieve the appropriate surface area and porosity, two methods of activation are commonly used: physical and chemical. Physical activation includes heating carbonized char to high temperatures (between 650 °C and 900 °C) with oxidizing gases such as air, carbon dioxide, and steam. Carbon dioxide is typically preferred due to its purity, easy handling, and regulated activation temperature of 800 °C. In comparison to steam, carbon dioxide activation can achieve high pore homogeneity. Steam, rather than carbon dioxide, is preferable for physical activation because AC with a comparatively large surface area may be formed. Water diffuses efficiently within the char structure due to its lower molecular size. Steam activation is two to three times more effective than carbon dioxide activation with the same degree of conversion.

The chemical technique, on the other hand, entails combining the precursor with activating chemicals (NaOH, KOH, $FeCl_3$, *etc.*). These activating chemicals work as both oxidants and dehydrators. Carbonization and activation are carried out concurrently in this method at lower temperatures of

300–500 °C than in the physical method. This has an impact on pyrolytic decomposition, which leads to the formation of a better porous structure and a higher carbon yield. The low-temperature requirement, high microporosity structures, large surface area, and reduced reaction completion time are all major advantages of the chemical approach over the physical approach. The superiority of chemical activation can be explained using a model provided by Kim *et al.*, according to which AC contains numerous spherical microdomains that are responsible for the production of micropores.[6] Mesopores, on the other hand, develop in the intermicrodomain areas. The authors used chemical (KOH) and physical (steam) activation to produce AC from phenol-based resin in the lab. The results showed that AC synthesized by KOH activation had a higher surface area of 2878 $m^2 g^{-1}$ than AC synthesized by steam activation, which had a surface area of 2213 $m^2 g^{-1}$. In addition, the pore size, surface area, micropore volume, and average pore width under KOH activation conditions were all shown to be superior to those in steam activation settings. The structural mechanistic model explanation for synthesis of AC by steam activation (C6S9) and KOH activation (C6K9) is shown in Figure 6.1.

AC can be classified into three types based on particle size and the processing method: powdered AC, granular AC, and bead AC. Powdered AC is made up of thin granules that are 1 mm in diameter and have an average diameter of 0.15–0.25 mm. Granular AC is greater in size but has a smaller external surface area. Depending on its dimension ratio, granular AC is employed for a variety of liquid- and gaseous-phase applications. The third type (bead AC) is often made from petroleum pitch and has diameters of

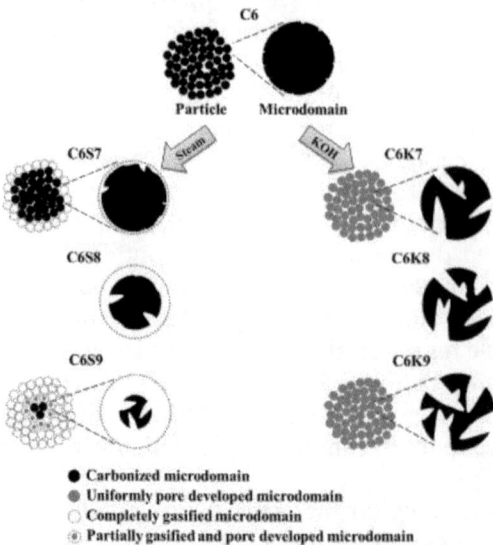

● Carbonized microdomain
● Uniformly pore developed microdomain
○ Completely gasified microdomain
⊛ Partially gasified and pore developed microdomain

Figure 6.1   Synthesis of AC by steam activation (C6S9) and KOH activation (C6K9). Reproduced from ref. 6 with permission from Elsevier, Copyright 2016.

0.35 to 0.8 mm. It has high mechanical strength and low dust content. Because of its spherical form, it is often used in fluidized bed applications such as water filtration.

6.2 Surface Chemistry of ACs

Active carbons are made up of highly disordered microstructures and are not graphitic or graphitizable. AC belongs to the family of microporous materials and can be thought of as a material with a very large surface area made up of millions of pores, similar to a molecular sponge. The surface is oxidized and any non-carbon impurities are eliminated. ACs are carbonaceous compounds that differ from elemental carbon in that the carbon atoms on the outer and inner surfaces are oxidized.[7] The internal pore structure of ACs is extremely amorphous, microcrystalline, and well developed. These materials have unusually large specific surface areas, well-developed porosity, and tunable surface-containing functional groups.[8,9] Agricultural and industrial waste materials with high carbon content and low inorganic content, such as coal, wood, lignin, and coconut shells, are suitable precursors for the manufacture of ACs.[10] Physical activation and chemical activation are the two basic processes in the production of ACs.[11] Controlling the activation process parameters can produce ACs with tailored characteristics for the adsorption of specific chemicals.[12] The nature of the raw material utilized, the nature of the activating agent, and the parameters of the carbonization and activation processes all have a significant impact on the final pore structure on the surface of AC.[13] ACs are employed in a variety of commercial applications as adsorbents for the removal of gaseous and liquid contaminants, as well as a variety of other applications, due to their high adsorption capacity.

6.3 Significance of Activation

ACs, which are porous and contain tiny holes and tunnels depending on the nature and manner of preparation, provide a suitably vast surface area for impurity adsorption (holding) and hence purging solutions. The prevalent holes in the active carbons can be filled with suitable materials to increase the sorption affinity towards hazardous ions, and therefore the surface area works as a substrate for the doped or loaded materials. The nature of porous carbon applications is determined by the specific surface area, pore structure, and chemical functional groups on the surface.[14,15]

All volatile substances are eliminated during the activation process, as carbon atoms are peeled away layer after layer, widening the internal pores and leaving behind a carbon skeleton. The internal surface area of the material is increased by reducing the number of carbon atoms. The physicochemical and catalytic characteristics of adsorbents are reported to be influenced by the surface chemical functional groups formed by the activation process.[16,17] As a result, many scientists have focused on how to modify and characterize the surface functional groups of carbon materials to improve or expand their practical uses.[16,18,19] Radovic *et al.* evaluated carbon materials as adsorbents in an aqueous solution and suggested that chemical and physical conditions might be controlled to develop carbon surfaces that are suitable for certain adsorption applications.[20]

6.4 Physical Structure

The physical structure of AC is responsible for its adsorption properties. The activation procedure entails carbonizing the raw material and then activating it at high temperatures between 800 °C and 1100 °C in the presence of oxidizing gases such as carbon dioxide and steam.[21,22] The pyrolytic decomposition of raw material eliminates most non-carbon components such as nitrogen, oxygen, and hydrogen as volatile (gaseous) species during the carbonization (physical activation) process. The remaining carbon atoms are organized into aromatic planes. These graphitic surfaces generate graphite-like crystallites. The angular orientations of the planes in AC, on the other hand, are arbitrary, whereas they are well organized in graphite. The aromatic sheets are arranged in an uneven pattern, leaving open interstices. Due to these pores, ACs are effective adsorbents. During the activation phase, the pore structure in carbonized materials further grows and strengthens with randomly distributed pores of varied sizes and shapes.[23] As a result, the AC has a large surface area. Because AC has a huge surface area compared to the size of the real carbon particle, it is easier to remove significant amounts of pollutants in a tiny contained region.

6.5 Chemical Structure

The surface of AC has a chemical structure in addition to its porous structure. The chemical structure of the AC surface has a significant impact on its adsorption properties. The presence of heteroatoms in the carbon matrix (*i.e.*, atoms present in the carbon structure which are not carbon, such as oxygen, hydrogen, and small amounts of nitrogen, which occur in the form of functional groups and/or atoms chemically bonded to the structure, giving rise to carbon–oxygen, carbon–hydrogen, and carbon–nitrogen surface compounds) modifies the surface characteristics of the carbon. The most abundant heteroatom in the carbon matrix is oxygen, which is present as functional groups such as carboxyls, carbonyls, phenols, lactones, quinones, and other groups linked to the margins of the graphite-like layers.[7,24] Various post-activation treatment procedures, the most notable of which is oxidation, can change the type and concentration of these groups. Chemical activation is a one-step procedure that involves both carbonization and activation at the same time. The precursor is combined with a chemical activating agent and then burned at high temperatures in chemical activation (oxidation).[25,26] Two principal oxidation techniques, dry and wet, are used to produce the surface oxygen functional groups of AC. Wet oxidation involves reactions between AC surfaces and oxidizing solutions such as aqueous nitric acid, sulphuric acid, orthophosphoric acid, hydrogen peroxide, zinc chloride, potassium permanganate, and potassium thiocyanate at low or reflux temperatures (about 100 °C), whereas dry oxidation involves reactions between AC surfaces and oxidizing gases such as steam, oxygen, carbon dioxide, and air at high temperatures. These carbon–oxygen surface functional groups have a profound impact on the unique adsorption capabilities of ACs.[27–35] These groups, which are mostly found on the basal plane's outer surface or edge, contribute to the carbon's chemical characteristics. Because these outer sites make up the majority of the adsorption surface, the amount of oxygen on the surface has a significant impact on the carbon's adsorption properties.[17,36–38] Cleaning out tar-clogging, generating new pores, and lastly increasing the surface areas are all part of the activation (oxidation) process. Chemical activation is preferable because it increases ACs' adsorptive qualities, such as large surface area, appropriate pore volume, and a wide range of pore size distributions (PSDs).[39,40] Nitric acid, ammonia, amines, and other nitrogen-containing reagents can be used to insert nitrogen functional groups on the surfaces of active carbons.[41–49] The nature of surface functional groups present in the inter-surface area and surface heterogeneity contribute to the active carbons' sorption nature.[50] The nature and number of functional groups are determined by the starting material and/or activation treatment procedures.[51,52] The acidic or basic character of the carbon structure is determined by its functional groups and delocalized electrons.[53]

6.6   Classification of the Surface of AC

The type of heteroatom present on AC's surface, *e.g.*, nitrogen, phosphorus, sulphur, and oxygen, has a significant impact on its surface chemistry. The nature of the activation process determines the type of heteroatom present. The surfaces of AC can be classified as acidic or basic depending on the type of group present.

6.6.1   Acidic AC

The reactivity of AC is determined by the groups present on the external surface. The acidic nature of AC is due to the presence of oxygen-containing molecules on its surface. Because the majority of the adsorbent area is made up of external sites, the concentration of such reactive groups has a significant impact on the adsorption capabilities of AC.[38,54–56] Carboxylic, quinone, carbonyl, and pyrone functional groups are examples of oxygen-containing acidic groups. Carboxylic anhydrides are another kind of carboxylic group.[57,58] The hydroxyl groups of carboxylic acid, lactone, and phenolic hydroxyl groups are the origins of surface acidity. The oxidation of gases and aqueous oxidants produces oxygen functionalities in AC.[58,59]

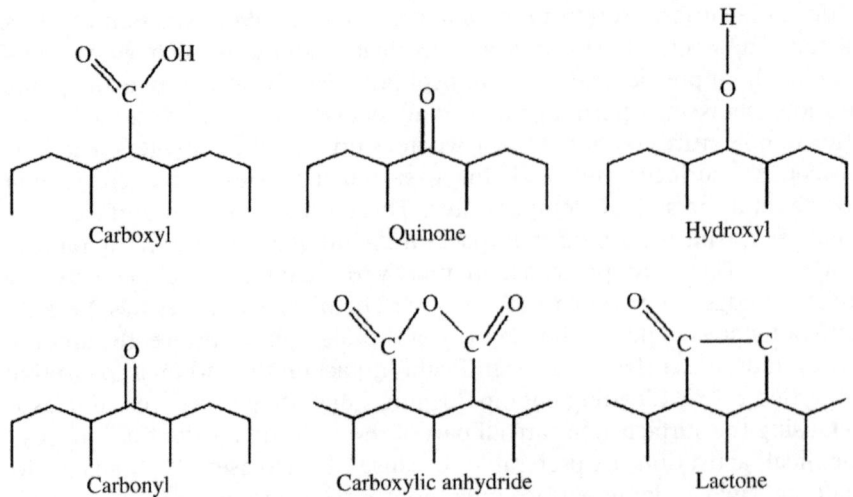

Figure 6.2 The bond formation between some acidic surface groups and aromatic rings on AC. Reproduced from ref. 19 with permission from Elsevier, Copyright 2006.

Oxygen, air, steam, and carbon dioxide can all be used in the gas-phase treatment. Low-temperature oxidations result in the formation of strongly acidic groups during these processes. High-temperature oxidations, on the other hand, produce weakly acidic groups.[31] Liquid-phase oxidation treatments can inject a greater amount of oxygen into the carbon moiety compared to gas-phase oxidation treatments. The oxidizing ability of a mixture of nitric acid and carboxylic acid allows for the introduction of substantial oxygenated acidic functional groups such as carboxylic, phenolic hydroxyl, and lactone onto the AC surface.[60,61] The results show that oxidation of AC in the gas phase is enhanced primarily due to the presence of hydroxyl and carbonyl groups on its surface; however, oxidation in the liquid phase can integrate a large amount of oxygen into the AC surface in the form of carboxylic and phenolic hydroxyl groups at much lower temperatures than oxidation in the gas phase. The bonding between acidic surface groups and aromatic rings on AC is shown in Figure 6.2.

6.6.2 Basic AC

The resonant electrons of aromatic carbon rings and basic surface functions (*e.g.*, nitrogen-containing groups) capable of interacting with protons are related to the basicity of AC. Basic AC can also be produced using oxygen-containing functions such as ketones, pyrones, and chromenes. Present electrons act as Lewis bases in basic AC. The ability of AC to adsorb carbon dioxide increases as the quantity of nitrogen-containing groups increases. By reacting with nitrogen-containing reagents (such as NH_3, nitric acid, and amines) or activating with nitrogen-containing precursors, nitrogen-containing functions can be hosted in AC. Dipole–dipole interactions, covalent bonding, and hydrogen bonding are the most common ways that basic AC interacts with acid molecules.

6.7   Synthesis Techniques of AC

The utilization of waste biomass as a precursor for porous ACs has emerged as an effective technique for the synthesis of AC. It has proven to be the ideal alternative to conventional raw resources such as petroleum wastes, coal, peat, and lignite, which are both expensive and non-renewable.[62] Pyrolysis carbonization and hydrothermal carbonization are the two main ways of generating carbon.[63] In general, pyrolysis carbonization is used to carbonize hard bio-waste materials, while hydrothermal carbonization is used to carbonize soft bio-based materials. Carbonization and activation are the two basic processes in the production of AC.

6.7.1   Pyrolysis Carbonization

Carbonization is the oldest method of transforming biomass into carbon material for human use. It is a slow and protracted process that converts organic material into carbon or a carbon-containing residue (biochar) by eliminating volatile, non-carbon species such as nitrogen, oxygen, and hydrogen from raw materials in a furnace under an inert gas atmosphere, strengthening the carbon content.[64] During the degasification process, the formation of narrow pore structures of precursors is followed by the elimination of residual compounds as the temperature increases. Furthermore, some residual substances may collide with this accumulation, causing the walls of the pores to collapse, resulting in hydro-cracking and carbon deposition.[65]

Temperature, the heating rate, the presence of an inert environment, and the processing time have the most notable effects in this process. A carbonization temperature of >600 °C usually results in a lower char yield and a faster liquid and gas release rate.[66] As the temperature increases, the amount of ash and fixed carbon in the atmosphere increases, while the amount of volatile stuff decreases. As a result, greater temperatures produce higher-quality biochar[67] but lower yields. The initial decomposition (de-volatilization) of biomass at a high temperature and the secondary decomposition (cracking) of biochar residue are assumed to be the causes of the lower yield. Pyrolysis carbonization is without a doubt the best process for producing large quantities of porous carbon products with good characteristics. This method, along with chemical activation, is now commonly employed to produce bio-based hierarchically porous ACs for commercial use.[68] The equipment for the fixed bed pyrolysis technique is shown in Figure 6.3.

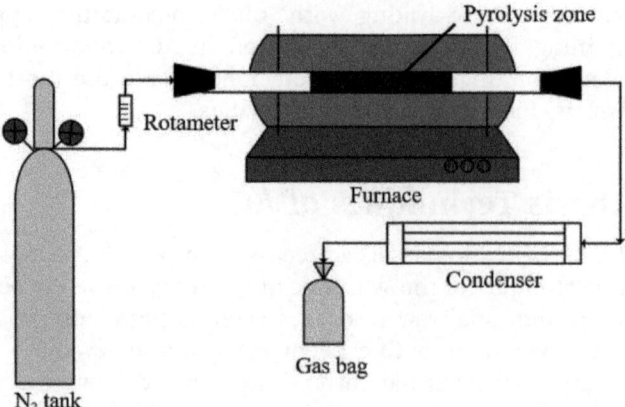

Figure 6.3   Equipment for the fixed bed pyrolysis technique. Reproduced from ref. 66 with permission from Elsevier, Copyright 2020.

6.7.2   Hydrothermal Carbonization

Hydrothermal carbonization is a promising thermochemical technology for converting biomass into ACs that is both cost-effective and environmentally friendly. Heating raw material dispersed in an aqueous solution and auto-claving it at temperatures between 150 °C and 350 °C for 2–24 hours under saturated pressure is known as hydrothermal carbonization. Water-soluble components and a carbon-rich hydrophilic solid known as 'hydrochar' are generated during this phase.[69,70] The hydrothermal carbonization method has few requirements in terms of biomass preparation and treatment; for example, unlike other thermal treatment procedures, it does not require pre-drying of wet waste, making it a cost-effective option. Different process parameters, such as reaction duration, pressure, and temperature, influence the final product formed.[71] This method is driven by exothermic de-hydration and decarboxylation processes, which make the process self-sustaining after activation. At subcritical temperatures, biomass is first hydrolyzed into monomers, then dehydrated and fragmented into soluble organics. The initial hydrolysis process (acid catalysis) is catalyzed by a decrease in pH, and the solution concentration increases due to the poly-merization and condensation of material, resulting in nucleation and growth.[72] Hydrochar has a chemical structure that is nearly identical to natural coal (core–shell structure with a hydrophobic core and a stabilizing hydrophilic shell) and contains oxygen functional groups such as hydroxyl, phenol, carbonyl, and carboxyl, making it an excellent adsorbent. The fun-damental advantage of this process over other thermochemical conversion technologies is that it can transform wet feedstock into a solid carbonaceous product (hydrochar) in high yields without dehydrating or drying it.[73] This technique can easily transform soft bio-based materials into biochar. Biochar can then be activated with various activating chemicals in a tube furnace for surface modification. Zhao *et al.* reported the construction of

ACs from tobacco rods (soft bio-waste), resulting in a hierarchically porous structure with micro-, meso-, and macropores coexisting.[74] Furthermore, the generated ACs have a large number of oxygen- and nitrogen-based functional groups with good electrical conductivity. The hydrothermal process is regarded as a basic and straightforward method for synthesizing innovative materials. Different metals or metal oxides can be doped onto the surfaces of ACs using this method. To make an effective photocatalyst from bio-waste materials, a three-step procedure involving pyrolysis carbonization, chemical activation, and hydrothermal treatment can be applied. Shrestha *et al.* reported the manufacture of ZnO rod-decorated ACs utilizing a hydrothermal technique.[61] Activation of phosphoric acid on the surfaces of ACs can provide functionalities that act as nucleation sites for the formation of ZnO nanorods on the surfaces of the porous ACs during hydrothermal treatment. The stable insertion of ZnO nanorods into ACs prevents the photocatalyst from being lost during recovery, allowing for successful water treatment. The hydrothermal synthesis technique to prepare the ZnO/AC composite is shown in Figure 6.4.

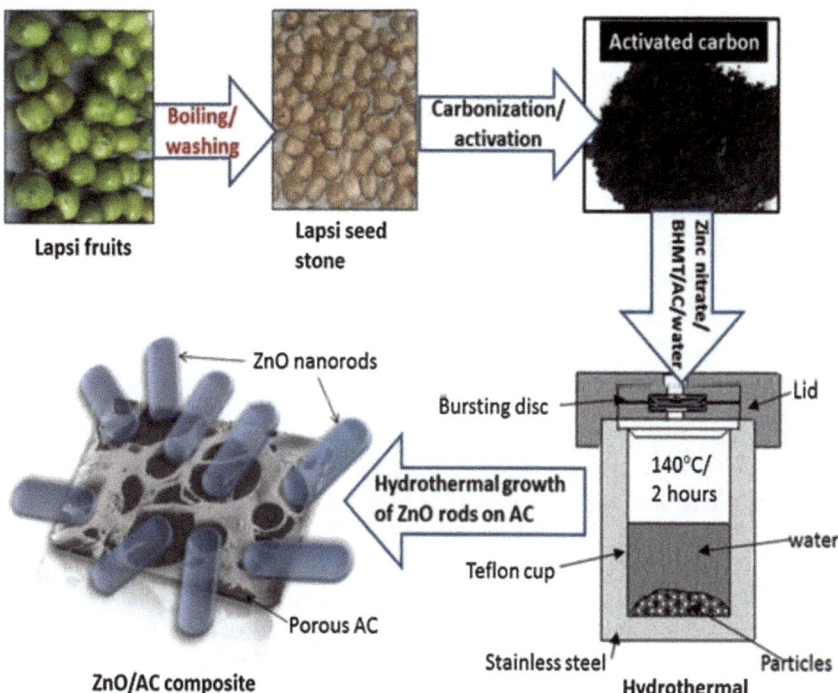

Figure 6.4   Synthesis of the ZnO/AC composite through a hydrothermal technique. Reproduced from ref. 61, https://doi.org/10.3390/ma13245667, under the terms of the CC BY 4.0 license https://creativecommons.org/licenses/by/4.0/.

6.8   Modification Methods of AC

Surface modification of ACs can lead to the development of appropriate physicochemical characteristics for use in environmental remediation. It is the process of tailoring the surfaces of ACs to generate desired physical and chemical properties by altering the physical, chemical, or biological characteristics on their surface.[75] Different activation procedures and activating substances might endow ACs with different physicochemical features.

6.8.1   Acid Modification

It's a common wet oxidation process used to make ACs from biomass. The porous carbon surfaces can remove mineral components and increase the surface's hydrophilic character during acid oxidation.[76] For this alteration, strong mineral acids and powerful oxidants such as HNO_3, H_2SO_4, HCl, H_3PO_4, H_2O_2, and HClO are used.[77] Weak organic acids, such as nitric acid and sulphuric acid, are rarely utilized due to their poor ionization capabilities. By lowering the minerals on the surfaces of ACs, acidification improves their acidic behaviour and surface hydrophilicity. Oxygen-containing functional groups such as hydroxyl, carbonyl, carboxyl, quinone, lactone, and carboxylic anhydride can be introduced into such adsorbent materials having an acidic surface. The presence of such functional groups on the outer surfaces or basal plane borders of ACs can have a significant influence on the chemical characteristics of the material.[78] Due to its ability to form metal complexes with anionic acid groups, acid-treated AC can develop a positive charge on its surface, boosting the adsorption of metal cations. This basic approach has been adopted by several research groups for the adsorption of heavy metals utilizing acid-treated AC with diverse precursors (produced from coal-based materials or biomass).

6.8.2   Alkaline Modification

A positive charge on the surface of AC is generated by a base (alkaline) treatment, which facilitates the adsorption of various negatively charged molecules. Treatment of ACs at high temperatures (400–900 °C) in an inert hydrogen or ammonia atmosphere is the simplest technique to improve the basic surface properties in porous carbons.[79,80] Alkali treatment can improve both the relative amount of alkali groups and the non-polarity of the surface. As a result of this approach, the adsorption capacity of ACs for non-polar compounds can be increased. Treatments with NaOH, KOH, Na_2CO_3, Na_2SiO_3, and oxides can change the surfaces of ACs. The basic nitrogen functionalities on the carbon surface arise when AC is treated with NH_3 at 400–900 °C.[48,81] Treatment with nitrogen precursors (such as ammonia and amines) or activation in a nitrogen-rich environment can be used to dope nitrogen functionalities.[41,82] The generated amide, imide, lactone, pyrrolic, and pyridinic groups usually give the appropriate property to enhance the dipole–dipole exchange, hydrogen, and covalent bonding interactions between adsorbents and acidic groups.

6.8.3 Impregnation

Impregnation is defined as the homogeneous distribution of other species over the surfaces of carbon materials (at the nano/microsize).[61,83] It is a method of using metal/metal oxides/chemicals to decorate the surfaces of porous materials. This technique can be employed using both dry and wet impregnation methods. A solvent can be used to pack the pores of the adsorbent in the dry impregnation procedure. However, after the pores have been filled, an excess solvent is introduced in the wet impregnation procedure. Impregnating materials are frequently metals or polymeric compounds that do not influence pH.

6.8.4 Ozone Treatment

The structure of carbon can be modified by treating it with ozone. The surface groups and chemical composition of activated carbon can also be modified through ozone treatment.[84,85] The interaction of ozone and ACs in a single process may be the best option for removing hazardous chemical compounds.[86] Previous research has found that, in addition to adsorption, other phenomena such as interactions between ozone, adsorbed organic matter, and produced free radicals can improve system efficiency.[87,88]

6.8.5 Plasma Treatment

Plasma treatment is quickly becoming a popular method for nanoscale surface modification of many materials. Strong electric fields, charged particles with adequate energy, powerful oxidizing and reducing species, UV radiation, ultrasound, electrohydraulic cavitations, and shock waves are used for the plasma treatment method.[89–91] The plasma oxidation technique is one of the most widely used methods for AC plasma treatment. The ACs can be exposed to plasma in the presence of controlled air or oxygen under vacuum or atmospheric pressure in this process. During plasma oxidation, oxygen is chemically added to carbon surfaces as oxygen free radicals, increasing surface acidity and dramatically altering the surface chemistry of ACs.

6.8.6 Biological Modification

Biological AC technology combines effective microbe degradation with an AC-based adsorption technique.[92] Bacteria are restricted within ACs for biological modification and begin to accumulate under a suitable temperature and nutrition for growth.[93] Biologically modified ACs have a distinct outer area and a well-developed pore structure according to the findings. As a result, they may efficiently adsorb dissolved oxygen and organic matter in water. Additionally, the attached microbes on the activated carbon convert the biomass, waste products, carbon dioxide, and other degradable components into useful materials using an adsorption process on the ACs. The biological AC system is based on interactions between AC particles, microorganisms, contaminants, dissolved oxygen, and other pollutants in wastewater that are soluble. Adsorption of microorganisms on AC may be beneficial in a variety of ways. In biologically modified ACs, for example, the carbon-bed life can be increased by converting a fraction of refractory organics to biodegradable organics. The biodegradable organics are then subsequently transformed by the microbes before this material may be adsorbed on the ACs. There will be a change in the surface charge density of the ACs due to the formation of a bio-film on the ACs. Hence, the adsorptivity of the AC will be improved.[94]

6.8.7 Microwave Treatment

Microwave treatment has recently been recommended as a promising strategy for AC modification due to its advantages of rapid temperature increase, homogeneous temperature distribution, and energy savings compared to standard heating methods.[95,96] Microwaves, which use dipole rotation and ionic conduction to provide energy to carbon particles, have been used in the manufacture and regeneration of ACs.[97] Microwave processing networks are also slightly flattened, convenient, long-lasting, and profitable, all of which can help ACs improve their adsorption capacity.

6.8.8 Grafting of Different Moieties

The modification of ACs to improve the selective adsorptive characteristics of ACs has garnered a lot of attention. Grafting is one of the promising approaches for modifying ACs among the many options. Grafting chemical agents onto AC surfaces aids the binding of functional groups from chemical agents to the porous carbon surface by forming new bonds as a result of the chemical reaction. Grafting terminates the functionalization of the carbon surface by evaporating it in chemical solvents, resulting in functional moieties that are eventually chemically linked to the carbon chain, providing the much-needed temporary stability.[98] The grasping of functional groups by bond formation is the most significant advantage of functionalization by grafting onto the surfaces of ACs.[99] However, because of the aversion resulting from the bond formation, grafting functional groups has the disadvantage of being less sensitive compared to free groups. When surface carbon has been completely utilized, this approach is also affected by limited chemical loading. The adsorption/desorption process for Cd(ii) ions onto the TATS@AC composite is shown in Figure 6.5.

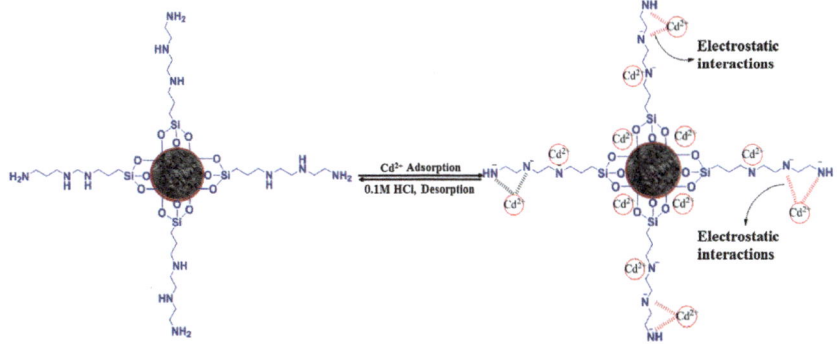

Figure 6.5 Process of adsorption/desorption for Cd(II) ions onto the TATS@AC composite. Reproduced from ref. 99 with permission from Elsevier, Copyright 2020.

6.9 Adsorption Process

Adsorption is a surface phenomenon that occurs when adsorbate molecules bind to an adsorbent. The variation in the surface energy of the material is the basic principle of the adsorption process. Interactions between the adsorbent surface and adsorbate molecules occur to stabilize the surface energy. Temperature, solution pH, adsorbate concentration, contact time, coexisting ions, and adsorbent surface attributes such as maximum accessible surface area, pore size, and pore distribution all influence the adsorption of contaminants onto the adsorbent. The green synthesis of iron oxide nanoparticles (IONPs) was carried out using *Excoecaria cochinchinensis* leaves. Both synthesis and the process of adsorption for rifampicin (RIF) onto IONPs are shown in Figure 6.6.

Based on the nature of interactions between adsorbate and adsorbent molecules, adsorption can be divided into two categories, which are physical and chemical. Chemical adsorption is defined by chemical interactions involving the exchange of electrons between the two phases, resulting in the formation of permanent bonds and an irreversible process. Physical adsorption, on the other hand, is a reversible process that involves electrostatic, π–π, van der Waals, or H-bonding interactions.[100,101] Two methods can be used to determine the electrical properties of an adsorbent surface which are based on the point of zero charges (PZC) and isoelectric point (IEP). The pH of a solution at which the surface charge equals zero is the PZC. It can be determined using many methods, such as the pH drift method or potentiometric titration. On the other hand, the pH at which the electrokinetic parameter at shear planes equals zero is by the IEP. The main distinction between the two is that the IEP only identifies the adsorbent's external surface charge, whereas the PZC determines both its internal and external surface charges. The internal surface charge of the adsorbent is determined by the difference between the two. The performance of an

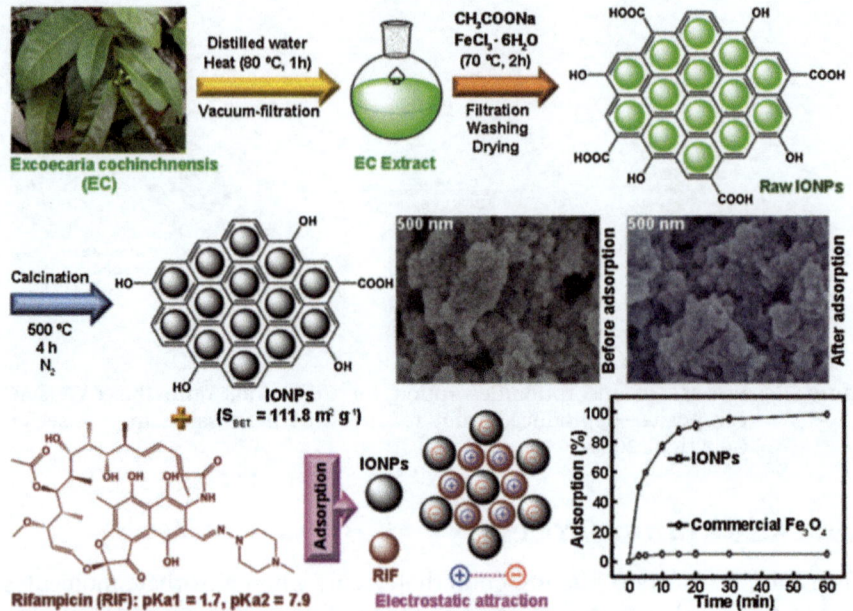

Figure 6.6 Green synthesis of IONPs using *Excoecaria cochinchinensis* leaves and the adsorption mechanism for RIF onto IONPs. Reproduced from ref. 101 with permission from Elsevier, Copyright 2021.

adsorbent is measured during adsorption trials in terms of its adsorption capacity, which can be estimated using eqn (6.1):

$$q_e = \frac{(C_0 - C_e)}{m} V \tag{6.1}$$

Here C_0, C_e, m, and V represent the initial adsorbate concentration and equilibrium adsorbate concentration in $mg\,L^{-1}$, the mass of the adsorbent in grams, and the volume of the whole reaction solution in liters, respectively. The solid-to-liquid ratio is defined as m/V. Furthermore, the adsorption percentage is used to indicate the entire amount of adsorption that happened during a process, which may be computed using eqn (6.2):

$$\% \text{ Adsorption} = \frac{(C_0 - C_t)}{C_t} \times 100 \tag{6.2}$$

where C_0 and C_t represent the initial adsorbate concentration in $mg\,L^{-1}$ and the adsorbate concentration in $mg\,L^{-1}$ at time t (in minutes), respectively. Because the adsorption percentage has some observational error, q_e is usually favoured over it.

Osmosis, photocatalysis, membrane filtration, and adsorption have all been proposed for environmental remediation purposes; however, adsorption is the most effective due to low-cost adsorbent production, high efficiency, simple design, and insensitivity to toxic pollutants, among other factors. At the end of the reaction, this approach does not even produce harmful by-products. Pesticides, organic dyes, metal ions, gases, and PPCPs are among the harmful contaminants that can be eliminated with this method.

6.9.1 Adsorption Isotherm

Adsorption isotherms aid in determining the mechanism of adsorbate molecule adsorption onto the adsorbent. Models also assist in establishing the adsorbent's maximal adsorption capacity. The sort of interaction between the adsorbate and adsorbent, which may be easily identified using the Langmuir and Freundlich isotherm models, respectively, defines whether adsorption is monolayered or multilayered. The Temkin, Redlich–Peterson, and Dubinin–Radushkevich models are some other isotherm models that are also often employed.

6.9.1.1 *Langmuir Isotherm Model*

The Langmuir isotherm model assumes that adsorbate molecules are adsorbed onto homogeneously distributed sites. It also assumes that the interaction takes place through monolayer adsorption onto the adsorbent surface.[102,103] The Langmuir isotherm model can be explained in the non-linear form by eqn (6.3):

$$q_e = \frac{q_m K_a C_e}{1 + K_a C_e} \tag{6.3}$$

where q_e, q_m, K_a, and C_e represent the adsorbent's equilibrium adsorption capacity in $mg\,g^{-1}$, the maximum adsorption capacity in $mg\,g^{-1}$, the Langmuir constant in $L\,mg^{-1}$, and the adsorbate's equilibrium concentration in $mg\,L^{-1}$, respectively.

6.9.1.2 *Freundlich Isotherm Model*

In contrast to the Langmuir isotherm model, the Freundlich isotherm model assumes that adsorbate molecules bind to heterogeneously distributed sites. This model explains the adsorbate's multilayered adsorption. The non-linear form (eqn (6.4)) for this model is as follows:

$$q_e = K_f C_e^{1/n} \tag{6.4}$$

The Freundlich isotherm model's linearized form is given by eqn (6.5):

$$\ln q_e = \ln K_f + \frac{1}{n} \ln C_e \tag{6.5}$$

where q_e, C_e, K_f, and n represent the quantity of metal ions adsorbed at equilibrium time ($mg\,g^{-1}$), the equilibrium concentration of dye in solution ($mg\,L^{-1}$), the adsorbent's capacity, and Freundlich's adsorption intensity constant, respectively. The K_f and n are calculated from the intercept and slope, respectively, using the plot of $\ln q_e$ vs. $l\ln C_e$.

6.9.2 Adsorption Kinetics

The adsorption kinetic models are generalized by a number of governing processes, including mass transfer coefficient, chemical reaction, and diffusion control. Kinetic studies aid in finding the ideal reaction conditions for conducting full-scale batch adsorption experiments. Adsorption kinetics provide insights into the rate of solute uptake and the mechanism by which this rate controls the adsorbate's residence time at the solution's interface. For assessing kinetics data, a variety of models can be used, including the pseudo-first-order kinetic model, the pseudo-second-order kinetic model, the intraparticle diffusion model, and the Elovich model.

6.9.2.1 Pseudo-first-order Kinetics

The adsorption at the solid–liquid interface is explained using a pseudo-first-order kinetic model. The linearized version of the pseudo-first-order kinetic model is given by eqn (6.6):[102]

$$\log(q_e - q_t) = \log q_e - \frac{k_1}{2.303} t \tag{6.6}$$

where q_e, q_t, k_1, and t represent the adsorption capacity of the adsorbent at equilibrium ($mg\,g^{-1}$), the adsorption capacity of the adsorbent at time t ($mg\,g^{-1}$), the pseudo-first-order rate constant, and t contact time (minutes) respectively. The values of k_1 and q_e can be estimated from the slope and intercept of a linear plot of $\log(q_e - q_t)$ vs. time.

6.9.2.2 Pseudo-second-order Kinetics

This model assumes that chemical interactions between adsorbate and adsorbent molecules regulate the rate-limiting phase in adsorption. Eqn (6.7) provides the differential form of this model:[103,104]

$$\frac{dq_1}{dt} = k_2(q_e - q_t)^2 \tag{6.7}$$

The pseudo-second-order rate constant ($g\,(mg\,min)^{-1}$) is represented by k_2. The linearized form of this model is given by the equation:[105]

$$\frac{t}{q_t} = \frac{1}{k_2 q_e^2} + \frac{1}{q_e} t \tag{6.8}$$

The values of q_e and k_2 can be derived from the slope and intercept of a linear plot of $\dfrac{t}{q_t}$ *versus* time.

6.9.2.3 Intraparticle Diffusion Method

The rate-limiting step in the intraparticle diffusion model is the transport of solute from solution into adsorbent pores *via* an intraparticle process. Weber and Smith proposed that the amount of adsorption varies proportionally with $t^{\frac{1}{2}}$ in spite of the contact duration for most adsorption processes, and this may be stated by eqn (6.9):[106]

$$q_t = k_{id}t^{0.5} \tag{6.9}$$

where q_t, $t^{0.5}$, and k_{id} represent the adsorption capacity at time t, the half-life period in seconds, and the intraparticle diffusion rate constant in $mg\,g^{-1}\,min^{-0.5}$. Plotting q_t *versus* $t^{0.5}$ yields a linear relationship, and the slope of the plot can be used to derive the rate constants.

6.9.3 Thermodynamic Studies

Thermodynamic studies aid in determining (a) the kind of adsorption, which can be either physical or chemical, (b) randomness, (c) exothermic or endothermic, and (d) the spontaneity of the adsorption reaction. All of these variables can be calculated using the following equations:[105,107]

$$\Delta G^0 = -RT\ln K_c \tag{6.10}$$

$$\Delta G^0 = \Delta H^0 - T\Delta S^0 \tag{6.11}$$

$$\ln K_c = \dfrac{\Delta H^0}{R} \times \dfrac{1}{T} + \dfrac{\Delta S^0}{R} \tag{6.12}$$

Here, T and R represent the operational temperature in Kelvin and the universal gas constant in $J\,mol^{-1}\,K^{-1}$. The value of ΔG^0 (Gibbs free energy change) can be derived directly from the aforementioned equation; however, the intercept and slope of the thermodynamic graph shown between $\ln K_c$ and $\dfrac{1}{T}$ are used to determine ΔS^0 (change in entropy) and ΔH^0 (change in enthalpy). In addition, the adsorption–isotherm constants are used to calculate K_c. ΔG^0 is measured in $J\,mol^{-1}$, and temperature is measured in K.

6.10 Mechanism of Adsorption onto AC

The mechanism of adsorption of various contaminants onto AC must be thoroughly investigated. Bulk diffusion, film diffusion, pore diffusion, and intraparticle diffusion are the four main processes involved. Bulk diffusion involves the immediate transfer of adsorbate molecules in the solution phase. Film diffusion refers to the movement of adsorbate molecules to the external surface of the adsorbent molecule through a hydrodynamic boundary. Pore diffusion is the process wherein adsorbate molecules are transferred into the pores of an adsorbent, lowering the total adsorption rate. Finally, intraparticle diffusion refers to the movement of adsorbate molecules from the outside of an adsorbent into its pores and along the pore-wall surfaces.[106–108] Bonding, electrostatic, π–π, and n–π interactions are only a few of the interactions that play a role in their adsorption.[109] Between the π-electrons of AC and the adsorbate molecules, there are π–π interactions, also known as π–π electron donor–acceptor interactions. The presence of electron-withdrawing groups at the AC's edges and surface causes a significant decline in electron density. The Cπ–cation interactions that occur between the π-electron cloud of AC and metal cations enhance the removal of metal cations by AC. Tran *et al.* reported the use of pyrolysis to produce biochar with characteristics similar to those of AC from orange peels. Cadmium ions adsorbed onto the biochar surface *via* Cπ–cation interactions, and a maximal adsorption capacity of 114.69 mg g^{-1} was achieved.[110] Bui *et al.* reported that the adsorption of metal cations is highly influenced by the pH of solution.[111] In addition to the interactions mentioned earlier, hydrogen bonding is significant in the adsorption of organic aromatic molecules onto AC. This interaction exists between the adsorbent and AC's functional groups, such as N–H in atrazine and O–H in paracetamol. According to one mechanism, the O–H groups on the surface of AC form hydrogen bonds with the functional groups in organic adsorbates. The magnitude of adsorption is controlled by the net strength of interactions. Some complexes between solvent molecules and surface oxides occur in this method, which may prevent solute molecules from migrating from the outside to the AC micropore structure.[112] These interactions are strongly temperature-dependent, as they disappear when the temperature increases due to the increase in adsorbed molecules' kinetic energy.

6.11 Environmental Applications of AC

6.11.1 Heavy Metals

Heavy metals are defined as metals with atomic weights of 63.5 to 200.6 g mol^{-1}.[113] Heavy metals are mostly introduced into the environment by the fertilizer, paper, pesticide, tannery, and battery industries, either directly or indirectly. Heavy metals are non-biodegradable in nature and accumulate in living beings. Most heavy metal ions, such as mercury, cobalt, cadmium, and lead, are poisonous or carcinogenic in nature.[114–116] Zinc is a transition metal with an atomic number of 30 that is extremely important for human health. Excess zinc consumption can result in major side effects such as skin irritations, stomach cramps, nausea, vomiting, and anemia. Mercury, a neurotoxin that damages the central nervous system, is another example. High mercury levels impede pulmonary and kidney function, as well as induce chest pain and dyspnea. Minamata disease is one of the most common diseases induced by mercury exposure.[117] Lead, in addition to mercury, can harm the central nervous system. It can also harm the kidneys, liver, and reproductive system, as well as disrupt brain functions. Anemia, headache, dizziness, sleeplessness, muscular weakness, and kidney impairment are among the hazardous side effects.[118] Chromium is found in the environment in two forms: Cr(III) and Cr(VI), the latter of which is very hazardous.[119] It affects human physiology, accumulates in the food chain, and produces a wide range of health problems, from minor skin irritation to lung cancer.[120] Through Cπ–cation interactions, metal ions interact with AC and are adsorbed onto its surface and into its pores. The elimination of lead ions from aqueous effluents was accomplished using AC made from rapeseed. FTIR analysis was performed to provide evidence for the participation of C–Pb(II) interactions. After the adsorption of Pb(II) ions, the typical peak of C=O at 1710 cm^{-1} vanished, indicating complexation Cπ–cation interactions.[121] In another experiment, AC made from coconut shells was used to adsorb Pb(II) ions from an aqueous solution. Adsorption was shown to be substantially dependent on solution pH, with a maximum adsorption capacity of 26.50 mg g^{-1} seen at pH 4.5. The reaction was found to be endothermic using thermodynamic analysis.[122] Chemically activated AC made from coirpith was used for the adsorption of Ni(II) ions from an aqueous solution. The adsorption rate was greatly influenced by the carbon/nickel ion concentration ratio. The highest adsorption of Ni(II) ions was found in a basic medium, and desorption in an acidic environment (HCl). Coirpith-based AC had a maximum adsorption capacity of 62.5 mg g^{-1}.[123] El-Hendawy investigated the adsorption of Pb(II) and Cd(II) ions onto AC that had been preactivated by physical (steam) and chemical (H_3PO_4 and KOH) treatments. The H_3PO_4 chemical treatment activation was shown to be the most successful of the three activation treatment methods, with a maximum adsorption capacity of 139 mg g^{-1} for Pb(II) and 129 mg g^{-1} for Cd(II). The adsorption isotherms of Pb^{2+} and Cd^{2+} on some ACs under different conditions are shown in Figure 6.7. Because of their lower solubility and lower pH needed for complexation, Pb(II) ions were removed or adsorbed in preference to Cd(II) ions.[124] Ameh *et al.* synthesized AC by treating Iraqi palm date pit samples with NaOH and then used it to adsorb Cu(II) and Cd(II) ions. Maximum adsorption capacities of 118.06 and 88.42 mg g^{-1} were observed for Cu(II) and Cd(II) ions, respectively, based on the Langmuir isotherm model. The adsorbent dosage was found to have a considerable influence on adsorption performance.[125]

6.11.2 Gas Adsorption

Adsorption of flue gases and by-product vapours has become crucial in industries, whether for the adsorption of global warming gases like carbon dioxide (CO_2) and methane, or poisonous gases like oxides of nitrogen (NO_x)

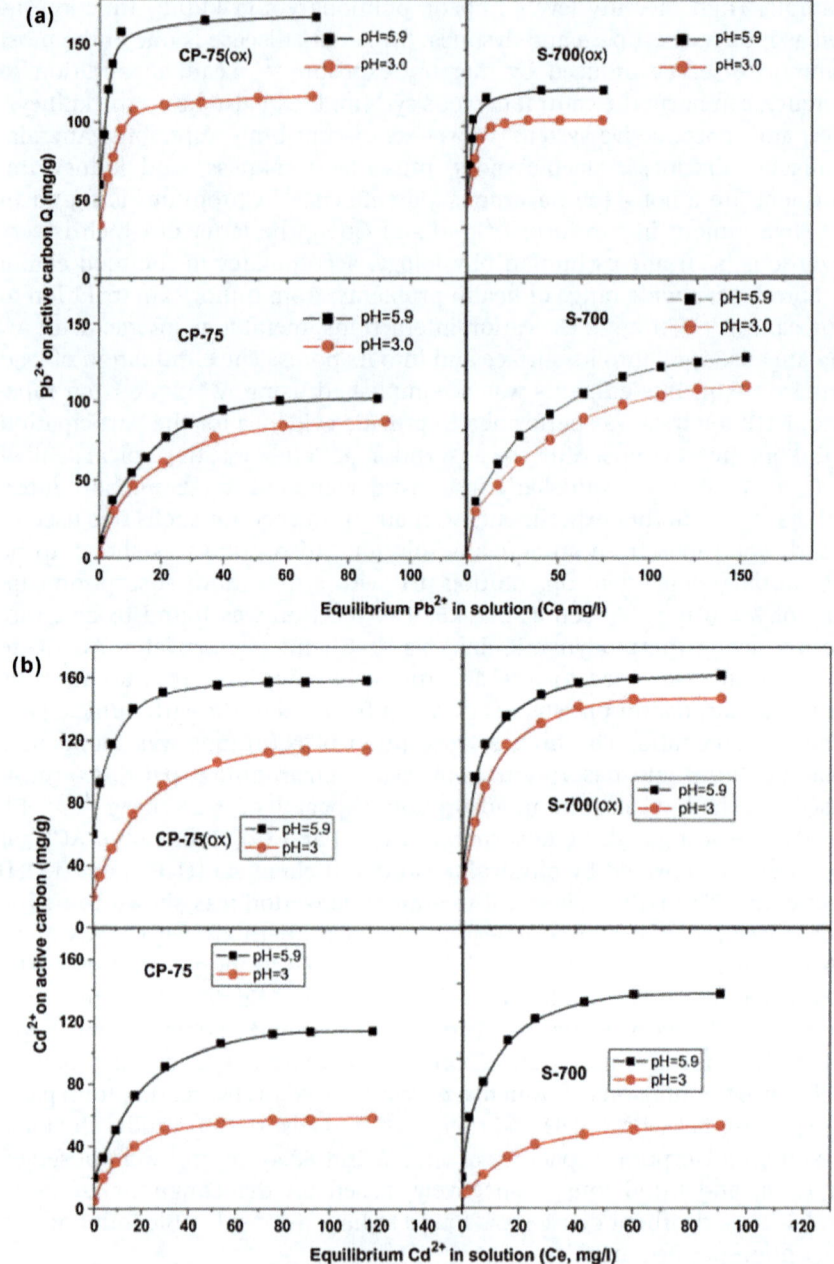

Figure 6.7 Adsorption isotherms of (a) Pb^{2+} and (b) Cd^{2+} on some ACs under different conditions. Reproduced from ref. 124 with permission from Elsevier, Copyright 2008.

and sulphur dioxide (SO_2). Carbon dioxide sequestration using appropriate adsorbents has become a major research topic in recent decades. Many carbon-based materials like activated biochar compounds, carbon nanotubes, metal–organic frameworks (MOFs), and AC composites have been investigated for gas sequestration *via* adsorption.[126]

Biochar is a viable option for producing AC. The thermal processes pyrolysis and carbonization can be used to treat a wide spectrum of biomass wastes. The substance can be activated, either physically or chemically, to produce biochar-AC. In terms of gas adsorption, AC from biochar was mostly used for carbon dioxide adsorption. Rice husk was pyrolyzed at 600 °C, then activated with a mix of nitrogen and ammonia, resulting in a carbon dioxide capture rate of up to 77.9 $mmol\,g^{-1}$. Because it is both cost-effective and environmentally friendly, activated biochar is chosen over ordinary AC. However, more research into the impact of feedstock composition, production parameters, and activation settings on biochar-AC performance is required.[127]

6.11.3 Pesticides

Pesticides are a class of chemical or biological chemicals with antibacterial and disinfecting capabilities that are used to control insects, weeds, birds, animals, and nematodes, among other microorganisms and pests. They aid in the production, processing, storage, and transportation of a wide range of foods.[128] Pesticides are distinguished by their remarkable chemical structures, which are designed to mimic and substitute certain molecules in targeted pests, resulting in a deadly disruption of expected biological responses.[129] Pesticides help to reduce the impact of undesired bacteria or other agents, as well as meet the ever-increasing food demands brought on by the world's growing population. Herbicides, insecticides, ovicides, rodenticides, fungicides, avicides, nematicides, and bactericides, among others, can be used in the agricultural sector.[130,131]

Use of pesticides as agricultural repellents, fumigants, defoliants, or sterilants has emerged in recent decades as one of the most difficult tasks, as a result of which it has garnered a lot of momentum and popularity around the world.[132] Precipitation, flotation, sedimentation, membrane processes, filtration, coagulation–flocculation, biological processes, adsorption, electrochemical techniques, chemical reactions, and ion exchange are some of the purification or treatment techniques that have been used with varying degrees of success in the scientific community.[133–136] Among them, the adsorption method has been proven to be the most successful and advantageous approach in wastewater treatment. 2,4-Dichlorophenoxyacetic acid (2,4-D) is a herbicide used to manage broadleaf weeds with care. Hameed *et al.* used AC treated with KOH/CO_2 for physiochemical activation to study its elimination from an aqueous solution. Various adsorption isotherms were used, with the Langmuir data fitting the best. The predicted maximum adsorption capacity was 238.10 $mg\,g^{-1}$. In comparison to solution pH and other parameters, the herbicide starting concentration had a stronger impact.[137] Chingombe *et al.* described a thermal treatment that increased the micropore concentration by annealing AC in a reactor, resulting in an increase in surface area from 790 to 960 $m^2 g^{-1}$ due to the reduction in oxygen-containing functional groups.[138] The PSD and adsorption process for F400 and F400AN are shown in Figure 6.8.

The uptake of 2,4-dichlorophenoxyacetic acid and benazolin onto its heat-treated AC was increased as a result of the enlarged surface area. Adam *et al.* studied the adsorption of propiconazole, a fungicide, under two distinct activation regimes, one basic (NaOH) and the other acidic (HNO$_3$ and H$_2$O$_2$). The acidic treatment of AC reduced its adsorption capacities, according to the findings. Acidic groups enhanced the oxygen-containing functionalities on the surface, which had a negative influence on propiconazole adsorption, resulting in a decrease in adsorption capacity.[139] Yoo *et al.* reported that the treatment of AC with nitric acid enhanced the concentration of carboxylic groups on its surface. This treatment, however, had no significant effect on

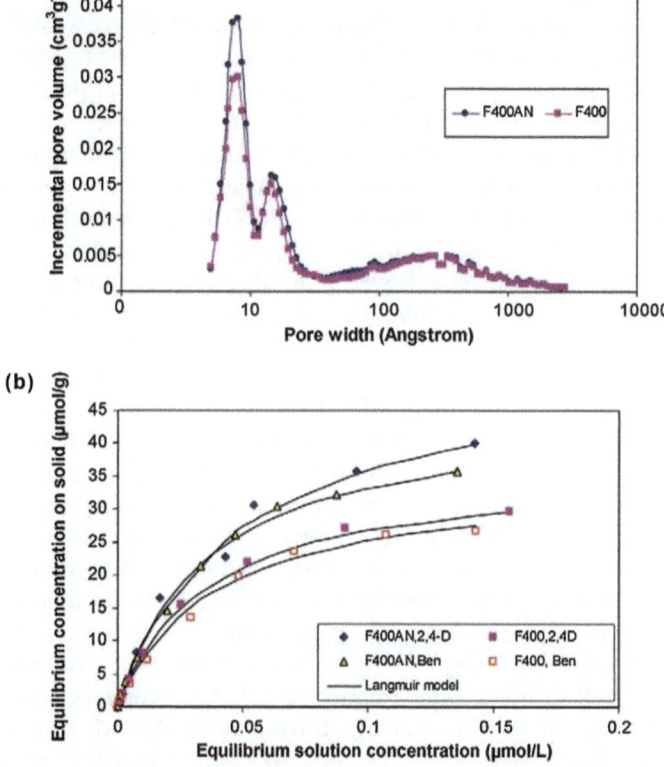

Figure 6.8   (a) PSD of F400 and F400AN and (b) adsorption of benazolin and 2,4-D onto F400 and F400AN. Reproduced from ref. 138 with permission from Elsevier, Copyright 2005.

the rate of oligosaccharide (carbohydrate component) adsorption from aqueous solutions.[140]

6.11.4 Pharmaceutical and Personal Care Products (PPCPs)

Several potential pathways have been proposed for identifying the causes of pharmaceutical pollution, which include incompletely metabolized and expelled medicines in the urine and faeces that reach the water system. Another possible cause is leftover or expired medications that are put into toilets. Few medications degrade quickly and thus persist in the water supply. Aquatic toxicity, genotoxicity, and endocrine disruption are some of the negative impacts of PPCPs. Furthermore, the existence of trace amounts of PPCPs in drinking water is a public health concern, as little is known about the long-term health effects of ingesting mixes of such chemicals through drinking water. Salman *et al.* investigated the elimination of insecticides bentazon and carbofuran from aqueous solutions. Date seed-based AC was synthesized and activated with potassium hydroxide and carbon dioxide.[141] In comparison to the Langmuir model, the Freundlich model fits the adsorption equilibrium data better for both pesticides. Bentazon and carbofuran had maximum adsorption capacities of 86.26 and 137.04 $mg\,g^{-1}$, respectively, based on Langmuir isotherms. Because carbofuran is smaller than bentazon, it easily penetrates the pores of AC, resulting in excellent removal efficiency. In comparison to the solution pH, the pesticide starting concentration was revealed to be the more relevant factor.

6.11.5 Dyes

Dyes are coloured substances that are used to give fabrics and paper their colour. These have a proclivity for adhering to any receptive material. These have been used for many ages. These were originally produced from natural sources such as plants, insects, and bacteria, but one of the major drawbacks of these dyes was their short-term applicability. These dyes faded under the effect of sunlight and temperature, necessitating the development of synthetic dyes. Synthetic dyes are laboratory-made dyes that were first invented by W. H. Perkins in 1856 to overcome the problem of dye colour persistence. The dyes created synthetically were structurally stable and so it was difficult to break them down.[142–144]

With their introduction, companies began to use them as a key component of their products, discarding them into the water system without treatment, which posed serious environmental and health risks. Although it appears that restarting the use of natural dyes is a better option, it was found to be worse than using synthetic dyes alone since a mordant is necessary to boost the bonding capabilities with the materials. Mordants are binders that aid in the adhesion of natural dye to various fabrics. These are far more hazardous and poisonous than synthetic colours. Dyes are divided into several classes depending on their structure, colour, application, and chemical makeup. Furthermore, these dyes can be classified as cationic, anionic, or non-ionic based on the particle charge gained after dissolution in aqueous media.[145–147] Dye-using enterprises dump dye effluents directly into water sources, with no treatment. Environmentalists are concerned about dye pollution in water because dyes are poisonous and harmful.[148,149] The introduction of concentrated dye effluents into acidic water bodies with high temperatures disrupts the oxygen transfer process and the ability of environmental water bodies to self-purify. When water bodies are mixed with dye effluents, turbidity develops. This enables the formation of a visible foam layer above the surface of the water. This process prevents sunlight from entering the system and limits the amount of sunlight available to aquatic animals for photosynthesis and respiration. Water bodies with dye effluents serve as breeding grounds for a variety of microbes such as bacteria and fungi, rendering the water unfit for human consumption. The International Dye Industry Wastewater Discharge Quality Standards recently developed legislation limiting the level of presence of these effluents, according to which enterprises must ensure to release the wastewater into water bodies.[150–152]

The surface characteristics of AC play a big role in dye adsorption. AC is an amphoteric substance that can exhibit both cationic and acidic properties. Its interaction with dye molecules is mostly impacted by the solution pH, which causes the surface to react differently depending on the pH of the solution. Malachite green was adsorbed from an aqueous solution using AC made from groundnut shells after chemical treatment with $ZnCl_2$.[153] Malachite green was adsorbed by the activated carbon due to its high surface area of 1200 $m^2 g^{-1}$. The maximum adsorption capacity was calculated to be 114 $mg g^{-1}$. Langmuir adsorption isotherms for malachite green dye adsorption from groundnut shell-based powdered AC (GSPAC) and commercially available powdered AC (CPAC) are shown in Figure 6.9.

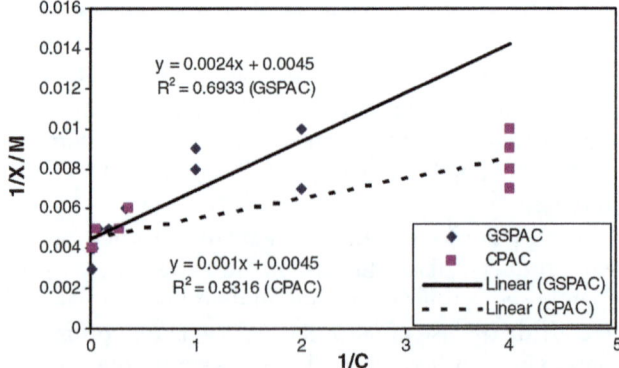

Figure 6.9   Langmuir adsorption isotherms for malachite green dye adsorption from groundnut shell-based powdered AC (GSPAC) and commercially available powdered AC (CPAC). Reproduced from ref. 153 with permission from Elsevier, Copyright 2006.

Cotton fibres were used as the carbon precursor, and $ZnCl_2$ was used as the activating agent in another study. Cotton fibre-AC (CFAC) was used to remove methylene blue from an aqueous solution as intended, and BET surface analysis revealed that CFAC had a large surface area of 2060 $m^2 g^{-1}$ and a maximum adsorption capacity of 597 $mg g^{-1}$. The Langmuir isotherms fit the adsorption data the best. The adsorption process was found to be encouraged by an acidic pH in a study. Chemical interactions promoted methylene blue adsorption on CFAC, according to the pseudo-second kinetic model.[154] Microorganisms such as *Enteromorpha prolifera*, in addition to agricultural products, were utilized to prepare carbon. It was chemically activated with $NaAlO_2$ to add functionalities to its surface and make it re-active. It was endowed with a high surface area of 1374 $m^2 g^{-1}$ after activation by $NaAlO_2$. It was effectively used to remove bright scarlet from an aqueous solution *via* adsorptive elimination. According to the Langmuir isotherm, a high adsorption capacity of 1000 $mg g^{-1}$ was obtained. *Albizia lebbeck*, hazelnut husk, macadamia nut endocarp, oil palm wood, olive stones, acacia fumosa seeds, vetiver roots, waste tea, and date stones are some of the AC precursors that have been reported for the removal of methylene blue.[154–164]

6.11.6 VOCs

Chemical substances known as VOCs are the primary sources of air pollution. Carcinogenic, mutagenic, and poisonous VOCs such as phenol, benzene, formaldehyde, ketone, butane, xylene, toluene, and chlorinated chemicals are among the most harmful. VOC emissions into the environment have increased as a result of increased industrialization. Transportation, construction, and natural emissions are all sources of VOCs in the environment. These are usually found in larger concentrations near petrochemical plants than in residential areas. Global warming, ozone depletion, acid rain, and photochemical smog are all caused by the presence of VOCs in the environment. They induce skin irritation, liver damage, liver malfunction, and asthma in humans. VOC concentrations in the environment have increased rapidly; the total amount of VOCs in China increased from 1.15 Tg in 1980 to 13.35 Tg in 2010.[165]

In the literature, a variety of approaches for removing VOCs have been proposed, including catalytic oxidation, photocatalytic oxidation, reverse osmosis, ion exchange condensation, thermal oxidation, biological therapies, and adsorption. The simplest, most efficient, and cost-effective solution is the adsorption method, with the following advantages: a high degree of operational stability, no harmful by-product production, no further treatment requirements, and cost-effectiveness. The surface area of AC made from avocado kernels and activated with carbon dioxide was 206 $m^2 g^{-1}$. It was used to remove phenol from the environment. The researchers found that a slightly acidic pH improved adsorption, with a maximum adsorption capacity of 90 $mg g^{-1}$.[166]

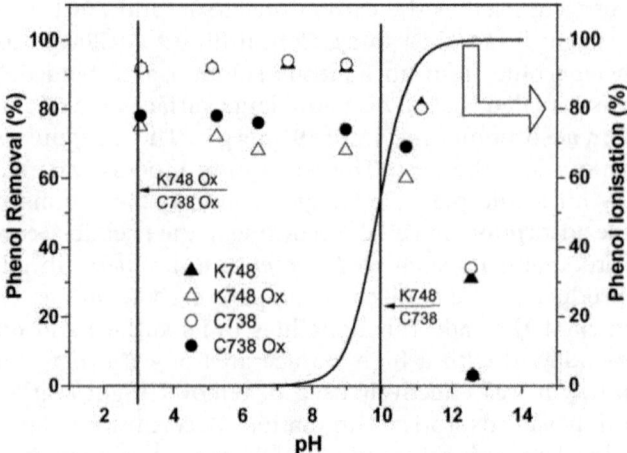

Figure 6.10 Phenol removal as a function of the initial solution pH from the ACs synthesized from kenaf and rapeseed. Reproduced from ref. 168 with permission from Elsevier, Copyright 2009.

According to a kinetic study, chemical interactions enhanced the adsorption of phenol onto AC. Durian shell-derived AC was activated with H_3PO_4 and had a high surface area of 1404 $m^2 g^{-1}$. It was utilized to remove toluene through adsorption, with a maximum adsorption capacity of 874 $mg g^{-1}$. In comparison to the Langmuir isotherm, isotherm investigations revealed that the Freundlich isotherm explained the adsorption process better.[167] As carbon precursors, kenaf and rapeseed were also used. Carbon dioxide was then used to activate it, resulting in a high surface area of 1112 $m^2 g^{-1}$. It was employed to remove phenol, with a maximum adsorption capacity of 84.1 $mg g^{-1}$ as measured by the Langmuir isotherm. The results revealed that adsorption was substantially influenced by the pH of the solution, with the highest adsorption occurring at pH 10 as shown in Figure 6.10.[168]

6.11.7 AC as a Catalyst Support

Any catalyst's overall activity and reactivity are determined by its chemical composition, surface area, and physical properties. Any active material or phase can be imprinted onto suitable support to increase its surface area and strengthen its mechanical properties. Furthermore, by depositing the highly dispersed active phase into the support's micropores, it can be stabilized. The active phase and support have a synergistic relationship that boosts its catalytic activity. It aids in the prevention of agglomeration. The most critical parameter to consider when building a catalyst is the correct degree of dispersion of the active phase onto the support, which is defined as the ratio of active phase surface atoms to total atoms present. A variety of materials can be used as supports, and one of them is AC, which is widely used because of its environmentally friendly nature, low cost, and easy synthetic pathways. It works as a sink for the created electron–hole pairs, which helps to reduce charge pair recombination. However, impurities, often known as catalyst poisons, such as copper, zinc, and iron, have a significant impact on their activity. AC's pore structure and internal surface may both be altered and this is a major benefit when AC is used as a catalyst support. It is made up of disorganized graphitic structural elements in a tiny area. Electron transport from the margins of carbon sheets determines their catalytic activity. As a result, the presence and spread of these active sites are critical. The inclusion of heteroatoms in the lattice, such as sulphur and nitrogen, can alter the catalytic activity of the graphite structure. The presence of acidic or basic surface oxide complexes affects its activity as well. AC has several advantages as a support material such as high resistance to acidic or basic media, tailored PSD for specific reactions, lower cost than silica and alumina supports, and good temperature stability. For chemical reactions like hydrolysis, esterification, and hydrogenation, AC and its composites can be used as catalysts or catalyst supports. The carbon catalyst, which has an amorphous shape and highly reactive groups like –COOH, –OH, and –SO$_3$H, is an effective catalyst for the hydrolysis of pure cellulose to glucose, which was previously impossible with catalysts like Amberlyst-15, Nafion, and H-mordenite. Because of its low activation energy need, the carbon catalyst increased the hydrolysis rate. The ability of the catalyst to adsorb 1,4-glucan contributes to its high performance.[169] Another branch in which carbon functions as a catalyst or as a support is the hydrodeoxygenation process. Such a reaction involve the conversion of carbohydrates to unsaturated compounds using a catalyst that favours C–O cleavage over C–C cleavage while using the least amount of H$_2$. MXenes supported using AC, such as Mo$_2$C, aid in the conversion of vegetable oils including rapeseed, olive, and soya bean oils to diesel-range hydrocarbons.[170,171] The treatment of substances with molecular hydrogen is known as hydrogenation. Hydrogenation on an industrial scale reduces nitro groups in nitroaromatic chemicals to decrease the possibility of other compounds such as aniline being produced. In 1977, the hydrogenation of nitrobenzene was investigated using a Pt catalyst supported on AC.[172] The results reveal that the impregnation of Pt onto AC increased its catalytic activity. The dispersion of Pt onto AC, the pore structure of AC, the chemical composition of the AC surface, and temperature were all found to have a significant impact on catalytic activity. It has been found that there is a linear relationship between activity and Pt dispersion (measured at equivalent reduction temperature). In the case of phenol oxidation, AC also serves as a support material. Shukla *et al.* synthesized a catalyst by impregnating AC with Co(II) ions, which were used to activate peroxymonosulphate to carry out phenol oxidation. The Co(II) ions were evenly dispersed across the AC surface. The results showed that Co/AC had high activity in oxidizing phenol with sulphate radicals, with 100% breakdown and 80% TOC elimination possible in 60 minutes as shown in Figure 6.11.[173]

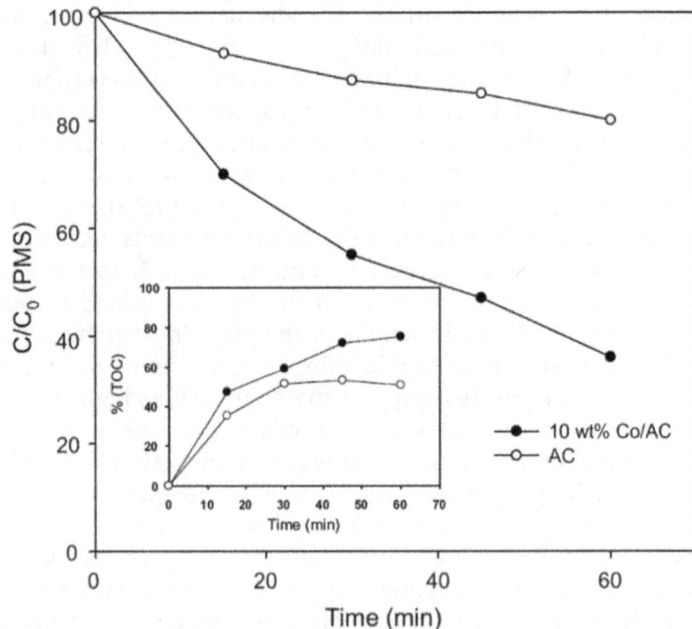

Figure 6.11   The degree of TOC removal from the solution in the presence of both AC and Co–AC. The TOC removal is associated with the amount of phenol removal. Reproduced from ref. 173 with permission from Elsevier, Copyright 2010.

6.11.8   AC in the Food and Pharmaceutical Industries

In the food and pharmaceutical industries, AC and its numerous compounds have been found to be useful. AC has a long history of usage as a pH regulator, chelating agent, and antioxidant synergist in the food and pharmaceutical industries. AC acts as an acidulant, assisting in the decolorization of maleic and adipic acids before conversion and crystallization. In the food, beverage, and pharmaceutical industries, sodium benzoate and sodium sorbate are employed as preservatives. Even at a concentration of less than 1%, granular AC can help decolorize and deodorize them during the production process.[174] AC has also been shown to be an effective adsorbent in pharmaceutical processing applications that require a highly efficient filtration system due to its excellent purity and efficacy. This protects soluble minerals with a minimal acid content, which reduces final product adulteration while having little effect on solution pH. Because of its unique adsorptive qualities and bio-origin, AC also has antitoxic properties, allowing it to swiftly reduce the effects of toxins or excessive stimulant consumption by assisting with digestion and sensitive gastrointestinal conditions. AC is widely used as a feed additive in the veterinary care industry. It works in the same way as human medicine by adsorbing unwanted or possibly hazardous compounds from the gastrointestinal system of animals, making it an efficient therapy for poisoning and diarrhoea.

6.11.9 AC as a Storage System for Gases

Increased industrialization has resulted in an increase in the emission of gases such as CO_2, SO_2, H_2, and CH_4 in the atmosphere (beyond their allowed limits). CO_2 is the primary cause of global warming, and it also contributes to the ozone layer's depletion. Industrial operations, fuel combustion, and deforestation are all major contributors of CO_2 to the atmosphere. CO_2 adsorption and sequestration from the place of origin have been proposed by researchers. Membrane separation, cryogenic techniques, liquid solvent absorption, temperature adsorption, solid sorbents, and pressure adsorption are some of the strategies proposed for CO_2 capture.[175] Because of its easy use, cost-effectiveness, and economic processing, adsorption is considered to be the most effective. CO_2 adsorption on AC needs the presence of basic functionalities on its surface, such as nitrogen-containing functional groups, that promote maximum CO_2 capture.[176] When ammonia is applied to carbon materials at room temperature, it decomposes into free radicals ($-NH_2$, $-NH$, and atomic hydrogen and nitrogen). These free radicals attack the carbon surface, forming functional groups that contain nitrogen.[177] Various techniques such as Fourier transform infrared spectroscopy (FTIR), X-ray photoelectron spectroscopy (XPS), and temperature-programmed desorption (TPD) can be used to successfully determine CO_2 capturing. Because of its abundant natural supply, relative safety (when compared to other fuels), cost-effectiveness, and minimal carbon emissions, methane has gained a lot of attention as a potential future clean energy source. For future transportation systems to be practical, proposed materials must have exceptional storage capacity, high storage and release kinetics, and long-term cycle efficiency under moderate thermodynamic conditions.[178] CH_4 storage utilizing AC made from polymers, hardwood, and coconut shells is similar to compressed natural gas at 250 bar, with a high methane absorption of roughly 200 v/v at 27 °C and pressures ranging from 34 to 48 bar[179] as shown in Figure 6.12.

Hydrogen, in addition to methane, is a superior fuel for future generations. The development of hydrogen infrastructure is critical to the commercialization of fuel cells. The US Department of Energy (DOE) determined that gravimetric systems require 5.5 wt% H_2 and volumetric systems require 40 g H_2 per L.[180] Storage containers with a high volumetric and gravimetric density are necessary for obtaining convenient hydrogen storage systems to operate fuel cells using H_2. Compression, cryogenic and cryo-compressed storage, metal hydrides, and AC are all common H_2 storage/transport systems today.

6.11.10 AC Cloths

AC cloths (ACCs) are gaining popularity due to their advantages over traditional forms and their potential for technological innovation in a variety of disciplines, including their prospective use in cell therapy as stem cell

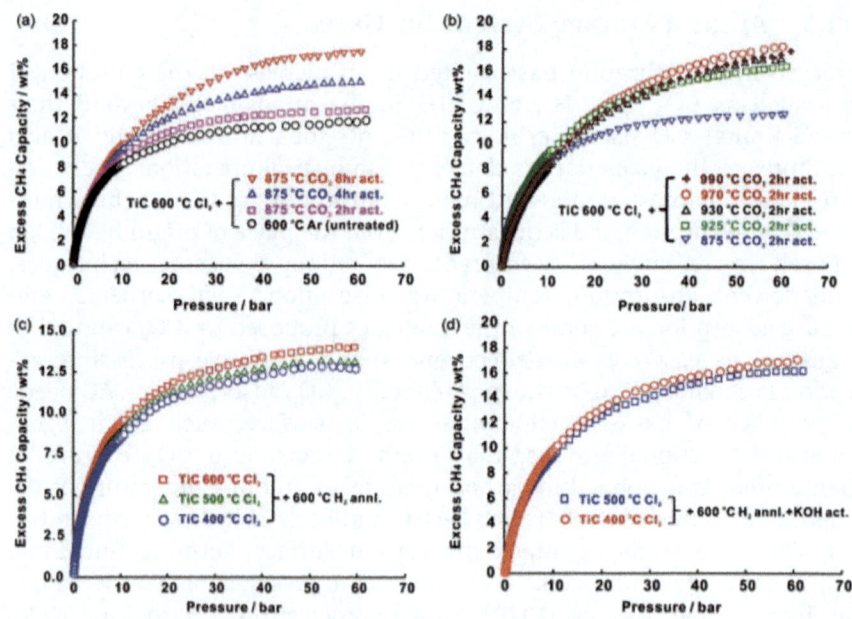

Figure 6.12 Methane storage capacity of the carbide-derived carbons activated with CO_2 or KOH. Reproduced from ref. 179 with permission from Elsevier, Copyright 2009.

growth supports. The tiny diameter of the fibres that make up ACC confers significant benefits. Because of the ACC's increased surface area and pore volume, it has faster adsorption kinetics, higher efficiency, and a bigger adsorption capacity. Micropores (2 nm) in ACC, in particular, are directly connected to the external surface area, lowering the heat and mass transfer resistances and pressure drops in flow units. ACCs are also light materials that can be assembled in a variety of stable patterns to create a continuous carbon form suitable for electrical and electrochemical applications.

Furthermore, the Joule heating effect of ACC provides further benefits in terms of desorption, *in situ* regeneration, and vapour recovery in VOC treatment processes. ACCs are encouraging a new notion in equipment design, favouring the creation of novel devices, such as microstructured reactors, for enhancing chemical process safety and intensification, due to their technological benefits. ACCs have been investigated as catalysts or catalyst/biocatalyst supports in a variety of processes, as electrode materials, and as potential substrates for nanotube formation. They've also been studied extensively for gaseous and liquid pollution abatement as well as biomolecule separation and purification. Gas storage and separation; H_2 recovery, purification, and/or sensing; and sour gas sweetening have all been examined. The two basic procedures used for the creation of AC textiles in traditional forms are physical or thermal activation and chemical activation. Briefly, physical activation entails pyrolysis of the precursor fabric, followed by gasification with an oxidizing gas, usually steam or CO_2, under carefully regulated conditions. The manufacture of AC fibres with supercritical fluids, water, and CO_2 has been explored in recent works. The properties of the AC textiles produced are strongly influenced by the precursor, activation method, and process parameters utilized. ACC development is more difficult than traditional forms because ACCs require a highly developed porous structure while maintaining the integrity of the constituent fibres, *i.e.*, they must be strong enough to avoid break-up and dusting.[181]

6.11.11   Energy Storage Applications of ACs

To meet contemporary society's expanding energy demand while avoiding resource depletion and pollution, high-performance, low-cost, and environmentally friendly energy storage and production systems must be developed. Because of their abundance, chemical and thermal stability, and processability, and the tunability of their textural and structural features to meet the requirements of specific applications, carbon-based materials have piqued interest in many energy-related applications. To meet evolving industrial–technological needs, ACs are also made accessible in other physical forms such as fibres, pellets, cloths, and felts. Controlling the textural features of ACs, as well as their diverse physical forms, has progressed significantly. Traditional ACs, on the other hand, have a broad PSD that spans the micropore to macropore range. Recent advancements in activation techniques and/or precursors have improved PSD control. These properties have expanded the range of uses for ACs, including catalysis/electrocatalysis, energy storage in supercapacitors and Li-ion batteries, CO_2 collection, and H_2 storage.

ACs remain the primary choice for the construction of electrodes for commercial supercapacitors despite recent advances in controlling the porosity of carbon materials through templating procedures and the development of novel materials such as graphene and carbon nanotubes due to their availability, cost, and simpler production methods. In the medium to long term, the scenario is likely to persist, albeit greater control over textural qualities (especially pore size) is required to maximize energy and power densities to meet future energy demands. Furthermore, significant cost reductions are required, and it is expected that low-cost, easily available, and renewable precursors such as biomass will play a vital role. ACs are being explored extensively for application in second-generation supercapacitors, which are promising in terms of increasing the amount of energy stored while maintaining electric double-layer capacitor (EDLC)-like cycling stability and power capability. These technologies are still in their early stages and will benefit from subsequent advances in materials science, such as porous carbon manufacturing and electrochemistry. The development of porous materials capable of storing enough hydrogen to meet the requirements for onboard applications (most notably storage temperature and pressure) is still a challenge. ACs are one of the most studied types of

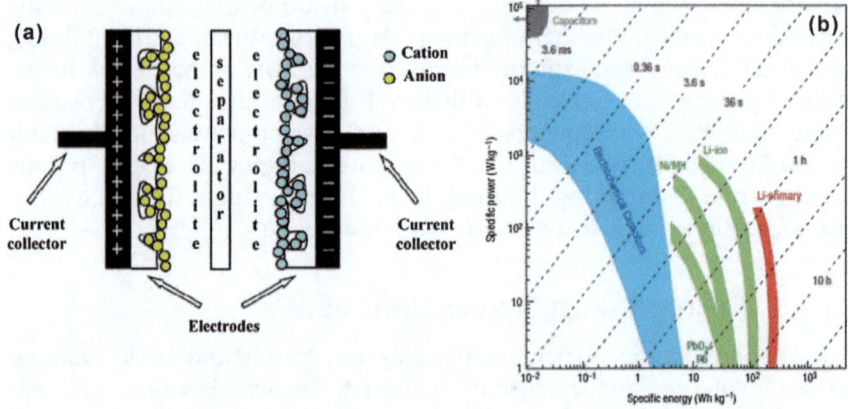

Figure 6.13   (a) Supercapacitor and (b) Ragone plots of electrical energy storage
systems. Reproduced from ref. 182 with permission from the Royal
Society of Chemistry.

materials in this particular field of research.[182] The supercapacitor and
Ragone plots for electrical energy storage systems are shown in Figure 6.13.

6.12   Regeneration of AC

Regeneration is critical since it extends the adsorbent's productive life and
improves its practical utility. When it comes to AC regeneration, there are
several options, some of which are listed as follows.

6.12.1   Thermal Regeneration

Temperature treatment is carried out utilizing fixed bed furnaces, fluidized
bed furnaces, and rotary kilns in this process. Adsorbent gases like carbon
dioxide, partially oxidized pollutants, and particulate matter can be extracted
from the adsorbent surface using this method. Gasification of biological
contaminants, drying at 105 °C, and pyrolysis in an inert atmosphere are all
part of the process. However, the adsorption effect of the adsorbent is
severely harmed by high-temperature conditions. Thermal regeneration is a
relatively easy regeneration process, but it is energy- and time-intensive, and
it also results in significant carbon loss. Furthermore, several hazardous
by-products can be formed as a result of the application of this approach.[183]

6.12.2   Chemical Regeneration

Certain chemicals are utilized to regenerate adsorbents by desorbing adsorbed molecules from the adsorbent surface in this regeneration approach.
The accuracy of this approach is highly dependent on the type of pollutant
found on or in the AC. This approach often results in 0% carbon attrition

and aids in the preservation of AC's true adsorption capacity. The extent of regeneration is determined by the affinity of adsorbed molecules with the acidic or alkaline reagents utilized. The solubility and reactivity of the adsorbents with the chemical reagents determine the efficacy of this type of regeneration.[184]

6.12.3   Electrochemical Regeneration

This method is very efficient because it uses a moderate temperature and does not require the use of chemicals. The cracking of organic impurities adsorbed on the surface of AC takes place *in situ* using this approach, while the structural integrity and properties of the carbon remain unchanged. Electro-oxidation, electro-reduction, and electro-sorption are the main mechanisms.[185]

6.12.4   Bioregeneration

Microbial agents are used to regenerate the adsorption capacity and surface activity of AC. When organic contaminant species may be easily desorbed, bioregeneration occurs. The bioregeneration is aided by the concentration gradient of organic pollutants between the adsorbent and the bulk liquid. In a closed-loop recirculating batch system, this two-step desorption and biological clearance of adsorbed organics has been verified.[186] As one proceeds from the carbon surface to the bulk solvent, it has been found that biodegradation occurs due to the presence of a pollutant concentration gradient. Many parameters influence the biogeneration of AC, including the carbon activation method employed, the porosity of AC, the concentration gradient and carbon saturation, the type of microorganism used, and the biomass concentration. Synthetic organic chemicals, explosive-contaminated groundwater, industrial effluent, and residual disinfectants are all major targets for adsorption bioregeneration.[187]

6.13   Conclusion

A thorough investigation based on several strategies for improving the adsorption capacity of AC made from biomass was done. Acid treatment, base treatment, salt treatment, ozone treatment, impregnation treatment, plasma treatment, and microwave treatment are some of the procedures that have been developed to create a new generation of surface functional groups in ACs. Treatments using chemical and physical methods were thoroughly discussed. The total activity of AC against different contaminants is affected by these treatments. The information provided emphasizes the necessity of essential functions on the surface of AC for the adsorption and storage of carbon dioxide. The surface area, pore size and volume, surface functions, and mineral matter concentration all affect the AC's adsorption ability. For the production of AC, a variety of precursors can be used, including lignocellulosic materials, agricultural waste, and fossil fuels. As powerful substitutes for ACs made from fossil fuels, ACs made from precursors of biobased origin are affordable and easily accessible. They are viable candidates for the adsorption of heavy metals, dyes, pesticides, and pharmaceutical effluents because of their appealing features. Additionally, carbon dioxide, methane, and hydrogen can all be stored with the aid of AC. There are various benefits to developing ACCs and using them as adsorbents in environmental applications, including faster adsorption kinetics, improved efficiency, and greater adsorption capacity. Traditionally utilized as catalyst supports or adsorbents, ACs—possibly the most researched class of porous carbons—are now increasingly used or have prospective uses in the production of supercapacitors and as hydrogen storage materials. Despite having such a wide range of uses, there is still a need for future research in areas like photocatalysis, hydrogen generation, and antimicrobials.

References

1. K. Y. Foo and B. H. Hameed, *Adv. Colloid Interface Sci.*, 2009, **149**, 27.
2. W. Y. Shi, H. B. Shao, H. Li, M. A. Shao and S. Du, *J. Hazard. Mater.*, 2009, **170**, 6.
3. K. Y. Foo and B. H. Hameed, *J. Hazard. Mater.*, 2009, **171**, 60.
4. J. Coronas, *Chem. Eng. J.*, 2010, **156**, 242.
5. Y. S. Tao, H. Kanoh, L. Abrams and K. Kaneko, *Chem. Rev.*, 2006, **106**, 910.
6. D. W. Kim, H. S. Kil, K. Nakabayashi, S. H. Yoon and J. Miyawaki, *Carbon*, 2017, **114**, 98.
7. J. S. Mattson and H. B. Mark Jr., *Activated Carbon: Surface Chemistry and Adsorption from Solution*, Marcel Dekker, New York, 1971.
8. F. S. Baker, C. E. Miller, A. J. Repic and E. D. Tolles, Activated Carbon, *Kirk-Othmer Encyclopaedia of Chemical Technology*, 1992, vol. 4.
9. Z. Jiang, Y. Liu, X. Sun, F. Tian, F. Sun, C. Liang, W. You, C. Han and C. Li, *Langmuir*, 2003, **9**, 731.
10. A. Bhatnagar, W. Hogland, M. Marques and M. Sillanpaa, *Chem. Eng. J.*, 2013, **219**, 499.
11. R. C. Bansal, J. B. Donnet and F. Stoeckli, *A Review of: Active Carbon*, Mercel Decker, New York, 1988.
12. M. V. Navarro, N. A. Seaton, A. M. Mastral and R. Murillo, *Carbon*, 2006, **44**, 2281.
13. J. Diaz-Teran, D. M. Nevskaia, A. J. Lopez-Peinado and A. Jerez, *Colloids Surf., A*, 2007, **167**, 187–188.
14. F. Rodriguez-Reinoso and M. Molina-Sabio, *Adv. Colloid Interface Sci.*, 1998, **271**, 76–77.
15. J. L. Figueiredo, M. F. R. Pereira, M. M. A. Freitas and J. J. M. Orfao, *Carbon*, 1999, **37**, 1379.
16. M. T. Izquierdo, B. Rubio, C. Mayoral and J. M. Andrés, *Appl. Catal., B*, 2001, **33**, 315.

17. G. S. Szymanski, Z. Karpinski, S. Biniak and A. Swiatkowski, *Carbon*, 2002, **40**, 2627.
18. J. W. Shim, S. J. Park and S. K. Ryu, *Carbon*, 2001, **39**, 1635.
19. C. Y. Yin, M. K. Aroua and W. A. W. Daud, *Sep. Purif. Technol.*, 2007, **52**, 403.
20. L. R. Radovic, C. Moreno-Castilla and J. Rivera-Utrilla, *Chemistry and Physics of Carbon*, Marcel Dekker, New York, 2001, vol. 27.
21. F. Rodriguez-Reinoso, A. Linares-Solano, M. Molinasabio and D. J. Lopez-Gonalez, *Adv. Sci. Technol.*, 1984, **1**, 513.
22. H. Marsh and F. Rodriguez-Reinoso, *Activated Carbon*, Elsevier Science and Technology Books, Oxford, 2006.
23. Y. Guo and A. C. Lua, *J. Therm. Anal. Calorim.*, 2000, **60**, 417.
24. L. B. McCarty, *Activated Charcoal for Pesticide Deactivation*, University of Florida Cooperative Extension Service, 2002.
25. B. S. Girgis and A. N. A. El-Hendawy, *Microporous Mesoporous Mater.*, 2002, **52**, 105.
26. J. Kong, Q. Yue, L. Huang, Y. Gao, Y. Sun, B. Gao, Q. Li and Y. Wang, *Chem. Eng. J.*, 2013, **221**, 62.
27. M. Smisek and S. Cerney, *Active Carbon: Manufacture, Properties and Applications*, Elsevier, Amsterdam, 1970.
28. B. R. Puri, *Advances in Chemistry Series*, American Chemical Society, Washington DC, 1983, vol. 77.
29. I. N. Ermolenko, I. P. Lyubliner and N. V. Gulko, *Chemically Modified Carbon Fibers and Their Applications*, VCH Publishers, New York, 1990.
30. K. Kutics and M. Suzuki, *Proceedings of the 2nd Korea-Japan Symposium on Separation Technology*, Seoul, 1990.
31. C. L. Mangun, K. R. Benak, M. A. Daley and J. Economy, *Chem. Mater.*, 1999, **11**, 3476.
32. H. M. Mozammel, O. Masahiro and S. Bhattacharya, *Biomass Bioenergy*, 2002, **22**, 397.
33. T. Garcia, R. Murillo, D. Cazorla-Amoros, A. M. Mastral and A. Linares- Solano, *Carbon*, 2004, **42**, 1683.
34. M. Santiago, F. Stuber, A. Fortuny, A. Fabregat and J. Font, *Carbon*, 2005, **43**, 2134.
35. Y. Onal, C. Akmil-Basar, C. Sarici-Ozdemir and S. Erdogan, *J. Hazard. Mater.*, 2007, **142**, 138.
36. B. R. Puri, *Chemistry and Physics of Carbon*, Marcel Dekker, New York, 1970.
37. R. C. Bansal, F. J. Vastola and P. L. Walker Jr., *Carbon*, 1974, **12**, 355.
38. T. Karanfil and J. E. Kilduff, *Environ. Sci. Technol.*, 1999, **33**, 3217.
39. M. Jagtoyen and F. Derbyshire, *Carbon*, 1998, **36**, 1085.
40. R. L. Tseng, *J. Hazard. Mater.*, 2007, **147**, 1020.
41. R. J. J. Jansen and H. Van Bekkum, *Carbon*, 1994, **32**, 1507.
42. P. Vinke, M. Van der Eijk, M. Verbree, A. F. Voskamp and H. Van Bekkum, *Carbon*, 1994, **32**, 675.

43. F. Stoeckli, T. A. Centeno, A. B. Fuertes and J. Muniz, *Carbon*, 1996, **34**, 1201.
44. C. U. Pittman Jr., G. R. He, B. Wu and S. D. Gardner, *Carbon*, 1997, **35**, 317.
45. J. Bimer, P. D. Salbut, S. Berlozecki, J. P. Boudou, E. Broniek and T. Siemieniewska, *Fuel*, 1998, **77**, 519.
46. M. C. Blanco-López, A. Martínez-Alonso and J. M. D. Tascón, *Microporous Mesoporous Mater.*, 2000, **34**, 171.
47. P. J. M. Carrott, J. M. V. Nabais, M. M. L. Ribeiro Carrott and J. A. Pajares, *Carbon*, 2001, **39**, 1543.
48. C. L. Mangun, K. R. Benak, J. Economy and K. L. Foster, *Carbon*, 2001, **39**, 1809.
49. T. C. Drage, A. Arenillas, K. M. Smith, C. Pevida, S. Piippo and C. E. Snape, *Fuel*, 2007, **86**, 22.
50. C. Pevida, M. G. Plaza, B. Arias, J. Fermoso, F. Rubiera and J. J. Pis, *Appl. Surf. Sci.*, 2008, **254**, 7165.
51. V. L. Snoeyink and W. J. Weber Jr., *Environ. Sci. Technol.*, 1967, **1**, 228.
52. Y. El-Sayed and T. J. Bandosz, *J. Colloid Interface Sci.*, 2004, **273**, 64.
53. K. Laszlo and A. Szucs, *Carbon*, 2001, **39**, 1945.
54. G. S. Szymański, Z. Karpiński, S. Biniak and A. Świątkowski, *Carbon*, 2002, **40**, 2627.
55. T. Karanfil, M. Kitis, J. E. Kilduff and A. Wigton, *Environ. Sci. Technol.*, 1999, **33**, 3225.
56. J. A. Menéndez, J. Phillips, B. Xia and L. R. Radovic, *Langmuir*, 1996, **12**, 4404.
57. M. S. Shafeeyan, W. M. A. W. Daud, A. Houshmand and A. Shamiri, *J. Anal. Appl. Pyrolysis*, 2010, **89**, 143.
58. H. Boehm, *Carbon*, 1994, **32**, 759.
59. M. J. McGuire and I. H. Suffet, *J. Environ. Eng.*, 1984, **110**, 629.
60. Y. Otake and R. G. Jenkins, *Carbon*, 1993, **31**, 109.
61. P. Shrestha, M. K. Jha, J. Ghimire, A. R. Koirala, R. M. Shrestha, R. K. Sharma, B. Pant, M. Park and H. R. Pant, *Materials*, 2020, **13**, 5667.
62. L. Yu, D. P. Gamliel, B. Markunas and J. A. Valla, *ACS Omega*, 2021, **6**, 8870.
63. X. Bai, Z. Wang, J. Luo, W. Wu, Y. Liang, X. Tong and Z. Zhao, *Nanoscale Res. Lett.*, 2020, **15**, 88.
64. N. Radenahmad, A. T. Azad, M. Saghir, J. Taweekun, M. S. A. Bakar, M. S. Reza and A. K. Azad, *Renewable Sustainable Energy Rev.*, 2020, **119**, 109560.
65. F. A. Lateef, H. O. Ogunsuyi and L. Jatropha Curcas, *Curr. Res. Green Sustainable Chem.*, 2021, **4**, 100142.
66. A. Promraksa and N. Rakmak, *Heliyon*, 2020, **6**, e04019.
67. M. S. Reza, S. N. Islam, S. Afroze, M. S. Abu Bakar, R. S. Sukri, S. Rahman and A. K. Azad, *Energy Ecol. Environ.*, 2020, **5**, 118.
68. D. N. K. P. Negara, T. G. T. Nindhia, I. W. Surata, F. Hidajat and M. Sucipta, *Surf. Interfaces*, 2019, **16**, 22.

69. F. T. You, G. W. Yu, Z. J. Xing, J. Li, S. Y. Xie, C. X. Li, G. Wang, H. Y. Ren and Y. Wang, *Appl. Surf. Sci.*, 2019, **471**, 633.

70. M. Usman, H. Chen, K. Chen, S. Ren, J. H. Clark, J. Fan, G. Luo and S. Zhang, *Green Chem.*, 2019, **21**, 1553.

71. H. E. Putra, E. Damanhuri, K. Dewi and A. D. Pasek, *IOP Conference Series: Earth and Environmental Science*, Indonesia, October, 2020.

72. M. P. Maniscalco, M. Volpe and A. Messineo, *Energies*, 2020, **13**, 4098.

73. N. Kumar, R. Weldon and J. G. Lynam, *Biocatal. Agric. Biotechnol.*, 2021, **36**, 102145.

74. Y. Q. Zhao, M. Lu, P. Y. Tao, Y. I. Zhang, X. T. Gong, Z. Yang, G. Q. Zhang and H. L. Li, *J. Power Sources*, 2016, **307**, 391.

75. S. M. Abegunde, K. S. Idowu, O. M. Adejuwon and T. Adeyemi-Adejolu, *Resour. Environ. Sustain.*, 2020, **1**, 100001.

76. S. Wenzhong, L. Zhijie and L. Yihong, *Recent Pat. Chem. Eng.*, 2008, **1**, 27.

77. J. Temuujin, T. Jadambaa, G. Burmaa, S. Erdenechimeg, J. Amarsanaa and K. J. D. MacKenzie, *Ceram. Int.*, 2004, **30**, 251.

78. H. Tamon and M. Okazaki, *Carbon*, 1996, **34**, 741.

79. J. A. Menéndez, J. Phillips, B. Xia and L. R. Radovic, *Langmuir*, 1996, **12**, 4404.

80. F. W. Shaarani and B. H. Hameed, *Chem. Eng. J.*, 2011, **169**, 180.

81. R. J. J. Jansen and H. van Bekkum, *Carbon*, 1995, **33**, 1021.

82. S. A. Dastgheib, T. Karanfil and W. Cheng, *Carbon*, 2004, **42**, 547.

83. H. R. Pant, S. P. Adhikari, B. Pant, M. K. Joshi, H. J. Kim, C. H. Park and C. S. Kim, *J. Colloid Interface Sci.*, 2015, **457**, 174.

84. I. Sutherland, E. Sheng, R. Bradley and P. Freakley, *J. Mater. Sci.*, 1996, **31**, 5651.

85. C. Zaror, G. Soto, H. Valdés and H. Mansilla, *Water Sci. Technol.*, 2001, **44**, 125.

86. C. A. Zaror, *J. Chem. Technol. Biotechnol.*, 1997, **70**, 21.

87. U. Jans and J. Hoigné, *Ozone: Sci. Eng.*, 1998, **20**, 67.

88. J. Rivera-Utrilla and M. Sánchez-Polo, *Appl. Catal., B*, 2002, **39**, 319.

89. A. A. Joshi, B. R. Locke, P. Arce and W. C. Finney, *J. Hazard. Mater.*, 1995, **41**, 3.

90. B. R. Locke, M. Sato, P. Sunka, M. R. Hoffmann and J. S. Chang, *Ind. Eng. Chem. Res.*, 2006, **45**, 882.

91. A. Bogaerts, E. Neyts, R. Gijbels and J. van der Mullen, *Spectrochim. Acta, Part B*, 2002, **57**, 609.

92. S. H. Kee, J. B. V. Chiongson, J. P. Saludes, S. Vigneswari, S. Ramakrishna and K. Bhubalan, *Environ. Pollut.*, 2021, **271**, 116311.

93. D. Wilcox, E. Chang, K. Dickson and K. Johansson, *Appl. Environ. Microbiol.*, 1983, **46**, 406.

94. J. Rivera-Utrilla, I. Bautista-Toledo, M. A. Ferro-García and C. Moreno-Castilla, *J. Chem. Technol. Biotechnol.*, 2001, **76**, 1209.

95. P. J. M. Carrott, J. M. V. Nabais, M. M. L. Ribeiro Carrott and J. A. Menéndez, *Microporous Mesoporous Mater.*, 2001, **47**, 243.

96. K. Y. Foo and B. H. Hameed, *Bioresour. Technol.*, 2012, **119**, 234.
97. E. Yagmur, M. Ozmak and Z. Aktas, *Fuel*, 2008, **87**, 3278.
98. A. A. Abdul Rasheed, A. A. Jalil, S. Triwahyono, M. A. A. Zaini, Y. Gambo and M. Ibrahim, *Renewable Sustainable Energy Rev.*, 2018, **94**, 1067.
99. M. Naushad, A. A. Alqadami and T. Ahamad, *Environ. Technol. Innovation*, 2020, **18**, 100686.
100. G. Sharma, B. Thakur, A. Kumar, S. Sharma, M. Naushad and F. J. Stadler, *Macromol. Mater. Eng.*, 2020, **305**, 2000274.
101. Z. Fallah, E. N. Zare, M. Ghomi, F. Ahmadijokani, M. Amini, M. Tajbakhsh, M. Arjmand, G. Sharma, H. Ali, A. Ahmad, P. Makvandi, E. Lichtfouse, M. Sillanpaa and R. S. Varma, *Chemosphere*, 2021, **275**, 130055.
102. M. D. G. de Luna, E. D. Flores, D. A. D. Genuino, C. M. Futalan and M. W. Wan, *J. Taiwan Inst. Chem. Eng.*, 2013, **4**, 646.
103. Y. S. Ho and G. McKay, *Process Biochem.*, 1999, **34**, 451.
104. G. Sharma, M. Naushad, A. Kumar, S. Rana, S. Sharma, A. Bhatnagar, F. J. Stadler, A. A. Ghfar and M. R. Khan, *Process Saf. Environ. Prot.*, 2017, **109**, 301.
105. G. Sharma, A. Kumar, C. Chauhan, A. Okram, S. Sharma, D. Pathania and S. Kalia, *Sustainable Chem. Pharm.*, 2017, **6**, 96.
106. W. J. Weber and E. H. Smith, *Environ. Sci. Technol.*, 1987, **21**, 1040.
107. O. Moradi and G. Sharma, *Environ. Res.*, 2021, **201**, 111534.
108. M. T. Yagub, T. K. Sen, S. Afroze and H. M. Ang, *Adv. Colloid Interface Sci.*, 2014, **209**, 172.
109. R. W. Coughlin and F. S. Ezra, *Environ. Sci. Technol.*, 1968, **2**, 291.
110. H. N. Tran, S. J. You and H. P. Chao, *Waste Manage. Res.*, 2016, **34**, 129.
111. T. X. Bui and H. Choi, *Environ. Sci. Technol.*, 2010, **44**, 4828.
112. Z. Jeirani, C. H. Niu and J. Soltan, *Rev. Chem. Eng.*, 2017, **33**, 491.
113. N. Srivastava and C. Majumder, *J. Hazard. Mater.*, 2008, **151**, 1.
114. M. Naushad, S. Vasudevan, G. Sharma, A. Kumar and Z. A. Alothman, *Desalin. Water Treat.*, 2016, **57**, 18551.
115. M. Naushad, Z. A. ALOthman and G. Sharma, *Ionics*, 2015, **21**, 1453.
116. G. Sharma, D. Pathania, M. Naushad and N. Kothiyal, *Chem. Eng. J.*, 2014, **251**, 413.
117. C. Namasivayam and K. Kadirvelu, *Carbon*, 1999, **37**, 79.
118. R. Naseem and S. Tahir, *Water Res.*, 2001, **35**, 3982.
119. G. Sharma, M. Naushad, A. H. Al-Muhtaseb, A. Kumar, M. R. Khan, S. Kalia, M. Bala and A. Sharma, *Int. J. Biol. Macromol.*, 2017, **95**, 484.
120. L. Khezami and R. Capart, *J. Hazard. Mater.*, 2005, **123**, 223.
121. I. Morosanu, C. Teodosiu, C. Paduraru, D. Ibanescu and L. Tofan, *New Biotechnol.*, 2017, **39**, 110.
122. M. Sekar, V. Sakthi and S. Rengaraj, *J. Colloid Interface Sci.*, 2004, **279**, 307.
123. K. Kadirvelu, K. Thamaraiselvi and C. Namasivayam, *Sep. Purif. Technol.*, 2001, **24**, 497.
124. A.-N. A. El-Hendawy, *J. Hazard. Mater.*, 2009, **167**, 260.

125. P. O. Ameh, *Int. J. Mod. Chem.*, 2013, **5**, 136.
126. K. Panchamoorthy Gopinath, D. V. N. Vo, D. G. Prakash, A. A. Joseph, S. Viswanathan and J. Arun, *Environ. Chem. Lett.*, 2021, **19**, 557.
127. X. F. Tan, S. B. Liu, Y. G. Liu, Y. L. Gu, G. M. Zeng, X. J. Hu, X. Wang, S. H. Liu and L. H. Jiang, *Bioresour. Technol.*, 2017, **227**, 359.
128. J. Dich, S. H. Zahm, A. Hanberg and H.-O. Adami, *Cancer, Causes Control*, 1997, **8**, 420.
129. M. Gavrilescu, *Eng. Life Sci.*, 2005, **5**, 497.
130. G. Sharma, A. Kumar, K. Devi, S. Sharma, M. Naushad, A. A. Ghfar, T. Ahamad and F. J. Stadler, *Int. J. Biol. Macromol.*, 2018, **114**, 295.
131. A. Hossain, K. Sakthipandi, A. A. Ullah and S. Roy, *Nano-Micro Lett.*, 2019, **11**, 1.
132. E. Ayranci and N. Hoda, *Chemosphere*, 2005, **60**, 1600.
133. A. L. Ahmad, L. Tan and S. A. Shukor, *J. Hazard. Mater.*, 2008, **151**, 71.
134. F. K. Yuen and B. Hameed, *Adv. Colloid Interface Sci.*, 2009, **149**, 19.
135. C.-F. Chang, C.-Y. Chang, K.-E. Hsu, S.-C. Lee and W. Höll, *J. Hazard. Mater.*, 2008, **155**, 295.
136. W. K. Lafi and Z. Al-Qodah, *J. Hazard. Mater.*, 2006, **137**, 489.
137. B. Hameed, J. Salman and A. Ahmad, *J. Hazard. Mater.*, 2009, **163**, 121.
138. P. Chingombe, B. Saha and R. Wakeman, *J. Colloid Interface Sci.*, 2006, **297**, 434.
139. O. Adam, M. Bitschené, G. Torri, F. De Giorgi, P. M. Badot and G. Crini, *Sep. Purif. Technol.*, 2005, **46**, 11.
140. J.-W. Yoo, T.-Y. Kim, S.-Y. Cho, S.-G. Rho and S.-J. Kim, *Adsorption*, 2005, **11**, 719.
141. J. Salman, V. Njoku and B. Hameed, *Chem. Eng. J.*, 2011, **173**, 361.
142. A. Ahmad, S. H. Mohd-Setapar, C. S. Chuong, A. Khatoon, W. A. Wani, R. Kumar and M. Rafatullah, *RSC Adv.*, 2015, **5**, 30801.
143. J. Abdi, M. Vossoughi, N. M. Mahmoodi and I. Alemzadeh, *Chem. Eng. J.*, 2017, **326**, 1145.
144. G. Sharma, A. Kumar, M. Naushad, A. García-Peñas, A. H. Al-Muhtaseb, A. A. Ghfar, V. Sharma, T. Ahamad and F. J. Stadler, *Carbohydr. Polym.*, 2018, **202**, 444.
145. G. Sharma, A. Khosla, A. Kumar, N. Kaushal, S. Sharma, M. Naushad, D. V. N. Vo, J. Iqbal and F. J. Stadler, *Chemosphere*, 2022, **289**, 133100.
146. M. Purkait, S. DasGupta and S. De, *J. Environ. Manage.*, 2005, **76**, 135.
147. A. Sharma, G. Sharma, M. Naushad, A. A. Ghfar and D. Pathania, *Environ. Technol.*, 2018, **39**, 917.
148. G. Sharma, A. Kumar, M. Naushad, B. Thakur, D. V. N. Vo, B. Gao, A. A. Al-Kahtani and F. J. Stadler, *J. Hazard. Mater.*, 2021, **416**, 125714.
149. G. Sharma, B. Thakur, M. Naushad, A. Kumar, F. J. Stadler, S. M. Alfadul and G. T. Mola, *Environ. Chem. Lett.*, 2018, **16**, 113.
150. S. A. Ganiyu, O. O. Ajumobi, S. A. Lateef, K. O. Sulaiman, I. A. Bakare, M. Qamaruddin and K. Alhooshani, *Chem. Eng. J.*, 2017, **321**, 651.
151. L. Liu, J. Zhang, Y. Tan, Y. Jiang, M. Hu, S. Li and Q. Zhai, *Chem. Eng. J.*, 2014, **244**, 9.

152. G. Sharma, M. Naushad, D. Pathania, A. Mittal and G. El-Desoky, *Desalin. Water Treat.*, 2015, **54**, 3114.
153. R. Malik, D. Ramteke and S. Wate, *Waste Manage.*, 2007, **27**, 1129.
154. K. L. Chiu and D. H. Ng, *Biomass Bioenergy*, 2012, **46**, 102.
155. Y. Gao, W. Zhang, Q. Yue, B. Gao, Y. Sun, J. Kong and P. Zhao, *J. Power Sources*, 2014, **270**, 403.
156. M. J. Ahmed and S. K. Theydan, *J. Anal. Appl. Pyrolysis*, 2014, **105**, 199.
157. G. Karaçetin, S. Sivrikaya and M. Imamoğlu, *J. Anal. Appl. Pyrolysis*, 2014, **110**, 270.
158. O. Pezoti Jr, A. L. Cazetta, R. C. Gomes, É. O. Barizão, I. P. A. F. Souza, A. C. Martins, T. Asefa and V. C. Almeida, *J. Anal. Appl. Pyrolysis*, 2014, **105**, 166.
159. A. Ahmad, M. Loh and J. Aziz, *Dyes Pigm.*, 2007, **75**, 263.
160. M. Berrios, M. Á. Martín and A. Martín, *J. Ind. Eng. Chem.*, 2012, **18**, 780.
161. M. Kumar and R. Tamilarasan, *J. Environ. Chem. Eng.*, 2013, **1**, 1108.
162. S. Altenor, B. Carene, E. Emmanuel, J. Lambert, J. J. Ehrhardt and S. Gaspard, *J. Hazard. Mater.*, 2009, **165**, 1029.
163. Y. Gokce and Z. Aktas, *Appl. Surf. Sci.*, 2014, **313**, 352.
164. M. J. Ahmed and S. K. Dhedan, *Fluid Phase Equilib.*, 2012, **317**, 9.
165. X. Zhang, B. Gao, A. E. Creamer, C. Cao and Y. Li, *J. Hazard. Mater.*, 2017, **338**, 102.
166. L. A. Rodrigues, M. L. C. P. Da Silva, M. O. Alvarez-Mendes, A. Dos Reis Coutinho and G. P. Thim, *Chem. Eng. J.*, 2011, **174**, 49.
167. Y. Tham, P. A. Latif, A. Abdullah, A. Shamala-Devi and Y. Taufiq-Yap, *Bioresour. Technol.*, 2011, **102**, 724.
168. J. V. Nabais, J. Gomes, P. Carrott, C. Laginhas and S. Roman, *J. Hazard. Mater.*, 2009, **167**, 904.
169. S. Suganuma, K. Nakajima, M. Kitano, D. Yamaguchi, H. Kato, S. Hayashi and M. Hara, *J. Am. Chem. Soc.*, 2008, **130**, 12787.
170. E. Lam and J. H. Luong, *ACS Catal.*, 2014, **4**, 3393.
171. J. Han, J. Duan, P. Chen, H. Lou and X. Zheng, *Adv. Synth. Catal.*, 2011, **353**, 2577.
172. M. M. Perez, C. S. M. de Lecea and A. L. Solano, *Appl. Catal., A*, 1997, **151**, 461.
173. P. R. Shukla, S. Wang, H. Sun, H. M. Ang and M. Tadé, *Appl. Catal., B*, 2010, **100**, 529.
174. A. Silem, A. Boualia, R. Kada and A. Mellah, *Can. J. Chem. Eng.*, 1992, **70**, 491.
175. F. Dong, H. Lou, A. Kodama, M. Goto and T. Hirose, *Sep. Purif. Technol.*, 1999, **16**, 159.
176. A. L. Chaffee, G. P. Knowles, Z. Liang, J. Zhang, P. Xiao and P. A. Webley, *Int. J. Greenhouse Gas Control*, 2007, **1**, 11.
177. B. Stöhr, H. Boehm and R. Schlögl, *Carbon*, 1991, **29**, 707.
178. M. Sabo, A. Henschel, H. Fröde, E. Klemm and S. Kaskel, *J. Mater. Chem.*, 2007, **17**, 3827.

179. S. H. Yeon, S. Osswald, Y. Gogotsi, J. P. Singer, J. M. Simmons, J. E. Fischer, M. A. Lillo-Ródenas and Á. Linares-Solano, *J. Power Sources*, 2009, **191**, 560.
180. H. T. Hwang and A. Varma, *Curr. Opin. Chem. Eng.*, 2014, **5**, 42.
181. A. L. Cukierman, *ISRN Chem. Eng.*, 2013, **2013**, 261523.
182. M. Sevilla and R. Mokaya, *Energy Environ. Sci.*, 2014, **7**, 1250.
183. G. San Miguel, S. Lambert and N. Graham, *Water Res.*, 2001, **35**, 2740.
184. Q. Li, Y. Qi and C. Gao, *J. Cleaner Prod.*, 2015, **86**, 424.
185. Y. Han, X. Quan, S. Chen, H. Zhao, C. Cui and Y. Zhao, *Sep. Purif. Technol.*, 2006, **50**, 365.
186. Ö. Aktaş and F. Çeçen, *J. Hazard. Mater.*, 2007, **141**, 769.
187. M. El Gamal, H. A. Mousa, M. H. El-Naas, R. Zacharia and S. J. S. Judd, *Sep. Purif. Technol.*, 2018, **197**, 345.

CHAPTER 7

Agricultural Applications of Activated Carbon

MOHAMED E. A. EL-SAYED(ⓘ 0000-0001-5547-0921),*[a]
HAYTHAM A. AYOUB(ⓘ 0000-0003-3886-7323)[b] AND
ISLAM A. ABDELHAFEEZ(ⓘ 0000-0001-8215-2267)[a,c]

[a] Soils, Water and Environment Research Institute, Agricultural Research Center, Giza 12112, Egypt; [b] Plant Protection Research Institute, Agricultural Research Center, Nadi El-Said Street – Dokki, Giza 12311, Egypt; [c] Department of Chemical and Environmental Engineering (ChEE), University of Cincinnati, Cincinnati, Ohio 45221, USA
*Email: eid1592003@yahoo.com

7.1 Introduction

Activated carbon (AC) is a well-developed porous carbonaceous material that has a large specific surface area. AC can be produced from different high-carbon-content precursors.[1] Biomass materials have emerged as a new platform for synthesizing functional porous carbon materials; preparing AC from biomass mostly involves physical and chemical methods which are mainly different in terms of activating agent and carbonization temperature.[1,2] Preparing AC *via* the physical activation process needs a very high temperature (>900 °C) in the presence of carbon dioxide (CO_2),[2] while chemical activation could be carried out at a relatively lower temperature.[3,4] Recently, AC has been applied in a wide variety of fields such as food preservation, the chemical industry, medicine, and agriculture.[5] Due to its large specific surface area, high porosity, and abundant surface functionalities, AC has been widely used as an alternative adsorbent.[6] In agriculture, AC has numerous applications

Activated Carbon: Progress and Applications
Edited by Chandrabhan Verma and Mumtaz A. Quraishi
© The Royal Society of Chemistry 2023
Published by the Royal Society of Chemistry, www.rsc.org

in soil remediation and fertility, water treatment (removal of heavy metals, organic pollutants, and pesticides), and biosensing.[7–9]

7.2 AC Applications in Agriculture

7.2.1 Fertilization Applications

Nowadays, more than two billion people suffer from various forms of micronutrient deficiencies, causing severe diseases such as anemia in women and blindness in children.[10] Fertilization has been used over for 50 years by farmers and has a crucial role in fostering food quality and production.[11] Nevertheless, conventional fertilizers can be leached and emitted (40–70%) from the soil to the environment due to their high solubility, resulting in environmental problems such as water pollution, eutrophication, and loss of soil fertility, and manifesting the impacts of global warming and climate change.[12,13]

Recently, AC has been used as a green and natural approach for enhanced fertilizers, slow-release micro fertilizer carriers, and pollutant removal to produce safe fertilizers.[14–16] Sultan *et al.* found that acidified AC is more efficient than biochar in improving soil fertility.[14] The results showed that sulfuric acid-acidified AC has the potential to reduce the alkaline soil pH and EC compared to the control in clay and sandy clay loam. Besides that, acidified AC increased the release and availability of micro- and macronutrients such as P, Zn, B, and Fe compared to biochar due to the decrease in soil pH. Activation of biochar is also a new technique to achieve slow-release fertilizers. H_3PO_4-activated biochar exhibited excellent slow release of urea into an aqueous medium with slow cumulative N leaching from soil columns and slow NH_3–N volatilization from the soil compared with the control (uncoated urea).[17] This could have originated from successful grafting of urea onto the surface of the activated biochar through carbonyl C=O groups and urea amine N groups, and hence it takes a longer time for urea to be released into the soil matrix. Use of organometallic-activated biochar is also another strategy to enhance foliar application for promoting plant growth.[18] Activated biochar was synthesized from pine wood:clay and wheat straw:bird manure. The aqueous extracts of activated biochar were used as foliar fertilizers and showed significant enhancement in lettuce growth and biomass compared to the corresponding chemical fertilizers. The reason for that could be directly related to the slow release of macro- and nanoscale organomineral particles, such as Si-rich complexes or aluminosilicate compounds, and agglomerates on leaf surfaces, leading to an increase in photosynthetic pigment concentrations.[18]

AC nanofibers (ACF) served as a micronutrient carrier of copper (Cu) nanoparticles (NPs) with a controlled release of the Cu NPs.[19] Cu NPs were effectively translocated from the root to the shoot *via* ACF. The Cu–ACF hybrid system enhanced the water uptake capacity, shoot and root lengths, germination rate, and protein and chlorophyll contents in the plant. Seed

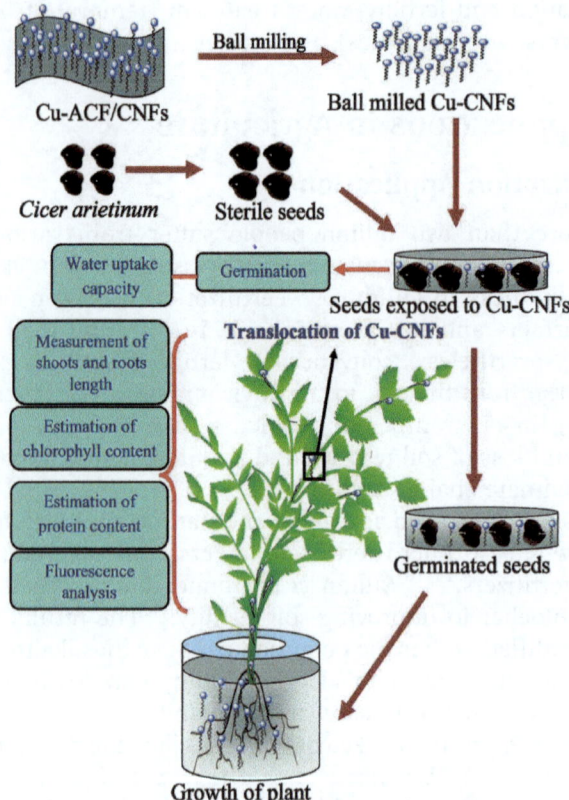

Figure 7.1   Seed germination and the uptake mechanism of Cu NPs in the plant *via* the Cu–ACF hybrid fertilizer.[19] Reproduced from ref. 19 with permission from the Royal Society of Chemistry.

germination and the uptake mechanism of Cu NPs during plant growth are illustrated in Figure 7.1. In another study, Prastiwi *et al.* investigated the role of activated charcoal in slow-release fertilizers after soaking in different solutions including $CuSO_4$, $FeSO_4$, and $ZnSO_4$.[20] The obtained results showed the good absorption of the micronutrients (Cu, Fe, and Zn) in the pores of activated charcoal, so they could be slowly released and could not easily leach out.

The application of AC to mitigate the environmental impacts or produce clean fertilizers has also been studied. The coupling of nitrification and AC adsorption has the potential to produce a safe fertilizer by minimizing the removal of beneficial nutrients and sufficiently decreasing the chemical concentrations. The addition of AC removed at least 90% of each of the twelve investigated chemicals at pH 6 and after 24 h contact time. In another study by Shin *et al.*, activated biochar pellet fertilizers were shown to have multifunctional applications including improvement of rice production and sequestration of greenhouse gas emissions.[21] After the addition of the activated pellets, the need for supplemental inorganic nitrogen (N) fertilizer was decreased by 60%. Furthermore, the lowest N_2O emission was observed (0.002 kg ha^{-1}) for activated pellets and greenhouse gas emissions were reduced by 10 kg ha^{-1} compared to the control during the crop season.

7.2.2 Environmental Applications

7.2.2.1 Soil Remediation

7.2.2.1.1 Soil Amendment. Pollution of the soil with a wide range of pollutants such as oil spills and organic and inorganic pollutants is often observed in municipal soils, around industrial plants and gas stations, and in areas where excavations are active.[22] Changes in soil properties due to such contamination can lead to water and oxygen deficit as well as a shortage of available forms of nitrogen and phosphorus, resulting in a negative impact on plant growth and yield, thereby leading to a threat to the food security and health of human beings.[23] Application of AC has significant potential for remediation of contaminated soils *via* immobilization of pollutants through which the migration of the contaminants will be retarded and the dissolved phase concentration will be decreased.[24] Retaining pollutants in the AC matrix provides longer residence time for contaminant degradation over reactive amendments, reducing the potential of contaminants to rebound and avoid groundwater pollution.[24]

Utilization of AC without other additives in contaminated soils shows a positive effect on reducing the impact of pollutants, improving the soil's physical properties, and enhancing the water-holding capacity.[25] Harwood *et al.* studied the impact of AC on bioaccumulation of dichlorodiphenyltrichloroethane (DDT) residues by earthworms. The results showed that AC amendment caused significant reductions (up to 92.94%) in DDT accumulation in earthworms, affirming the ability of AC to reduce the toxic bioavailability in contaminated sites.[26] Coupling AC with other additives will further enhance the remediation process. For example, combining (Triton X-100) surfactant-enhanced washing with AC as an adsorbent improved the removal efficiency of organochlorine pesticides, such as chlordane and DDT, in contaminated soils, with further recycling of the surfactant for use in the remediation.[27]

On the other hand, integrating advanced oxidation processes (AOPs) with ACs as heterogeneous catalysts has been investigated as an alternative strategy for the cleanup of organic-contaminated soil systems. Annamalai *et al.* reported a novel and low-cost colloidal AC/persulfate (PS) system.[28] The catalytic system exhibited excellent degradation performance for phenol in the field-contaminated soil with 65.7% phenol degraded within 24 h. The reason for such degradation performance could be attributed to the formation of radical scavengers such as sulfate radicals ($SO_4^{\bullet-}$), hydroxyl radicals ($^{\bullet}OH$), and superoxide radicals ($O_2^{\bullet}$), and non-radical species such as singlet oxygen ($^1O_2$). In another study, Liu *et al.* studied the effect of the combination of thermal activation of peroxymonosulfate (PMS) and AC on the degradation of perfluorooctanoic acid (PFOA).[29] The results showed that 76% of PFOA was removed by the hybrid system at 60 °C merely in 12 h. That could be attributed to the promotion of the production of $SO_4^{\bullet-}$ radicals, causing the fast diminution of PFOA in soil.

Introducing the electrokinetic method with AC hybrid materials has been widely examined as a substitutional technique for the remediation of inorganic-contaminated soil. Wang *et al.* tested a hybrid system of iron-doped AC to enhance the Cr(vi) contaminated soil remediation using microbial fuel cells (MFCs).[30] The spiking of AC–Fe particles into the contaminated soil enhanced the MFC system with a high Cr(vi) removal efficiency of $84.2 \pm 1.2\%$ in 24 h, and the maximum power generation of the MFC was 11.5 mW m^{-2}. Here, the AC–Fe composite system worked as a simultaneous adsorbent and reducing agent for Cr(vi) to Cr(iii), and at the same time, Fe(ii) loaded onto the surface of the AC oxidized to Fe(iii). Besides that, zero-valent iron/AC (ZVI/AC) particles, as a permeable reactive barrier, showed a significant reduction of uranium(vi) (U(vi)) in artificially contaminated soil by Fe/C micro-electrolysis, and the maximum removal efficiency was obtained after 120 h ($80.58 \pm 0.99\%$).[31] The addition of citric acid (CA) mixed with ferric chloride (FeCl$_3$), as a composite electrolyte for electrokinetic remediation, promoted the dissolution and leaching of U(vi) in the soil. On the other hand, Fe and AC worked as an anode and a cathode, respectively, where Fe lost electrons to reduce U(vi) to the insoluble U(iv) and AC can boost the conversion of Fe(iii) to Fe(ii) by obtaining electrons from ZVI, as shown in Figure 7.2. In another study, coupling of AC with CA enhanced the electrokinetic remediation of lead-contaminated soil. CA can accelerate the dissolution and desorption of Pb from the soil matrix by forming soluble complexes, while AC regulated the pH distribution in the cathode area in the handling cell. Such regulated distribution prevented Pb from precipitation and enhanced the ionic mobility and removal of Pb ions in porous media.[32]

7.2.2.1.2 Bioremediation. Bioremediation is an alternative strategy for the remediation of contaminated soil, relying on the utilization of appropriate microorganisms for the degradation of specific pollutants such as total petroleum hydrocarbons (TPH) and pesticides, due to their low cost and low harmful impact on nature. Although bioremediation is a green approach and exhibits high efficiency for pollutant degradation, its long-term nature (1–3 years) with its slow treatment process, strong inhibition at high contaminant concentrations, and high probability of secondary pollution of groundwater hinder its applicability.[33]

Use of AC has been investigated as a green strategy to enhance microorganisms' biodegradation of organic pollutants in the soil due to its enriched functional groups on its surface, and immobilize the pollutant's segments to the microbes and plants during the bioremediation process.[34] Vasilyeva *et al.* observed that the addition of AC to soil contaminated with organic pollutants, including 3,4-dichloroaniline (DCA), 2,4,6-trinitrotoluene (TNT), and polychlorinated biphenyls (PCB), enhanced the bioremediation process and

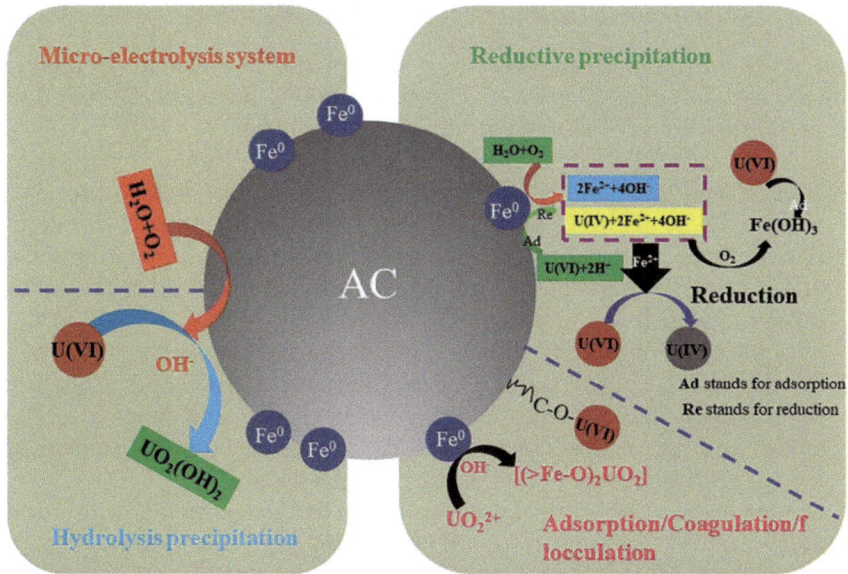

Figure 7.2   Schematic diagrams of uranium removal mechanisms in the ZVI/AC system.[31] Reproduced from ref. 31 with permission from Elsevier, Copyright 2020.

reduced their toxicity by transferring to less toxic fractions.[35] Furthermore, amendment with AC led to a sharp decrease of pollutants extractable and promoted the counterpart-degrading strains. The amendment of diesel-contaminated soil with activated charcoal and a bio-preparation considerably reduces the bio- and phytotoxicity of TPH. The introduction of activated charcoal during the bioremediation increased the remediation and localized the hydrocarbon pollutants twice more than unsupplemented soil which may decrease the risk of their penetration to groundwater during the bioremediation process.[36]

To further improve the AC performance, some additives may be useful to enhance the bioremediation process. The combination of diatomite and AC added in petroleum-contaminated soil resulted in a significant reduction and degradation of TPH by 9–10% and 5–8% at the end of the first and second years, respectively.[33] Besides that, the count of petroleum-degrading microorganisms was significantly increased. In another report, Kondrashina *et al.* investigated the same mixture with a bio-preparation for bioremediation of oil-contaminated soil.[37] The best results were observed for the bio-mixture compared to each component alone where the TPH content was reduced by 90% and 92% at 2 and 13 months, respectively. The addition of the AC bio-mixture did not only result in enhanced petroleum-degrading strains and plant growth, due to the reversible adsorption of toxic hydrocarbon components and metabolites, but also increased the soil porosity and water-holding capacity and enhanced plant resistance to abiotic stress.

7.2.2.2   Agricultural Drainage Water Treatment

7.2.2.2.1   Removal of Heavy Metals.   Heavy metals are harmful to plants, animals, and humans. There are several sources of heavy metal pollution that are available in the agro-environment, including:[38]

- Natural sources: land clusters enriched with heavy metals are considered to be natural sources of heavy metal pollution in the agro-environment. Several sources such as natural rocks, volcanic eruptions, and wind-blown dust particles are major contributors to soil and water pollution with heavy metals.
- Agriculture: inorganic fertilizers are a major contributor of heavy metals, including Cd, Cr, Ni, Pb, and Zn, in agricultural soil and water. The other sources include the wide range of pesticides and fungicides. Besides, land application of sewage sludge is one of the major sources of heavy metal pollution in soil and irrigated wastewater.
- Domestic sources: some irresponsible activities particularly in rural areas such as the burning of bio-wastes and mismanagement of wastewater effluents are the largest contributor to the high concentration of heavy metals in soil and water and cause air pollution in agroecological zones. In addition, misuse of inorganic and organic wastes and failure to recycle e-wastes discharged into the water cause huge environmental pollution.
- Industrial sources: the rapid pace of industrial activities' growth such as refinement and mining highly influences the water and air quality *via* volatized heavy metals dispersed by wind or dissolved metals runoff the water surface. These practices affect consequently soil health. Besides that, discharges from the chemical industry especially in nearby urban cities are also responsible for soil and water contamination with heavy metals.
- Other sources: there are various sources that contribute to agroecological heavy metals pollution, including landfill areas or open dumps, traffic or transportation emissions, residues and emissions of incineration, and medical wastes.

AC has been thoroughly and efficiently investigated for heavy metal removal from water matrixes. There are several mechanisms for the adsorption of heavy metals from aqueous solutions, which are summarized in Figure 7.3, and they include the following:

(1) Physical adsorption: in this part, the adsorption process does not rely on chemical bonds, but on the diffusion and desorption of heavy metal ions in the pores of AC. That is, it depends on the pore size distribution and surface area of the adsorbent. The more the mesopores and micropores increase, the greater the increase in surface area; thus the diffusion of more heavy metal into the pores then promotes the physical adsorption

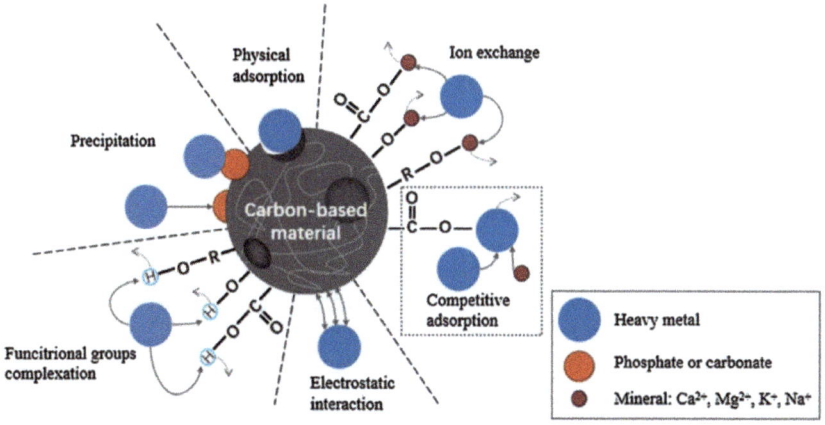

Figure 7.3 Possible mechanisms of heavy metal removal by AC.[39] Reproduced from ref. 39 with permission from Elsevier, Copyright 2020.

and accelerates the adsorption kinetics.[39] For instance, Kharrazi *et al.* studied the adsorption activity of lignocellulosic-based AC prepared physically *via* thermal tension by washing with 80 °C, 95 °C, and subsequently 2 °C (cold) DI water.[40] The surface area increased by 133.33% and then the adsorption capacity was enhanced by 51.14% and 160.48% for Cr(vi) and Pb(ii), respectively.

(2) Electrostatic interaction: the interaction obtained between heavy metals and AC relies on the interaction between the opposite ions.[39] AC synthesized by pre-incubating with H_3PO_4 and then activated in a muffle furnace showed high adsorption of Pb and Cd by electrostatic attraction. This is due to the strong electrostatic interaction and complexing adsorption of Cd^{2+} and Pb^{2+} with the negatively charged surface (caused by deprotonation of –OH) of AC.[41]

(3) Ion exchange: ion exchange occurs between the protons of oxygen-containing functional groups (–OH, –COOH) on the surface of AC and divalent or trivalent metal cations (M^{2+} or M^{3+}) and can be indicated by cation exchange capacity (CEC).[42] Rivera-Utrilla and Sánchez-Polo found that the adsorption capacity of AC by ion exchange could be improved with the increase in oxygenated acid groups on the surface of AC. The cationic exchange was achieved between Cr(iii) and protons in the $-C\pi-H_3O^+$ interaction.[43]

(4) Surface complexation: enriched functional groups on the surface of AC (such as –O–, –OH, –COOH, and –C–NH–) react with heavy metal ions or complexes to form multiatom structures *via* surface complexation.[39] Bohli *et al.* found that Cd(ii), Ni(ii), and Cu(ii) can be adsorbed and form binary complexes with carbonyl, phenol, and carboxylic active groups on the surface of microporous AC.[44] Also, the modification of the AC surface can further enhance the surface complexation. For instance, sulfur-containing ligands doped on the

surface of the AC can significantly improve the adsorption kinetics of Cd(II), which was 2–10 times higher than that of the pristine AC.[45] The modification of the AC surface with cysteine and thiourea enhanced the Cd(II) adsorption capacity *via* complexation, which was twice that of unmodified AC.

(5) Precipitation/coprecipitation: heavy metal ions can dissociate from water by forming solid precipitates with others groups and ions on the adsorbent surface.[39] Suo *et al.* reported that several ions such as Pb(II), Cu(II), Zn(II), and Cd(II) mostly precipitated on the surface of AC at higher pH due to the production of more hydroxide precipitates.[46]

7.2.2.2.2 Removal of Pesticides. Pesticides are (natural or synthetic) agrochemicals used in different agricultural activities for preventing, destroying, repelling, or mitigating pests. According to the target organisms and their biological function, pesticides include herbicides, insecticides, fungicides, rodenticides, and nematicides.[47,48] The inadequate utilization of pesticides results in the hazardous contamination of soil, air, surface water, and food products.[49] Thus, the removal of pesticides from water has been one of the peak areas of research in recent years.[50]

Removal of pesticides *via* adsorption on AC for water reuse is considered to be the most convenient compared to other techniques due to the initial cost, flexibility, simplicity of design, easy operation, and insensitivity to toxic pollutants.[51] 2,4-Dichlorophenoxyacetic acid (2,4-D) is one of the extensively applied herbicides. 2,4-D is applied against various weed species for numerous crops in different countries in the world. Due to its low biodegradability and high mobility, 2,4-D poses a threat to human health and farm animals.[52] Amiri *et al.* prepared mesoporous and microporous AC derived from canola stalk and evaluated its ability to remove 2,4-D from synthetic and natural water. The results showed that the maximum uptake was 135.8 mg g^{-1} at an initial concentration of 150 mg and pH 2.[7] AC derived from waste tangerine seeds was prepared by a one-step pyrolysis method and was utilized to remove carbamate pesticides from complex solutions (Figure 7.4). Batch adsorption studies confirmed that the adsorption behavior was well fitted with the Langmuir isotherm model and pseudo-second-order kinetics model. Further, van der Waals' force and hydrogen bonding were the main forces in a spontaneous and exothermic adsorption process for carbamate pesticides on the surface of AC.[8]

Use of the AC derived from sewage sludge holds promise as a low-cost and sustainable solution to manage this waste; sludge from a pharmaceutical industry waste was utilized as a precursor for the synthesis of ACs. Different activating agents ($ZnCl_2$, $FeCl_3·6H_2O$, $Fe(NO_3)_3·9H_2O$, and $Fe(SO_4)_3·H_2O$) were used in the synthesis process. The fabricated ACs were employed in the removal of three neonicotinoid insecticides acetamiprid (ACT), thiamethoxam (THM), and imidacloprid (IMD) by adsorption. AC prepared *via* $ZnCl_2$ displayed a multilayer profile adsorption isotherm with the highest adsorption capacities (q_e) of 128.9, 126.8, and 166.1 mg g^{-1} for ACT, THM, and IMD, respectively.[9]

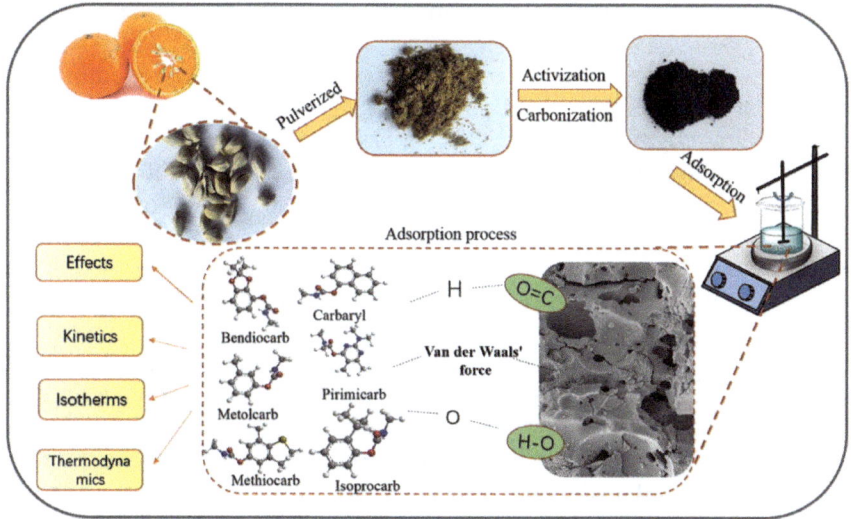

Figure 7.4   Schematic representation of preparing AC from waste tangerine seeds by a one-step pyrolysis method, and its role in removing carbamate pesticides.[8] Reproduced from ref. 8 with permission from Elsevier, Copyright 2020.

Poly(cyclodextrin (CD)–AC composite materials were synthesized, which enabled using an amphiphilic material (CD) with a highly efficient material (AC) with a high capacity to interact and remove the cymoxanil and imidacloprid pesticides from polluted water. Composites of poly(cyclodextrin) with 5 and 10 wt% of AC showed particularly good stability, high removal efficiencies (>75%), and pesticide sorption capacities up to 50 mg g^{-1}.[53] Micro-/mesoporous AC derived from tangerine peels (TPAC) was synthesized *via* chemical treatment with phosphoric acid. The synthesized TPAC was utilized to remove the acetamiprid insecticide from an aqueous solution. The adsorption process followed the Langmuir isotherm and fitted the pseudo-second-order kinetics model with a maximum adsorption capacity of 35.7 mg g^{-1}.[54]

7.2.3   Desalination

Water scarcity as a serious global crisis is threatening food security worldwide, especially in developing countries. Therefore, using low-quality water such as drainage water, saline groundwater, and treated wastewater has become an important consideration and is a new policy that is being applied.[55] The high salinity of agricultural drainage water (ADW) is a serious issue that results in real risks to soil, plant, and public health when used in crop irrigation. For instance, the average salinity of ADW in Egypt is estimated to be 565 ppm and can reach up to 6000 ppm on the northern coast of the Mediterranean.[56] At high salinity concentrations, the soil water becomes less available for the plant to accommodate its transpiration due to the high osmotic value of the saline solution.

This phenomenon affects directly the primary production, in a similar way to the drought.[57]

AC has been widely investigated for the desalination of ions. Separation of the ions from water using membranes is the traditional way of desalination. Bian *et al.* investigated the role of AC fibers (ACF) compared with stainless steel fibers (SSF) and filter paper (FP) for the desalination of NaCl.[58] The highest desalination rate was obtained for ACF (513.4 $mg\,L^{-1}\,h^{-1}$), followed by SSF (374.1 $mg\,L^{-1}\,h^{-1}$) and FP (297.9 $mg\,L^{-1}\,h^{-1}$). This enhancement of the ACF separator was attributed to the decrease in the internal resistance of membrane capacitive deionization upon the packing of ACF.

Capacitive deionization (CDI) is an alternative and promising ion-separation technology for water treatment of brackish water with low or moderate salt concentrations (below 10 $g\,L^{-1}$), relying on applying an electrical potential difference over two electrodes.[59] AC as an active porous carbon has significant potential for water desalination. Electrospun-AC fiber (ACF) showed an enhanced electrosorption capacity of 10.53 $mg\,g^{-1}$ at 1.6 V, ascribed to the supplementary interfiber pores with high specific surface area, large pore volume, and enhanced mesoporosity.[60] Such additional interfiber pores served as ion buffering reservoirs in the electrosorption process. On the other hand, super-AC with ultra-high pore volume was synthesized from the coconut shell immersed in H_3PO_4 and then kept in a CO_2 atmosphere at 865 °C for 2 h.[61] This super-AC exhibited a superior salt adsorption capacity of 27.2 $mg\,g^{-1}$ at 1.2 V due to the large total pore volume caused by the abundant presence of both micropores and mesopores in the material. The different mechanisms of important electrochemical reactions and processes in CDI electrodes are summarized in Figure 7.5.

The combination of carbon nanotubes (CNTs) with AC further enhances CDI. Le *et al.* reported the efficiency of CNT/AC hybrid electrodes for water desalination. The electron transfer between the AC particles was improved by enhancing the electrical bridging of CNT fibers. The salt adsorption capacity was 11.97 $mg\,g^{-1}$ at 1.2 V, with stable performance for 100 cycles with an energy consumption of 0.284 $kW\,h\,m^{-3}$.[62] Matching with this finding, Chen *et al.* investigated the role of CNTs combined with AC for desalination of real saline wastewater.[63] Addition of CNTs promoted the charge transfer of the hybrid electrode by the bridging effect and significantly improved the average salt removal rate (ASRR) of 1.59 $\mu mol\,cm^{-2}\,min^{-1}$. Besides, the flow-electrode capacitive deionization (FCDI) for desalination of wastewater exhibited a high charge efficiency (>99%), a high decrease in conductivity (94%), and low energy consumption (0.034 $kW\,h\,mol^{-1}$), demonstrating that using CNTs is a feasible approach for sustainable wastewater desalination.[63]

7.2.4 AC-based Electrochemical Sensors for Agro-applications

Electrochemical sensors are based on the electrocatalytic reaction of active electrode materials toward the electrochemical behavior of analytes (toxic

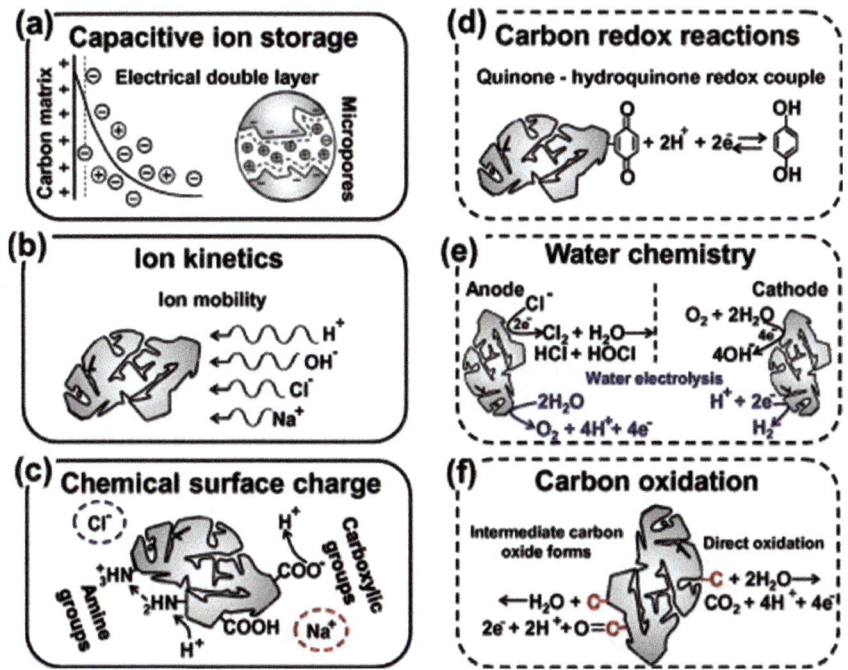

Figure 7.5 Mechanisms of important electrochemical reactions and processes in CDI electrodes.[73] Reproduced from ref. 73 with permission from Elsevier, Copyright 2013.

substances); therefore, for any toxic substance to be detected, it must be electrochemically active on the electrode material.[64]

7.2.4.1 Pesticides

Disproportionate use of pesticides in modern agriculture to balance the increasing demand for food results in water contamination. Thus, monitoring water quality for detecting pesticide residues is of great importance.[65] An electrochemical sensor for indirect determination of carbofuran was fabricated (PET-AC/GCE), and a glassy carbon electrode (GCE) was modified with the AC derived from post-consumer polyethene terephthalate (PET) bottles.[66] PET-AC/GCE provides high sensitivity with a low limit of detection (0.03 μM) as an amperometric sensor (Figure 7.6).

7.2.4.2 Organic Pollutants

Dihydroxybenzene isomers (hydroquinone, HQ, and catechol, CC) have a wide range of industrial applications as developing agents, despite their toxic effects on the environment. Thus, monitoring the trace amounts of hydroquinone and catechol has particular importance. An electrochemical sensor based on a

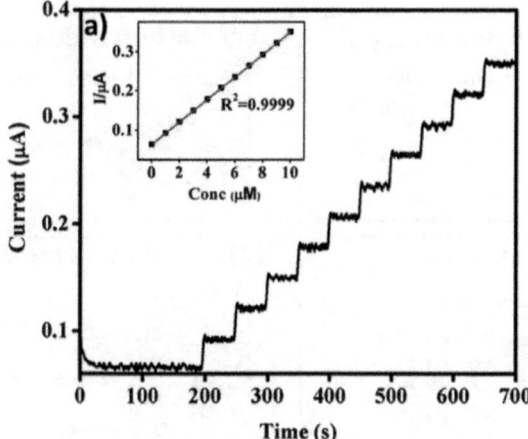

Figure 7.6 Amperometric *i–t* response of PET-AC/GCE upon successive addition (1 μM) of carbofuran–phenol in 0.1 M PBS (pH 7.0), inset: plot of the response current (μA) *vs.* carbofuran–phenol concentration (μM).[66] Reproduced from ref. 66, https://doi.org/10.1038/s41598-018-31627-8, under the terms of the CC BY 4.0 license https://creativecommons.org/licenses/by/4.0/.

glassy carbon electrode (GCE) modified with a sensitive material of bamboo-AC (MCPBAC) was fabricated. The modified sensor (MCPBAC/GCE) can distinguish and sensitively measure HQ and CC by using differential pulse voltammetry (DPV) with a low detection limit of 0.2 mM for both (Figure 7.7).[67]

7.2.4.3 Heavy Metals

Detection of heavy metals at trace levels using AC based on modified electrodes has been recently recognized as an economic and eco-friendly electrochemical sensing technique.[68] For example, a palladium (Pd) nanoparticle-modified porous AC electrode exhibited highly sensitive and selective detection of heavy metal ions such as Cd^{2+}, Pb^{2+}, Cu^{2+}, and Hg^{2+} simultaneously and individually.[69,70] A nitrogen-doped activated nano-porous carbon-modified GCE showed also excellent sensitivity towards Pb^{2+} ions with DPV, and the peak current showed a linear response in the Pb^{2+} concentration range of 2.0 to 120 mg L^{-1}.[71] In another report, kelp-derived porous AC exhibited high sensitivity in the simultaneous detection of Pb^{2+} and Cd^{2+} ions by using square wave anodic stripping voltammetry (SWASV): 53.4 μA μM^{-1} and 26.5 μA μM^{-1}, respectively.[72]

7.3 Future Perspectives

Numerous approaches might be utilized to improve the efficiency of AC and make it more generally available such as nano-AC or coating the surface of

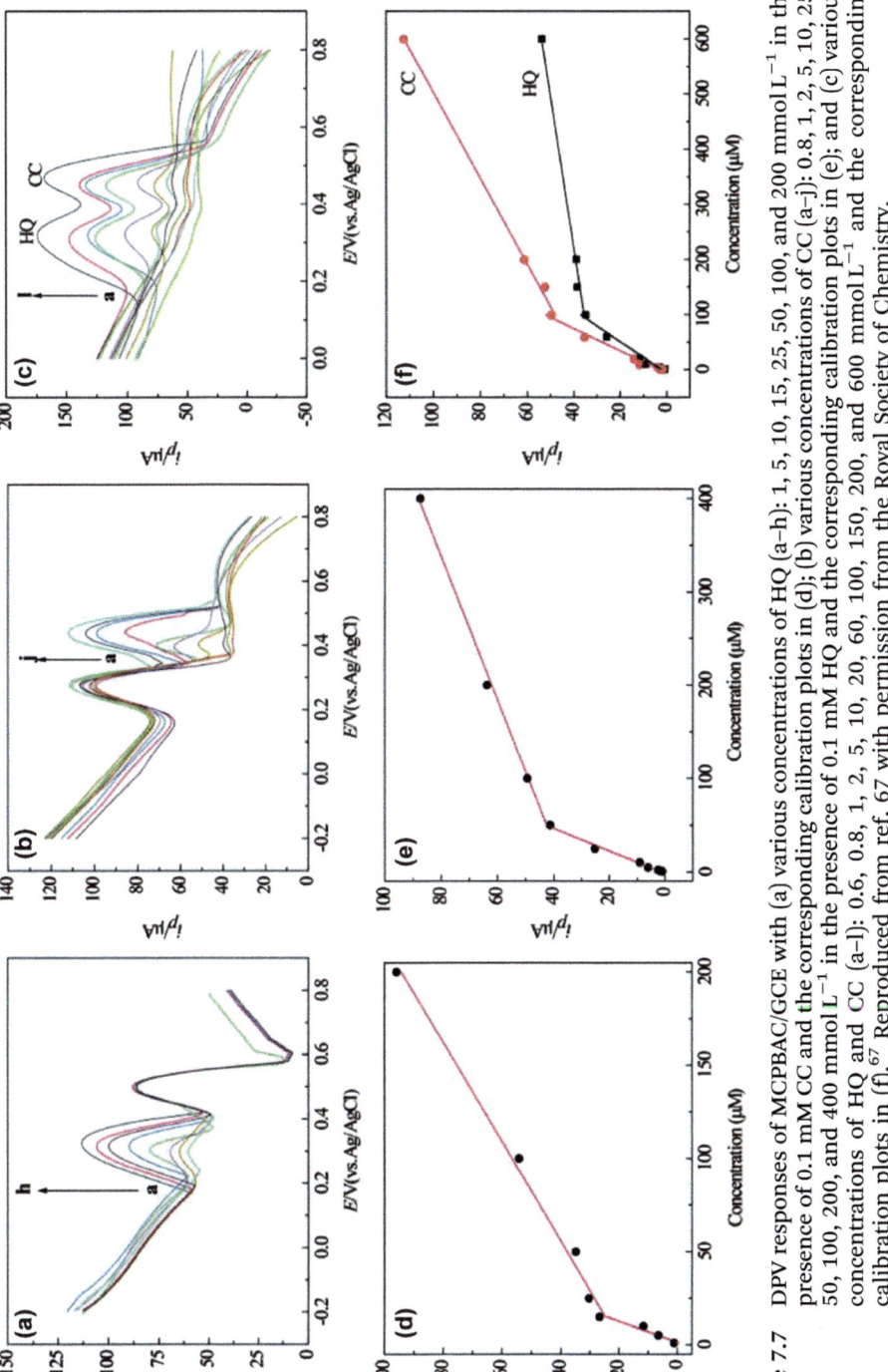

Figure 7.7 DPV responses of MCPBAC/GCE with (a) various concentrations of HQ (a–h): 1, 5, 10, 15, 25, 50, 100, and 200 mmol L^{-1} in the presence of 0.1 mM CC and the corresponding calibration plots in (d); (b) various concentrations of CC (a–j): 0.8, 1, 2, 5, 10, 25, 50, 100, 200, and 400 mmol L^{-1} in the presence of 0.1 mM HQ and the corresponding calibration plots in (e); and (c) various concentrations of HQ and CC (a–l): 0.6, 0.8, 1, 2, 5, 10, 20, 60, 100, 150, 200, and 600 mmol L^{-1} and the corresponding calibration plots in (f).[67] Reproduced from ref. 67 with permission from the Royal Society of Chemistry.

AC with additional active compounds. These strategies improve the unique features of AC, making it more useful in agricultural settings; nonetheless, additional research into the downsides or risk evaluation is needed, particularly with nanomaterial structures.

References

1. W. Du, J. Sun, Y. Zan, Z. Zhang, J. Ji, M. Dou and F. Wang, *RSC Adv.*, 2017, **7**, 46629–46635.
2. Y. Fu, Y. Shen, Z. Zhang, X. Ge and M. Chen, *Sci. Total Environ.*, 2019, **646**, 1567–1577.
3. A. Zubrik, M. Matik, S. Hredzák, M. Lovás, Z. Danková, M. Kováčová and J. Briančin, *J. Cleaner Prod.*, 2017, **143**, 643–653.
4. H. M. al Saidi, O. A. Farghaly, M. E. A. El-sayed, M. A. Elmottaleb, A. Taha and M. G. Z. Ahmed, *Life Sci. J.*, 2020, **1**, 17.
5. A. Wang, K. Sun, R. Xu, Y. Sun and J. Jiang, *J. Cleaner Prod.*, 2021, **283**, 125385.
6. W. S. Chai, J. Y. Cheun, P. S. Kumar, M. Mubashir, Z. Majeed, F. Banat, S. H. Ho and P. L. Show, *J. Cleaner Prod.*, 2021, 296.
7. M. J. Amiri, R. Roohi, M. Arshadi and A. Abbaspourrad, *Environ. Sci. Pollut. Res.*, 2020, **27**, 16983–16997.
8. Y. Wang, S. Ling Wang, T. Xie and J. Cao, *Bioresour. Technol.*, 2020, **316**, 123929.
9. E. Sanz-Santos, S. Álvarez-Torrellas, L. Ceballos, M. Larriba, V. I. Águeda and J. García, *Appl. Sci.*, 2021, **11**, 308.
10. Food and Agriculture Organization of the United Nations (FAO), *The future of food and agriculture – Alternative pathways to 2050*, 2018.
11. H. Rose, L. Benzon, M. Rosnah, U. Rubenecia, V. U. Ultra and S. C. Lee, *Int. J. Agron. Agric. Res.*, 2015, **7**, 2223–7054.
12. M. A. Iqbal, *Nano-Fertilizers for Sustainable Crop Production under Changing Climate: A Global Perspective*, IntechOpen, 2016.
13. C. Tarafder, M. Daizy, M. M. Alam, M. R. Ali, M. J. Islam, R. Islam, M. S. Ahommed, M. Aly Saad Aly and M. Z. H. Khan, *ACS Omega*, 2020, **5**, 23960–23966.
14. H. Sultan, N. Ahmed, M. Mubashir and S. Danish, *Sci. Rep.*, 2020, **10**, 595.
15. P. Priyadi and W. Mangiring, *Sains Tanah – J. Soil Sci. Agroclimatol.*, 2019, **16**, 147.
16. B. D. Özel Duygan, K. M. Udert, A. Remmele and C. S. McArdell, *Resour., Conserv. Recycl.*, 2021, **166**, 105341.
17. S. Bakshi, C. Banik, D. A. Laird, R. Smith and R. C. Brown, *ACS Sustainable Chem. Eng.*, 2021, **9**, 8222–8231.
18. A. Kumar, S. Joseph, E. R. Graber, S. Taherysoosavi, D. R. G. Mitchell, P. Munroe, L. Tsechansky, O. Lerdahl, W. Aker and M. Sæbø, *Chem. Biol. Technol. Agric.*, 2021, **8**, 21.

19. M. Ashfaq, N. Verma and S. Khan, *Environ. Sci.: Nano*, 2017, **4**, 138–148.
20. D. A. Prastiwi, B. Sumawinata, Iskandar and G. Pari, in *IOP Conference Series: Earth and Environmental Science*, Institute of Physics Publishing, 2019, vol. 359.
21. J. du Shin, D. G. Park, S. G. Hong, C. Jeong, H. Kim and W. Chung, *Environ. Pollut.*, 2021, **285**, 117457.
22. C. J. Clark, *Environ. Forensics*, 2003, **4**, 167–173.
23. J. Wyszkowska and J. Kucharski, Biochemical Properties of Soil Contaminated by Petrol, *Pol. J. Environ. Stud.*, 2000, **9**, 479–485.
24. United States Environmental Protection Agency (U.S. EPA), https://clu-in.org/s.focus/c/pub/i/2727, 2018, 9.
25. I. Zamulina, D. Pinsky, M. Burachevskaya, T. Bauer and A. Pshenichnaya, *E3S Web Conf.*, 2020, **175**, 3–8.
26. A. D. Harwood, S. A. Nutile and A. M. Simpson, *Environ. Pollut.*, 2022, **295**, 118687.
27. S. Zhang, Y. He, L. Wu, J. Wan, M. Ye, T. Long, Z. Yan, X. Jiang, Y. Lin and X. Lu, *Pedosphere*, 2019, **29**, 400–408.
28. S. Annamalai, A. Septian, J. Choi and W. S. Shin, *J. Environ. Manage.*, 2022, **310**, 114709.
29. G. Liu, J. Qian, Y. Zhang, K. Lin and F. Liu, *J. Environ. Chem. Eng.*, 2022, **10**, 107475.
30. H. Wang, J. Liu, C. Gui, Q. Yan, L. Wang, S. Wang and J. Li, *Environ. Res.*, 2022, **208**, 112707.
31. J. Xiao, Z. Pang, S. Zhou, L. Chu, L. Rong, Y. Liu, J. Li and L. Tian, *Sep. Purif. Technol.*, 2020, **244**, 116667.
32. N. Xie, Z. Chen, H. Wang and C. You, *J. Cleaner Prod.*, 2021, **308**, 127433.
33. G. Vasilyeva, V. Kondrashina, E. Strijakova and J. J. Ortega-Calvo, *Sci. Total Environ.*, 2020, **706**, 135739.
34. P. Meynet, S. E. Hale, R. J. Davenport, G. Cornelissen, G. D. Breedveld and D. Werner, *Environ. Sci. Technol.*, 2012, **46**, 5057–5066.
35. I. Twardowska, H. E. Allen, M. M. Häggblom and S. Stefaniak, *Soil and Water Pollution Monitoring, Protection and Remediation*, NATO Science Series: IV, volume 69, Springer, 2006, ISBN: 978-1-4020-4726-8.
36. N. N. Semenyuk, V. S. Yatsenko, E. R. Strijakova, A. E. Filonov, K. V. Petrikov, Y. A. Zavgorodnyaya and G. K. Vasilyeva, *Microbiology*, 2014, **83**, 589–598.
37. V. Kondrashina, E. Strijakova, L. Zinnatshina, E. Bocharnikova and G. Vasilyeva, *Soil Sci.*, 2018, **183**, 150–158.
38. V. Srivastava, A. Sarkar, S. Singh, P. Singh, A. S. F. de Araujo and R. P. Singh, *Front. Environ. Sci.*, 2017, **5**, 1–19.
39. C. Duan, T. Ma, J. Wang and Y. Zhou, *J. Water Process Eng.*, 2020, **37**, 101339.
40. S. M. Kharrazi, N. Mirghaffari, M. M. Dastgerdi and M. Soleimani, *Powder Technol.*, 2020, **366**, 358–368.

41. C. Tang, Y. Shu, R. Zhang, X. Li, J. Song, B. Li, Y. Zhang and D. Ou, *RSC Adv.*, 2017, **7**, 16092–16103.
42. E. A. Deliyanni, G. Z. Kyzas, K. S. Triantafyllidis and K. A. Matis, *Open Chem.*, 2015, **13**, 699–708.
43. J. Rivera-Utrilla and M. Sánchez-Polo, *Water Res.*, 2003, **37**, 3335–3340.
44. T. Bohli, A. Ouederni and I. Villaescusa, *EuroMediterr. J. Environ. Integr.*, 2017, **2**, 19.
45. M. Fronczak, K. Pyrzyńska, A. Bhattarai, P. Pietrowski and M. Bystrzejewski, *Int. J. Environ. Sci. Technol.*, 2019, **16**, 7921–7932.
46. C. Suo, K. Du, R. Yuan, H. Chen, F. Wang and B. Zhou, *Desalin. Water Treat.*, 2020, **183**, 315–324.
47. W. Aktar, D. Sengupta and A. Chowdhury, *Interdiscip. Toxicol.*, 2009, **2**, 1–12.
48. P. C. Abhilash and N. Singh, *J. Hazard. Mater.*, 2009, **165**, 1–12.
49. X. Chang, M. T. Meyer, X. Liu, Q. Zhao, H. Chen, J. an Chen, Z. Qiu, L. Yang, J. Cao and W. Shu, *Environ. Pollut.*, 2010, **158**, 1444–1450.
50. G. Liu, L. Li, X. Huang, S. Zheng, X. Xu, Z. Liu, Y. Zhang, J. Wang, H. Lin and D. Xu, *J. Mater. Sci.*, 2018, **53**, 10772–10783.
51. K. Y. Foo and B. H. Hameed, *J. Hazard. Mater.*, 2010, **175**, 1–11.
52. L. Ding, X. Lu, H. Deng and X. Zhang, *Ind. Eng. Chem. Res.*, 2012, **51**, 11226–11235.
53. G. Utzeri, L. Verissimo, D. Murtinho, A. A. C. C. Pais, F. X. Perrin, F. Ziarelli, T. V. Iordache, A. Sarbu and A. J. M. Valente, *Molecules*, 2021, **26**, 1426.
54. S. G. Mohammad, S. M. Ahmed, A. E. G. E. Amr and A. H. Kamel, *Molecules*, 2020, **25**, 2339.
55. J. Wang, M. E. A. El-sayed and I. A. Abdelhafeez, *Sustainability*, 2021, 1–12.
56. M. Shaban, *Irrig. Drain.*, 2020, **69**, 788–805.
57. S. E. A. T. M. van der Zee, S. F. Stofberg, X. Yang, Y. Liu, Md. N. Islam and Y. F. Hu, in *Current Perspective on Irrigation and Drainage*, 2017, pp. 1–22.
58. Y. Bian, P. Liang, X. Yang, Y. Jiang, C. Zhang and X. Huang, *Desalination*, 2016, **381**, 95–99.
59. S. Porada, R. Zhao, A. Van Der Wal, V. Presser and P. M. Biesheuvel, *Prog. Mater. Sci.*, 2013, **58**, 1388–1442.
60. N. L. Liu, L. I. Chen, S. W. Tsai and C. H. Hou, *Environ. Sci.*, 2020, **6**, 312–320.
61. Z. Zhang, Y. Zhang, C. Jiang, D. Li, Z. Zhang, K. Wang, W. Liu, X. Jiang, Y. Rao, C. Xu, X. Chen and N. Meng, *Desalination*, 2022, **529**, 115653.
62. V. H. Le, L. T. N. Huynh, T. N. Tran, T. T. N. Ho, M. N. Hoang and T. H. Nguyen, *J. Appl. Electrochem.*, 2021, **51**, 1313–1322.
63. K. Y. Chen, Y. Y. Shen, D. M. Wang and C. H. Hou, *Desalination*, 2022, **522**, 115440.
64. G. Hernandez-Vargas, J. E. Sosa-Hernández, S. Saldarriaga-Hernandez, A. M. Villalba-Rodríguez, R. Parra-Saldivar and H. M. N. Iqbal, *Biosensors*, 2018, **8**, 29.

65. N. Lezi and A. Economou, *Electroanalysis*, 2015, **27**, 2313–2321.
66. S. Ayyalusamy, S. Mishra and V. Suryanarayanan, *Sci. Rep.*, 2018, **8**, 13151.
67. X. Yang, C. He, Y. Lin, Y. Qiu, P. Li, Y. Chen, B. Huang and X. Zheng, *Anal. Methods*, 2022, **14**, 34–43.
68. L. G. Djemmoe, E. Njanja, F. M. M. Tchieno, D. T. Ndinteh, P. G. Ndungu and I. K. Tonle, *Int. J. Environ. Anal. Chem.*, 2020, **100**, 1429–1445.
69. P. Veerakumar, V. Veeramani, S. M. Chen, R. Madhu and S. Bin Liu, *ACS Appl. Mater. Interfaces*, 2016, **8**, 1319–1326.
70. T. Zhang, H. Jin, Y. Fang, J. B. Guan, S. J. Ma, Y. Pan, M. Zhang, H. Zhu, X. D. Liu and M. L. Du, *Mater. Chem. Phys.*, 2019, **225**, 433–442.
71. Y. Baikeli, X. Mamat, N. Yalikun, Y. Wang, M. Qiao, Y. Li and G. Hu, *RSC Adv.*, 2019, **9**, 23678–23685.
72. J. Guan, Y. Fang, T. Zhang, L. Wang, H. Zhu, M. Du and M. Zhang, *Electrocatalysis*, 2020, **11**, 59–67.
73. S. Porada, R. Zhao, A. van der Wal, V. Presser and P. M. Biesheuvel, *Prog. Mater. Sci.*, 2013, **58**, 1388–1442.

CHAPTER 8

Alcoholic Beverage Purification Applications of Activated Carbon

ABHINAY THAKUR,[a] ASHISH KUMAR*[b] AND RENHUI ZHANG*[c]

[a] Department of Chemistry, School of Chemical Engineering and Physical Sciences, Lovely Professional University, Phagwara, Punjab, India; [b] NCE, Department of Science and Technology, Government of Bihar, India; [c] NCE, School of Materials Science and Engineering, East China JiaoTong University, Nanchang 330013, China
*Emails: drashishchemlpu@gmail.com; 3067@ecjtu.edu.cn

8.1 Introduction

Alcoholic beverages (ABs) are among the most well-liked products around the globe. According to reports, the mean amount of sheer alcohol consumed by adults over the age of 15 in 2016 was 6.4 L per person; however, there are significant regional differences.[1-3] Non-distilled products are often made using liquid fermentation rather than high-temperature distillation. Because the alcohol percentage is typically moderate and the flavor of non-distilled spirits substantially preserves the fragrances of the basic ingredients utilized, the end product is appropriate for the majority of customers. Contrarily, distilled spirits are made by fermenting raw materials and then distilling them at extreme temperatures. Since these products are typically made *via* liquid- or solid-state fermentation, they have significant alcohol concentration. Additionally, following high-temperature distillation, these

Activated Carbon: Progress and Applications
Edited by Chandrabhan Verma and Mumtaz A. Quraishi
© The Royal Society of Chemistry 2023
Published by the Royal Society of Chemistry, www.rsc.org

beverages are rich in volatile flavor compounds as a result of their complex fermentation process. For instance, in Chinese Baijiu, more than 2000 trace chemicals have been found in the final yield of alcohol. Consumers frequently use ABs' flavor as the primary indicator of their excellence.[4–7] Extremely varied flavors of ABs are a result of a variety of factors, including various source ingredients, fermentation apparatus, saccharification and/or fermentation reagents, fermentation and/or distillation methods, manufacturing and preservation environments, and mixing.[8–15] It is noteworthy that off-flavor, which is an unbalanced flavor that deviates beyond the alcohol's and beverage's natural scent, flavor, and taste, has garnered increasing attention when it pertains to flavor. Off-flavor refers to any comparative aberrant impulses or aberrant pattern mixtures that flavor and smell sensors can detect in particular circumstances. Off-flavors share several traits in general, including a lower detection limit, challenging characterization, and a significant impact on the entire flavor of ABs. These impurities have unfavorable behavioral effects on perception. Such contamination generation and financial penalties of ABs are rather significant. For example, according to statistics, China produced 0.372 million kiloliters of grape wine from January to November in the year 2019, a 12.1% decline from the production of the year before (China Alcoholic Drinks Association, 2019).[16–20] The causes that may explain this are as follows. Initially, there was fierce rivalry between imported wine by countries in Europe, the United States, and other regions, and there was a poor production of grape juice with minimal sugar concentration. Secondly, cork smells and other issues linked to bad flavors could cause a significant drop in alcohol (especially wine) output in China. One of the biggest issues on the market today is off-tastes, which can occur in both distilled and non-distilled spirits. Numerous volatile substances have been found in ABs, including esters, alcohols, ketones, ethers, hydrocarbons, fumarates, pyrrolines, pyridines, and sulfur compounds. However, there are limited comprehensive observations on such effluents. According to reports, the taste and smell of soil mold and cork are the two most prevalent off-flavors in grape wine. Pit mud, moldy, rice husk, iron, tallowy, sour, rubbery, and burnt odors and tastes have all been noted in Baijiu. While molecules like unsaturated alcohols, volatile phenolic compounds, aldehydes, and ketones often contribute to the fragrance of stables, drugs, mushrooms, and other things, certain chemicals only have an unpleasant smell whenever they upset the harmony of other fragrance elements. It is difficult to categorize these components as off-tastes. For instance, several single ingredients, including dimethyl disulfide, 3-methylindole, and dimethyl trisulfide, initially enhance the flavor of ABs characteristic at reduced amounts but become unfavorable while prevalent at proportions greater than a specific critical level.[21–23] A comprehensive study from the perspectives of the off-flavors and other contributing elements is essential to maximize productivity and reduce the business costs of ABs brought on by off-flavors.[11,24] Brewing is the collective name for the processes used to produce alcohol (beer), and Figure 8.1 depicts a summary of these operations.[25]

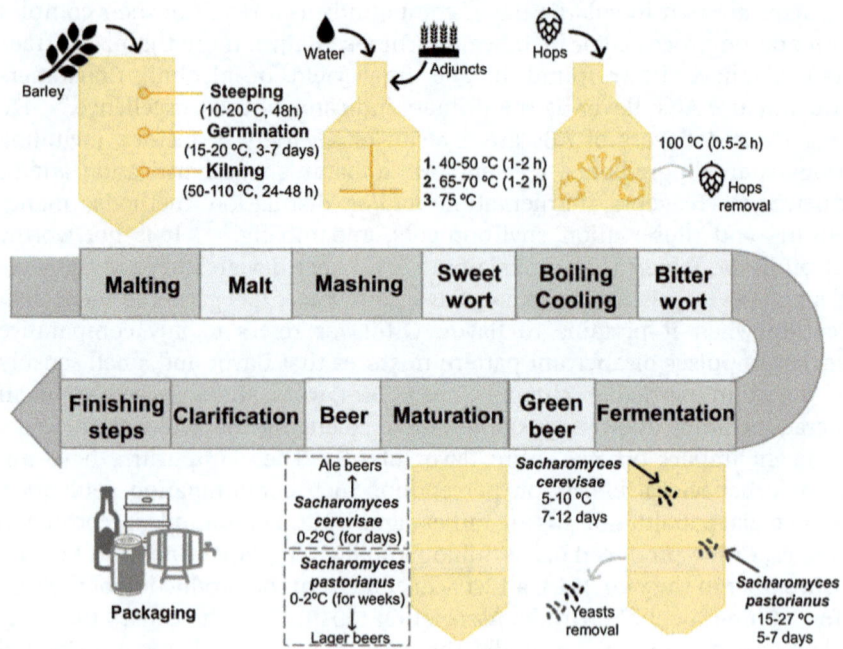

Figure 8.1 Illustration of the brewing process. Reproduced from ref. 25, https://doi.org/10.3390/molecules26216460, under the terms of the CC BY 4.0 license https://creativecommons.org/licenses/by/4.0/.

Carbon adsorbents (active and activated carbons (ACs)) are special and excellent sorption substances which enable resolving a multitude of problems linked to maintaining the biological and chemical wellbeing of people, the ecosystem, and infrastructures owing to their physicochemical characteristics. When fabricated from carbon-containing sources, ACs are extremely porous materials with expanded internal surfaces that can exceed 2500 $m^2 g^{-1}$ and have excellent absorption properties for contaminants in the medium to be cleaned. Each kind of organic micro-impurity is absorbed by the porous configuration of coals owing to adsorption factors (surface interaction forces). Such adsorbents simultaneously accomplish several activities to enhance the purity of drinkable water and ABs, especially vodka. For instance, active coal BAU-A, which is manufactured from charcoal produced by the pyrolysis of thick wood varieties, is commonly employed in distilleries in Russia (birch, oak, beech). AC adsorbents, which remove practically any organic contaminants, are employed to purify alcohols and several other beverages in the food sector. To produce β-galactosidase using *Lactobacillus bulgaricus* and generate purified alcohols, Duan *et al.*[26] developed an integrated use of dairy whey with AC. This research developed an integrated method for using whey to support the development of probiotics, produce enzymes, and synthesize galactooligosaccharides (GOS) (Figure 8.2).

To substitute the MRS medium for growing *L. bulgaricus* L3 to manufacture β-galactosidase, a unique, inexpensive whey-based medium was created. For the very first time, the lysozyme-treated crude *L. bulgaricus* L3 β-galactosidase without the exclusion of cell debris served as a highly effective catalyst for the synthesis of GOS utilizing less costly whey as the substrate for the creation of highly purified liquor.

Continuing into the research mentioned previously, a novel, reasonably priced whey-based medium blended with AC was prepared for cultivating *L. bulgaricus* to manufacture β-galactosidase. Involving several empirical models, including the Plackett–Burman model, the steepest ascent test, and the central composite design, response surface methodology (RSM) was used to optimize the medium. The β-galactosidase activity of *L. bulgaricus* achieved was 2034 U L^{-1} in the optimized medium, which was double that produced in the conventional MRS medium. After being removed from the whey-based medium, the *L. bulgaricus* cells were next given lysozyme treatment. By employing whey (200 g L^{-1}) as the substrate, the resultant crude enzyme was employed as an effective catalyst, which catalyzed the production of the prebiotic galactooligosaccharides (GOS) at a significant yield of 44.7%. Activated charcoal adsorption was used to thoroughly clean the sugar mixture, producing a high-purity GOS with a purity level of 77.6%.

Similarly, chitosan's effectiveness as an adsorbent for extracting copper(II) from cachaça was assessed by Santos *et al.*[27] FTIR and SEM were used to assess the morphological properties of the produced chitosan and the impact of absorbed copper. Specifically, SEM was used to acquire backscattered electron images (BEIs) of the cross-section of the sample (Figure 8.3). It was feasible to see that the morphological alterations were facilitated by the deacetylation processes. Qi exhibited a consistent, hard morphology with structured fibers in photomicrographs (Figure 8.3A and B), which were

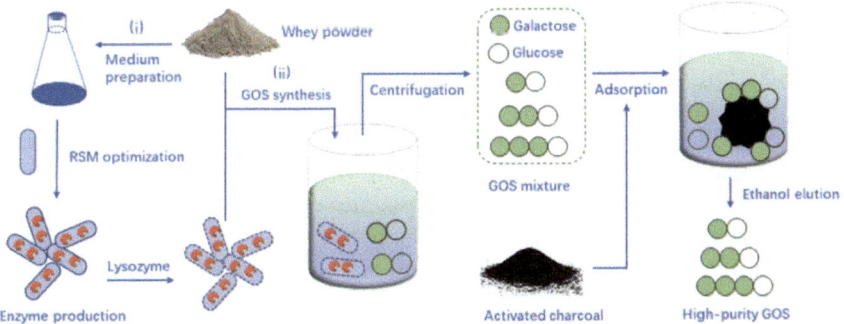

Figure 8.2 A description of the integrated use of dairy whey in the production of GOS and generation of β-galactosidase. Reproduced from ref. 26, https://doi.org/10.3390/catal11060658, under the terms of the CC BY 4.0 license https://creativecommons.org/licenses/by/4.0/.

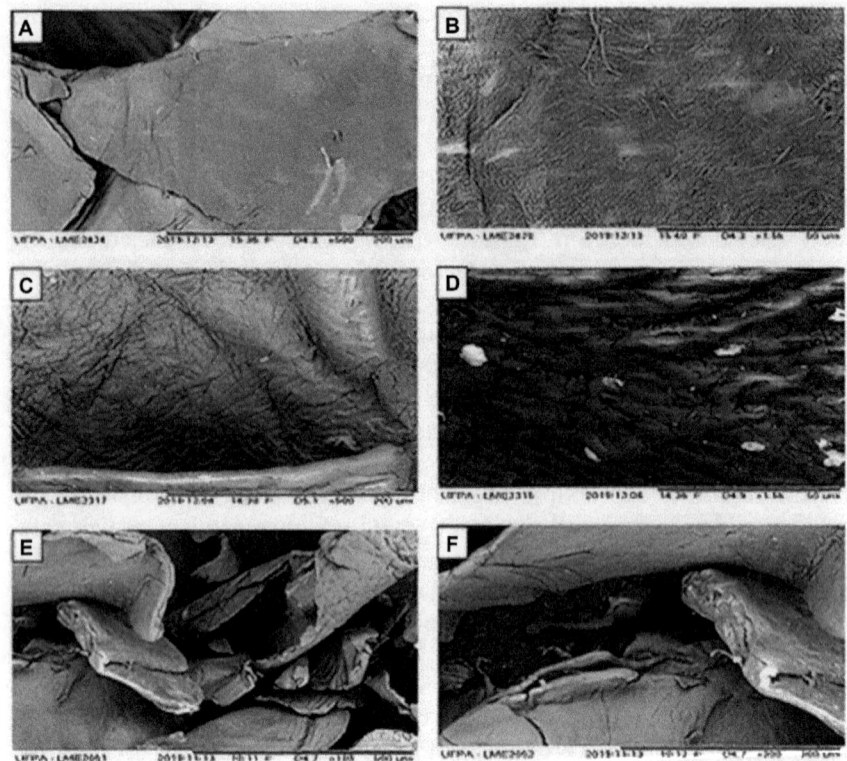

Figure 8.3   ERE photomicrographs for Qi (A and B) and Qt amplified by 500 and 1500
times (C and D), and 180 and 300 times the original size for Qc (E and F).
Reproduced from ref. 27, https://doi.org/10.3390/polym14030573, under
the terms of the CC BY 4.0 license https://creativecommons.org/licenses/
by/4.0/.

typical of chitin granules. In contrast, the structures of Qt (Figure 8.3C and D)
had a rougher surface made up of finer, irregularly shaped granules than the
structures of Qi. Another research study revealed a comparable set of traits.
The utilization of chitosan for particular purposes, including its use as a
bio-adsorbent, was closely tied to variations in the degree of deacetylation
and the molar mass in the polymer framework. Photomicrographs
Figure 8.3(E and F) of Qc showed a substrate with a brighter and highly vivid
appearance, which was similar to the appearance of Cu(II) in chitosan. The
illumination level in BEI is inversely related to the atomic number of the
element. Deacetylation of chitin (shrimp shell) produced chitosan, which
had an approximate molecular mass of 162.96 kDa, with a deacetylation
degree of 88.9% (potentiometric titration) and 86.9% (FTIR). By using
spectrophotometric titration, the copper elimination rate was estimated to
be 84.09%. The kinetic study demonstrated a superior fit of the data by the
Elovich equations, indicating that the chemosorption method regulated the
kinetic phase. The amine sites of chitosan showed a propensity for

adsorbing copper ions. The findings indicate that chitosan may enhance the cachaça's flavor and efficiency.

Similarly, by examining major volatile compounds at various phases of distillation, Gantumur *et al.*[28] aimed to improve the understanding of the desirable sensory attributes of drinkable whey-based spirits made using mid-supplemented, non-supplemented, and highly-supplemented whey samples with the aid of AC. The outcomes revealed that commercial *Saccharomyces cerevisiae* strains in lactase-hydrolyzed whey produced ethanol efficiently and quickly, up to 28.4, 41.9, and 55.7 gL^{-1} in about 26, 31, and 37 hours, respectively. During the fermentation, the differences in specific gravity, titratable acidity, pH, residual protein, sugar concentration, and alcohol yield were examined. The total content of volatile compounds declined noticeably from the head (2087–2549 mgL^{-1}) to the tail whey spirits (889–1398 mgL^{-1}). The most common dominant chemicals in the whey spirit were 3-methyl-1-butanol, 2-methyl-1-butanol, 1-propanol, acetaldehyde, 2-methyl-1-propanol, and ethyl acetate, which together accounted for the majority of the total volatile components. The number of volatile chemicals found was much lower than the permitted legal level. The findings imply that AC/lactose-supplemented whey could be fermented with *S. cerevisiae* cells to produce drinkable spirits with excellent sensory attributes.

Liu *et al.*[29] used the fluorescent reactant 4-((aminooxy)methyl)-7-hydroxycoumarin (AOHC) to label low-molecular-mass aldehydes (LMMAs) *via* a reductive oxymination reaction, producing single *N,O*-substituted oxyamine derivative products at ambient temperature with up to 96.8% derivatization efficacy. 12 LMMAs, such as aromatic aldehydes, furfurals, and aliphatic aldehydes, were baseline-separated on an ODS column, with LODs of 0.2–50 nM and decent accuracy (the intraday relative standard deviations [RSDs] were 2.40–4.68%, and the interday RSDs were 4.65–8.91%) in the subsequent high-performance liquid chromatography with fluorescence detection. Following this, nine LMMAs with LODs in the range of 0.28–798.16 μM were effectively detected with outstanding precision (the recoveries were 92.2–106.2%) in the analysis of six ABs and five dairy items using this method. The outcomes were then quantitatively examined and evaluated. Excellent sensitivity, room-temperature labelling, and the elimination of additional extraction and/or enrichment processes are just a few benefits of the suggested technique, which shows how useful it is for tracking LMMAs in a variety of complicated matrices.

8.2   Synthesis Routes of AC

The four primary mechanisms of AC production, as documented by numerous scientists, are pyrolysis, physical and chemical activation, carbonization, and steam/thermal activation. Steam and CO_2 are the most utilized agents for the physical activation process, which involves carbonization and activation, and they have a considerable impact on the AC's permeability.[30–35] Chemical reagents including phosphoric acid (H_3PO_4), potassium hydroxide (KOH), and zinc chloride ($ZnCl_2$) may easily be utilized at ambient temperature in a single step of the chemical activation method of producing AC. Contaminants including zinc (Zn) and phosphorus (P) could be detected in AC products, based on the chemical reagents employed; however, doing so could increase the cost of operations because more chemicals must be utilized.

8.2.1 Pyrolysis Process

Pyrolysis is the terminology for the thermochemical process that transforms organic material into gaseous or/and liquid fuels at exceptionally elevated temperatures in the absence of halogens (or mostly oxygen). Pyrolysis is an irreversible process which concurrently alters the chemical structure and physical state of substances. Most often, pyrolysis occurs when substances are heated to higher temperatures.[36–40] Several parameters such as temperature, initial concentration and nitrogen flow rate exert significant impact on the pyrolysis process. Usually, increasing the temperature of the process results in a decrease in the formation of AC and char, whereas increasing the degree of pyrolysis results in a decrease in the yield of particulates and an improvement in the production of both gas and liquid fractions. On either side, increasing the temperature causes a decrease in volatile materials while increasing ash and AC%. As a result, at increasing temperatures, AC of better grade is produced. Additionally, increasing the temperature may produce a reduction in char production due to the sheer substantial breakdown of biomass at high temperatures or as a result of the pyrolysis of char remnants. However, higher secondary pyrolysis temperatures may result in certain gaseous byproducts that are not condensable, eventually enhancing gas production.

For the adsorption of nimesulide in aqueous wastewater, Raupp *et al.*[41] focused on producing and analyzing activated charcoal using olive pomace (OP), an agro-industrial byproduct, and assessed the adsorption behavior and equilibria utilizing an experimental method. The raw product was dehydrated in an oven at 105 °C for 24 hours, pulverized, and chemically activated using OP, $ZnCl_2$, and CaOH in a ratio of $1:0.8:0.2$, and thermal decomposition by pyrolysis in a hardened steel furnace at 550 °C for 30 min. OP, which is composed of pulp and a shell, was roasted in an oven for 24 hours at 105 °C. The processed sample was rinsed with water to neutral and consistent pH after acid leaching with HCl (6 mol L^{-1}). The material was then dried at 105 °C for 24 hours before being kept in a dry environment. ACOP will be used to refer to this material moving forward. The reported surface area was $650.9 \text{ m}^{-2} \text{g}^{-1}$. The PSO and Freundlich models were the mathematical isothermal and kinetic methods that effectively captured the adsorption kinetics, and the greatest adsorption capacity measured was 353.27 mg g^{-1}. The findings revealed that the ACOP performed well as an adsorbent material and had a lot of ability to get rid of emerging pollutants like nimesulide.

DES (choline chloride/urea)/H_3PO_4 was used as the activating agent by Pam *et al.*[42] in a new process for producing palm kernel shell AC. The AC's adsorption capacities and pore characteristics were investigated. The ability of the AC samples to adsorb Pb(II) in the aqueous phase at two different pyrolysis temperatures (500 and 600 °C) was examined. The findings showed that the pyrolysis temperature and the ratio of the precursor to the activating agent had a substantial impact on the formation of AC and its adsorptive capabilities. With a total specific surface area of 1413 $m^2 g^{-1}$ and a total porosity of 0.6181 $cm^3 g^{-1}$, DES/H_3PO_4 AC showed effective Pb(II) extraction. Even when Pb(II) was adsorbed from an aqueous medium with all examined AC samples, each material's adsorption capacity varied depending on its unique characteristics. As compared to the precursor concentration, the pyrolysis temperature had less impact on the Pb(II) adsorption on the AC samples. The porous structure might have contributed to their superior desorption capability.

8.2.2 Physical Activation Process

Physical activation proceeds in two steps. Before activating the ensuing char at extreme temperatures under the influence of CO_2, water, atmosphere, or oxidizing gases, carbonaceous substances must initially be carbonized. Since it is simple to operate and hygienic and shows slow reaction kinetics at a temperature of about 800 °C, CO_2 is frequently employed as an activation gas. This property makes it easier to manage the activation procedure. The carbonization temperature was determined to be around 400 and 850 °C, though it could occasionally reach up to 1000 °C, and the activation temperature was determined to be around 600 and 900 °C. The physical activation process used to create the AC has significant drawbacks that prevent it from being used as a filter or as an adsorbent. Several agricultural biomass byproducts, including mango pits, rice hulls, sawdust, sunflower shells, husks, olive shafts, pine cones, wood scraps, and peanut hulls, could be used for physical activation.

Chang *et al.*[43] synthesized lignosulfonate-based AC fibers (ACF) using the electrospinning method, physical activation, and carbonization. The obtained lignosulfonate-based ACF mostly had mesopores and a negligible quantity of micropores in their porous structure. Furthermore, after CO_2 activation, inadequate carbonization damaged the fibers. Materials with a carbonization temperature of 700 °C were of a higher grade than those with lower temperatures. As the carbonization duration increased, so did the rate of weight loss and specific surface area. A reduced weight loss rate, simplified manufacturing, and a large surface area were all benefits of the two-step carbonization method, which also produced fibers of a higher grade. The lignin fibers prepared in this work that were carbonized at 700 °C and activated at 800 °C for 30 min demonstrated the potential to manufacture lignin-based ACF for other uses. The findings were also analogous to some other investigations which also utilized a physical activation process even

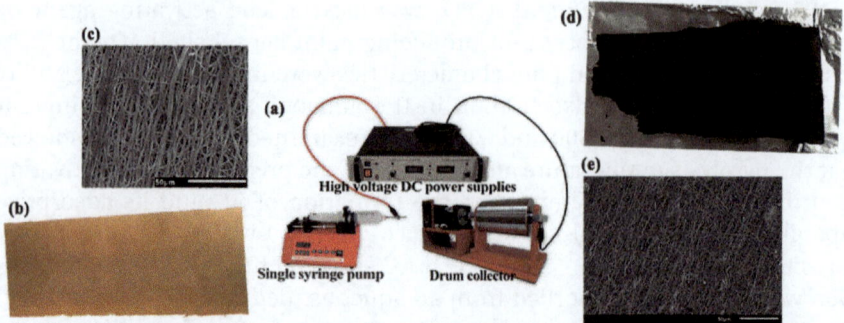

Figure 8.4   (a) Electrospinning instrument; (b) synthesized lignosulfonate fibers; (c) synthesized lignosulfonate fibers' SEM image; (d) AC fibers; and (e) AC fibers' SEM image. Reproduced from ref. 43, https://doi.org/10.3390/ma11101877, under the terms of the CC BY 4.0 license https://creativecommons.org/licenses/by/4.0/.

though the values were not as high as those of other similar studies that subjected ACF to chemical activation. The electrospinning process was carried out longitudinally (Figure 8.4a). The prepared hardwood lignosulfonic acid sodium salt (HLS) solution was placed into a syringe and charged with a power source for electrospinning. The applied voltage was 15 kV, and the syringe needle and a collector were attached to the power supply's positive terminal and ground, respectively. The flow rate was 0.03 mL min^{-1}, the syringe to collector distance was 20 cm, the collector was rotated at 720 rpm, and the needle gauge was 18 G. To prepare lignin fiber mats, the electrospun HLS fibers were gathered on a surface. The mean width and diameter of the fiber mat were 43.70 μm ± 3.09 μm and around 1611 nm ± 351 nm, respectively (Figure 8.4b and c). Figure 8.4d and e, respectively, displays the finalized LACF output and images obtained using a scanning electron microscope (SEM). Additional research into the uses of lignosulfonate-based ACF, which could be employed as very effective adsorption and filtering substances, would be beneficial.

8.2.3   Chemical Activation Process

The chemical activation technique entails numerous consecutive phases in which the oxidizing agents and dehydrates from chemical activating reagents react with the precursor. ACs with improved porosity frameworks are produced when activation and carbonization are carried out concurrently during the chemical activation method at reduced temperatures, even if environmental safety concerns might restrict the usage of chemical activators. Additionally, KOH, trihydroxidooxidophosphorus, H_3PO_4, $ZnCl_2$, and potassium carbonate (K_2CO_3) are among the substances which are frequently utilized as activation agents.

By chemically activating the rubber fruit shell (RFS) using H_3PO_4, accompanied by a straightforward hydrothermal method at ambient temperature, without the need for a vacuum, and without a gas catalyst, Suhdi *et al.*[44] generated carbon nanofibers (CNFs). The material produced was an amorphous graphite deposition, according to XRD and Raman investigations. H_3PO_4 85% by mass solution in water provided by Fisher Scientific was used to saturate the 3 g RFS charcoal (RFSC) powder. The weight percentage for implantation of the activating reagent into charcoal was 1:4 wt%. A magnetic stirrer was used to agitate the slurry for 1.5 hours at 300 rpm. Thereafter, the homogenous slurry was processed in an oven at 110 °C. Then, the thick, dark slurries were placed in a ceramic crucible and heated to 500 °C at a rate of 10 °C min^{-1} in a furnace. The activation period ranged from 60 minutes and took place in a self-created atmosphere that had been warmed to ambient temperature. The samples of RFS AC (RFSAC) were centrifuged continuously for 30 minutes at 1000 rpm to eliminate the surplus activating agent until a pH of about 7 was attained. The samples were rinsed with distilled water combined with a NaOH solution. Following this procedure, the hydrothermal procedure was carried out. The supernatant was submerged in 40 mL of distilled water, agitated at 60 °C for 3 hours at 300 rpm, and left in the air for a day. In the final step, the liquid was dried for about three days at 90 °C in a dry oven to achieve total dryness. The steps are depicted in Figure 8.5.

The surface of the RFSAC had numerous large CNFs with varied diameters, as revealed by SEM and TEM analysis, where the mean diameter distribution was close to 172 nm. The findings of the Raman spectroscopy investigation showed that the CNF properties were identical to those of graphite, with intensity peaks at about 1340 and 1582 cm^{-1} (D and G bands, respectively). The XRD study showed consistent observations, revealing the formation of an amorphous graphitic carbon framework. The TGA and BET findings showed that the CNFs had excellent thermal

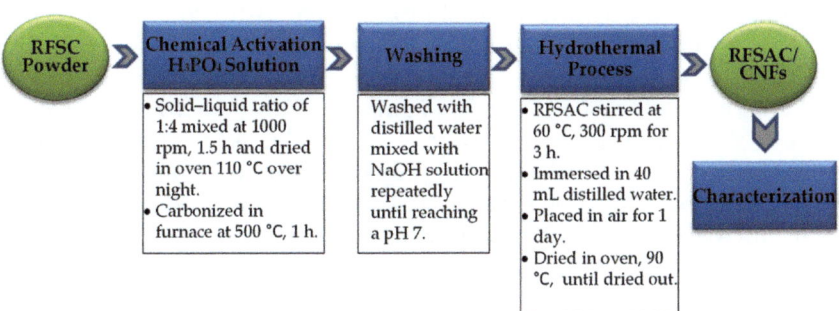

Figure 8.5 The chemical activation and hydrothermal methodology. Reproduced from ref. 44, https://doi.org/10.3390/nano11082038, under the terms of the CC BY 4.0 license https://creativecommons.org/licenses/by/4.0/.

conductivity, a specific surface area of 63 $m^2 g^{-1}$, and a mean pore diameter for adsorption of 2.11 nm. As a consequence, depending on the findings of the research, the RFS is regarded as a renewable biomass carbon resource substitute for the synthesis of CNFs, with the biomass being transformed into amorphous carbon by chemical activation and the hydrothermal method.

8.2.4 Steam Pyrolysis

In this procedure, basic agricultural wastes are alternatively warmed at lower temperatures of 500–700 °C or maximum temperatures of 700–800 °C under a flowing stream of pure steam. Numerous kinds of agricultural biomass waste, including agricultural residues, Ziziphus seeds, sawdust, tropical wood waste, banana peels, mangosteen shells, *etc.*, have been effectively processed using steam pyrolysis. Excluding those made from cherry stones, apricot stones, and almond shells, steam pyrolysis-produced AC was frequently recognized to be less efficacious. To research the role of sulfur in the formation of AC, Shalaby *et al.*[45] selected three types of apricot stones with varying sulfur contents due to various drying methods. The influence of processing parameters including the activation temperature, immersion duration, and particle diameter range on these samples was investigated. The activation temperature and testing period were 650–850 °C and 1–4 hours, respectively. The adsorption (iodine number), chemical (elemental content), and surface (BET surface area (SBET), mercury porosimetry) characteristics of the ACs were assessed. Through thermogravimetric measurement, the apricot stones' carbonization behavior was examined. The outcomes of the investigation showed that the sulfur concentration affects the pore architectures and adsorption properties of carbons produced under identical activation conditions. The low sulfur concentration (0.04%) apricot stone with a particle size distribution of 1–3.35 mm and an activation temperature of 800 °C for 4 hours produced the maximum SBET carbon (1092 $m^2 g^{-1}$). The investigation's findings demonstrated that Turkey is capable of producing porous ACs from Malatya apricot stones for economic purposes.

To turn palm kernel shells into AC, Lam *et al.*[46] used microwave pyrolysis alongside steam activation. The AC was then evaluated as an adsorbent to adsorb the herbicide 2,4-dichlorophenoxyacetic acid (commonly known as "2,4-D") that polluted the surface water in farmland. With steam activation, microwave pyrolysis produced AC in a relatively quick 65 minute procedure using a rapid heating rate of 20 °C min^{-1}. With a high specific surface area of up to 419 $m^2 g^{-1}$, extremely porous AC in an 83 wt% yield was produced. The AC demonstrated the best 2,4-D adsorption capacity of 11 $mg g^{-1}$ of AC. Therefore, the pyrolysis method shows promise for preparing AC for use as an adsorbent to adsorb herbicides polluting the surface water in farmland.

8.3 Applications of AC for AB Purification

The availability of numerous high-end brands of renowned ABs on the global market today inspires malicious motives regarding the possibility of additives or contamination by the addition of water, low-grade alcohol, or other cheap substances. Quality assurance of ABs is therefore crucial for managing the alcohol industry and safeguarding consumer health. To do this, one must be able to distinguish a regulated and purified alcoholic drink from another. It must be mentioned that the chemical composition of ABs is extremely complex. Diverse tannins, smells, and flavors emerge during the many processes that culminate in the creation of different types of liquor. These volatile compositions have a particular purpose and could vary based on the kinds and manufacturers of ABs. Owing to its distinctive optical characteristics, high biocompatibility, nontoxicity, high water stability, and simple manufacturing, AC has received a lot of interest recently. Taking into consideration the usage of ACs, Krochmalny *et al.*[47] demonstrated how the shape and adsorption ability of brewers' spent grain (BSG) were affected by hydrothermal carbonization, and how further cold atmospheric plasma jet treatment with argon and helium affected the yield. Herein, a cold atmospheric plasma jet was utilized to treat AC for comparison. The volumetric approach under moderate pressure was used to determine how activation affected the porosity of the substances (N_2, 77 K). Various theoretical approaches were used to calculate the specific surface area, total pore capacity, mean pore diameter, and pore size distribution (PSD). PSDs based on two separate simulations were characterized to discover how the volume varied along the spectrum of distinct pore sizes. The DFT simulation was used to establish that the plasma treatment had an impact on the smaller pores (Figure 8.6A). No pores smaller than 1.3 nm in size were found, and the volume expansion in pores between 1.3 and 15.0 nm was rather significant. Following exposure to argon plasma, the total pore volume grew from 0.15 $mm^3 g^{-1}$ (BSG-HC) to 3.87 $mm^3 g^{-1}$. As per the mesopore-specific BJH concept, plasma treatment increased the volume of the pores in the size range of 1.9 nm to 164 nm in BSG-HC substances (Figure 8.6B). While there was also a sizable improvement in total pore volume, this metric was only 7.6 $mm^3 g^{-1}$ (BSG-HC/Ar). Consequently, the pore size ranged from 8.7 to 11.7 nm.

Compared to hydrochars before cold atmospheric plasma jet treatment, hydrochars treated with cold atmospheric plasma jet showed a significant enhancement in their sorption capability: 7.5% (utilizing He) and 11.6% (utilizing Ar). Increases in specific surface area were also observed: 5-fold (for He) and 15-fold (for Ar). Such a significant shift was achieved only with the usage of AC after plasma activation with the specific surface areas increasing by 1.1–1.3.

Similarly, after extracting tea, Bhusari *et al.*[48] utilized the spent (processed) tea trash to make a sulfonated AC catalyst. In this experiment, SEM, FTIR, and BET studies were used to analyze the surface geometry of

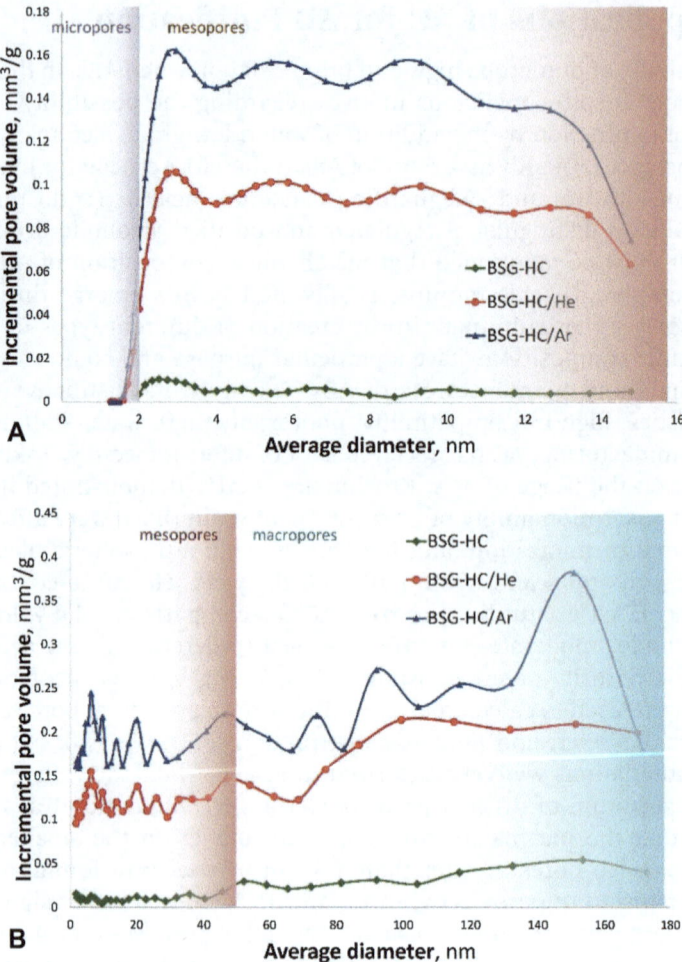

Figure 8.6 Pore size distribution of BSG-HC: (A) as per the DFT analysis and (B) as per the BJH framework. Reproduced from ref. 47, https://doi.org/10.3390/en15124396, under the terms of the CC BY 4.0 license https://creativecommons.org/licenses/by/4.0/.

the sulfonated tea waste carbon catalyst (STWCC). The catalyst's structure was found to be porous and to have outstanding catalytic capability. The STWCC's surface area was calculated to be 381 $m^2 g^{-1}$. Additionally, the catalyst contained a variety of functional components, the majority of which were related to electron withdrawing and used to prevent the catalyst from leaching. Ethyl alcohol and propionic acid were esterified using the STWCC at temperatures of 328 to 348 K and a catalyst ratio of 1 : 3 (wt/wt%). It was found that when the catalyst (AC) dosage and process temperature increased, the reaction rate and the conversion of propionic acid also increased. The tortuosity factor and effectiveness factor were also evaluated;

they were determined to be 0.99 and 0.2, respectively. At various temperatures, the efficient diffusivity of the STWCC was evaluated. By using the Weisz–Prater test, it was concluded that there were no internal and external diffusion limitations. Simply stated, the amount of tea waste that is available, legal considerations, price, and end product usage are cited as the bases for the argument that the current work is innovative. Earlier studies estimated that the price of tea waste from manufacturing enterprises ranged between INR 25 and INR 30 per kilogram. The residual tea waste indicated in the current article, however, was free of cost. Additionally, manufacturers of industrial tea waste were required to adhere to different waste treatment regulations. Additionally, they observed limited yields and surface areas for the ACs. Another conclusion that may be made is that all past research concentrated on creating adsorbents for the filtration of wastewater, alcohol, *etc.* Therefore, they asserted that their work is innovative in creating a catalyst from residual tea waste and then using it to esterify alcohols and beverages.

8.3.1 Detoxification Using AC to Remove Microbial Fermentation Inhibitors

Proteins and starch are mostly broken down into amino acids and sugars during the mashing process of ABs like whiskey. During the distillation and fermentation processes, ethanol is produced from sugars using yeast, and the resultant spirit is aged in oak. During these activities, several taste compounds are formed with no major reduction in their quantities.[49–53] In the malting process, sulfur, heterocyclic, and phenolic chemicals are produced. Besides producing ethanol, yeast also participates in the fermentation process by producing several volatile chemicals as byproducts, notably esters, sulfur compounds, aldehydes, and lactones. Distillation results in the production of ketones, as well as the formation and reduction of sulfur compounds. Finally, during oak ripening, a variety of flavoring substances are removed, including aromatic phenols, lactones, hetero cyclic compounds, and tannins. As these have nasty smells and tastes, and eventually undesired properties, the quantity, concentration, and toxicity of some volatile compounds such as sulfur must be decreased during this procedure with the aid of ACs. In an experiment, Tavares *et al.*[54] investigated alcohol synthesis using cardoon hemicellulosic hydrolysates, which were produced through a pretreatment step involving dilute sulfuric acid hydrolysis. Utilizing both AC adsorption and a nanofiltration membrane technology, a detoxification phase to eliminate liberated microbial fermentation inhibitors was tested. The fermentation efficiencies of the transgenic *Scheffersomyces stipitis* CBS5773 yeast and *Escherichia coli* MS04 were tested under various experimental conditions. The positive outcomes of employing *E. coli* resulted in a bioethanol volumetric output of 0.30 g $L^{-1}h^{-1}$, with a conversion efficacy of 94.5%, employing detoxified cardoon by membrane nanofiltration. Concerning *S. stipitis*, a volumetric output of 0.091 g L^{-1} h^{-1} with a conversion efficacy of 64.9% was recorded under comparable fermentation conditions. In conclusion, detoxifying hemicellulosic cardoon hydrolysate to produce bioethanol is a viable approach to producing 2nd-generation bioethanol, particularly while employing engineered *E. coli*.

With the help of the designated fungal strain *Rhizopus oryzae* K20, Boondaeng *et al.*[55] investigated the viability of steam-exploded oil palm empty fruit bunches (EFBs) for the manufacture of fumaric acid, a food and alcohol additive and detoxifier commonly used for flavor and preservation of alcohols. Herein, an effective two-stage precipitation process was used, accompanied by an AC-mediated purification process to eliminate impurities, to extract and purify fumaric acid using fermented oil palm EFBs. A recovery efficiency of 81.2 %and a purity of 83.5% were reached after these two procedures were finished. Additionally, Zhou *et al.*[56] studied ethanol's unrecognized impact on the adsorption of phthalate acid esters (PAEs) in an alcoholic medium. The dibutyl phthalate (DBP) adsorption capacities on AC in liquids containing ethanol levels of 30, 50, 70, and 100 vol% were only 59, 43, 19, and 10% of what they were (16.39 mg g^{-1}) in water, respectively. With the production of water clusters, the ethanol concentration increased from 50% to 100%, dramatically reducing the adsorption capacity from 13.99 mg g^{-1} to 2.34 mg g^{-1}. When the ethanol level soared from 0 to 100%, the molecular dynamics simulation revealed that the DBP evolved to be spread further out from the AC (the statistics indicated that the distance increased from 5.25 Å to 15.3 Å). The inclusion of ethanol had a greater impact on PAEs with fewer chains than those with larger chains. The adsorption capacity of the AC in ethanol (0.41 mg g^{-1}) for DBP as an example was just 2.2% of that in water (18.21 mg g^{-1}). The sorption of PAEs on the AC had a significant influence on the tastes of the real Baijiu samples, according to usage data. To analyze the dispersion of DBP at various ethanol levels, five adsorption simulations were developed. Screenshots of the simulations following equilibration are shown in Figure 8.7. Figure 8.7a–e shows how DBP progressively drifts apart from the AC as the ethanol level increases. This indicates that ethanol lessens the impact of the AC's adsorption of DBP. Figure 8.7b–e shows the co-solvent impact by demonstrating that the DBP is constantly accompanied by ethanol.

Alcaraz *et al.*[57] evaluated a method that used AC generated from spent coffee grounds (SCG) to extract terbium from liquid solutions. SCG was hydro-alcoholized to produce AC, which was then physically activated to produce the final substance. Figure 8.8 depicts the steps to produce AC. The corresponding nitrogen adsorption–desorption isotherms and SEM photographs demonstrated the AC's highly porous and microporous frameworks. Using the micro-Raman measurement technique, a specific graphitic feature was also found. This AC was used to study terbium adsorption and assess the effects of media, pH, temperature, and adsorbent quantity on terbium

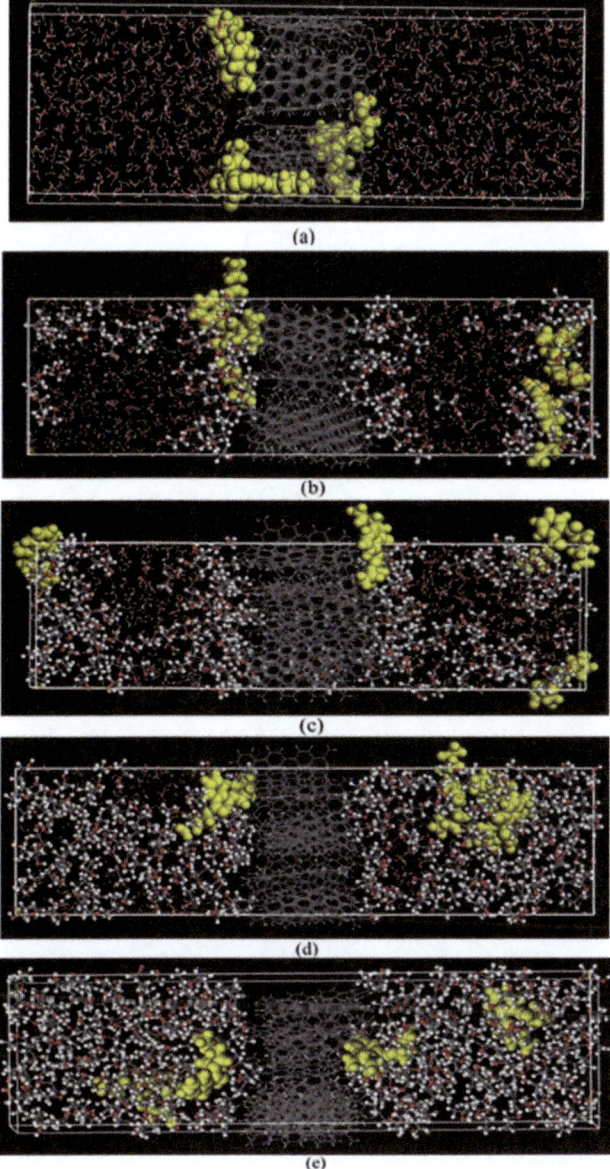

Figure 8.7 Analysis framework of media with distinct ethanol concentrations: (a) 0 v%; (b) 30 v%; (c) 50 v%; (d) 70 v%; and (e) 100 v%. Reproduced from ref. 56, https://doi.org/10.3390/foods11142114, under the terms of the CC BY 4.0 license https://creativecommons.org/licenses/by/4.0/.

absorption. Kinetic investigations and adsorption isotherm studies were also performed. The type 1 Langmuir isotherm and pseudo-second-order kinetics theory showed the best fit. Terbium adsorption is a spontaneous and endothermic reaction, according to thermodynamics research. Terbium

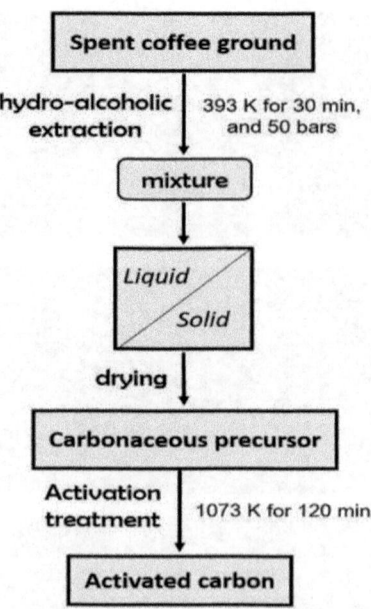

Figure 8.8   Process flow diagram to produce the relevant AC. Reproduced from ref. 57, https://doi.org/10.3390/met11040630, under the terms of the CC BY 4.0 license https://creativecommons.org/licenses/by/4.0/.

desorption was evaluated using acidic media. This study showed that the current AC can be used to recover this precious metal from a liquid medium.

Azis *et al.*[58] investigated a sand filtration–AC adsorption method to get rid of the antimicrobials in water that have undergone biological treatment to attain a high yield of alcohol. Although sand filtering made a small contribution to the reduction of toxic components, the purifying operation was primarily carried out on the AC column. For the tertiary treatment of biologically processed wastewater with residual quantities of imazalil and fludioxonil, a laboratory-scale filtration system and an AC column with a 5 L working capacity each were constructed and operated in sequence (Figure 8.9). The cylindrical, opaque plexiglass columns holding the sand and AC were manufactured in this way. Every column had a usable height of 30 cm and an internal diameter of 14 cm. There were 4 cm of pebbles at the bottom and 26 cm of quartz sand on top of the gravel stratum in the sand filter. Coarse gravel, AC, and quartz sand were all added to the AC column in that order, from the bottom to the top (23 cm in height). The cleaned filtrate was routed from the bottom to the top of the AC column, while the secondary wastewater was transported downwards in the sand column.

The tertiary treatment system, which operated in a batch system for 25 bed volumes, detected total COD levels of <50 mg L^{-1} in the AC column filtrate. The total and soluble COD extraction efficiencies were $76.5 \pm 1.5\%$ and $88.2 \pm 1.3\%$, respectively. Following AC adsorption, a sizable pH rise and a corresponding electrical conductivity (EC) decline also took place. The total

Figure 8.9 The laboratory-scale sand-AC apparatus for the tertiary treatment and detoxification of biologically processed fungicide-containing wastewater is constructed of quartz sand (A) and granulated AC (B). Reproduced from ref. 58, https://doi.org/10.3390/pr9071223, under the terms of the CC BY 4.0 license https://creativecommons.org/licenses/by/4.0/.

and ammonium nitrogen in the AC permeates decreased considerably to 2.44 ± 0.02 mg L^{-1} and 0.93 ± 0.19 mg L^{-1}, respectively. Considering this, the initial imazalil content in the biologically processed wastewater was higher than the fludioxonil level (*i.e.*, 41.26 ± 0.04 mg L^{-1} *versus* 7.35 ± 0.43 mg L^{-1}, respectively). During AC adsorption, imazalil was eliminated, although fludioxonil was still present in small amounts. The biologically processed fungicide-containing effluent was substantially detoxified by AC processing, which led to an increase in germination index of either 68% or 47% following 1 : 1 v/v dilution of the wastewater.

8.3.2 Dyes Integrated with AC to Attain or Lose the Optimized Color

Numerous novel compounds that are relatively inexpensive and accessible analogues of natural products are emerging as a result of the food industry's fast evolution.[59–62] In an experiment, Shungite, a native nanomaterial based on AC, was employed by Alham *et al.*[63] as a solid-phase extraction (SPE) absorbent and demonstrated a superb capability for the absorption of red synthetic dye. NMSh, an amalgam of several carbon allotropes, is employed globally in scholastic and food investigation as an affordable and efficient sorbent. Shungite has been considered as an enticing substance for the advancement of nanotechnology, which is of considerable intrigue for the advancement of science and technology. NMSh was not just amorphous carbon, but a combination of different carbon allotropes. This information led to the decision to utilize shungite to produce the sorbents which were utilized to prepare samples for assessment. An SPE approach using NMSh packed cartridge coupled with HPLC-UV monitoring was created to identify the red synthetic dye Ponceau 4R in samples of juice and wine. A comparison of activated and non-activated NMSh for usage as sorbents was made after studying the particle shape and weight of the NMSh sorbent. The relative standard deviations (RSDs) throughout the day were 0.18% for juice and 0.15% for wine, with the analyte extraction rates of 99.3% for wine and 94.3% for juice. The findings demonstrated that this technology is particularly accurate and efficient for identifying food dyes at low concentrations and very complex matrices.

On using ACs, Aissa *et al.*[64] effectively discolored low-grade dark maple syrup for the production of ABs. The size of the particles (25, 50, 75 m), agitation speed (200, 400, 600 rpm), combining duration (20, 40, 60 min), AC content (0.1, 0.3, 0.5 g/100 mL), grade of AC (I, II, III), and temperature (40, 60, 80 °C) were all investigated as empirical factors. The collected data showed that the discoloration was best achieved by using parameters like a 40 minute swirling period and 0.3 g/100 mL of type III AC. Based on these factors, the syrup had an extremely clear glass light transmittance of $83.70 \pm 0.21\%$. According to the findings, the adsorption on the type III carbon adhered to the Freundlich, Langmuir, and Langmuir–Freundlich isotherms amongst the studied carbons. The collected data indicated that the best adsorption efficiency was achieved for an median particle size of 25 m using a swirling frequency of 200 rpm and an operating temperature of 80 °C. The Freundlich and Langmuir theories performed well in the optimized settings. The discoloration procedure utilizing type III AC followed pseudo-second-order kinetics.

Lopes *et al.*[65] prepared AC using malt bagasse as a precursor by physically activating it with steam. Rotatable central composite design, RSM, and desirability function were used to analyze and optimize the process parameters, including temperature, activation time, and steam flow. The SBET and output of the AC were significantly affected by the parameters. The AC produced under ideal circumstances (ACMop) of 841 °C, 82 minutes, and 164 cm^{-3} min^{-1} steam flow displayed an SBET of 917 m^{-2} g^{-1} and an output of 9.45%, in contrast to surface acid groups and pH_{pzc} 3.92. To remove the food color sunset yellow (SY) from the aqueous phase, the ACMop was used. The ACMop showed quick adsorption kinetics and adsorption equilibrium about 240 minutes. The fit of the Morris and Weber model demonstrated that intraparticle dispersion is not the rate-limiting phase, but the Elovich model provided a better explanation for the kinetic data. Given that the g value was close to 1, the Redlich–Peterson isotherm model offered a superior match to the empirical observations and indicated that the Langmuir hypotheses were better suited to explain the adsorption of SY onto the ACMop. Furthermore, at a temperature of 55 °C, the ACMop's monolayer adsorption capacity was discovered to be 199.7 mg g^{-1}. The thermodynamic analysis revealed that the adsorption mechanism was endothermic but spontaneous, with an enhancement in unpredictability at the solid/liquid interface.

Actinobacillus succinogenes fermented succinic acid (SA) was extracted by Omwene *et al.*[66] using a combined technique that included reactive extraction, ultrafiltration, and vacuum distillation in the pretreatment of fermentation of alcohol. For both liquid–liquid (LLE) extraction and supported liquid membrane (SLM) separation, tri-*n*-octylamine (TOA) using 1-octanol as a diluent was utilized as the extractant. The yield and SA titer of the generated substance were 11.16 g L^{-1} and 0.44 g g^{-1}, respectively. Using 10% (w/v) of granular ACs, the fermentation broth was completely decolored, and the steady flow ultrafiltration permeate flux varied from 31.18 to 33.42 L m^{-2} h^{-1}. The LLE had a 51.5% extraction efficacy, while the SLM recovered 57.3%. SA showed a permeability and transfer rate of 0.08605 cm h^{-1} and 0.00697 h^{-1} ($R^2 > 0.92$), respectively. As the aqueous pH increased from 2 to 5, the amount of SA that could be extracted decreased significantly. Initially, the SA flux in SLM was estimated to be 9.65 g m^{-2} h^{-1}, which was twice as much as lactic acid flux. Although selective extraction of just SA was not successful, residual biological matter and macromolecules were successfully eliminated. Here, the authors conclusively showed that reactive extraction with process integration is a viable technique for recovering SA from fermentation broth.

8.3.3 Detection of Organic Compounds in ABs

To understand the existence of both favorable and unfavorable contributing chemicals in the end product, the identification of organic compounds in alcohols is of great importance.[66–70] To examine how bacterial growth affects the breakdown of organic matter, Habiba *et al.*[71] studied an anaerobic sludge technique for the detection and remediation of industrial beverage effluent that is inexpensive, effective, and ecologically benign and is frequently utilized for the preparation of ABs. In this experiment, microbial kinetics was also used to assess the process, linkages among the microbes' proliferation and operational parameters of an anaerobic batch reactor, and the effectiveness of the wastewater remediation process. A 10 liter anaerobic batch reactor 50 cm in length and 15 cm in diameter was used in this method, as depicted in Figure 8.10. In the batch reactor, a procedure known as anaerobic sludge was conducted. The procedure involved three phases: the first phase involved adding 5 L of beverage industry wastewater discharge to a single batch reactor, and the second phase required adding

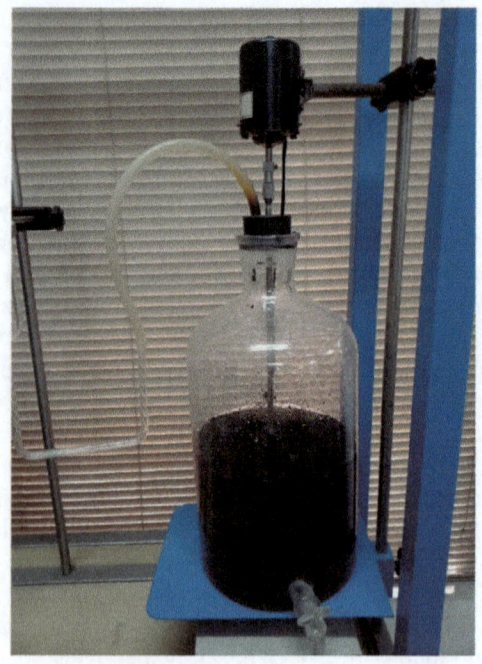

Figure 8.10 Laboratory-scale anaerobic reactor for pretreatment of commercial beverage wastewater using 3 mg L^{-1} ACs. Reproduced from ref. 71, https://doi.org/10.3390/separations8040043, under the terms of the CC BY 4.0 license https://creativecommons.org/licenses/by/4.0/.

3 mg L^{-1} of AC powder (200 mesh, Ash 6% (max)) and stirring it for three hours to eliminate odor and contaminants. 100 mL of the sample were taken during the settling procedure every two days until the overall hydraulic retention time (HRT) of 13 days was achieved. The seven samples that had been obtained were then examined utilizing ASTM standards to produce data for substrate removal characteristics. The final phase involved discharging the flocs from the batch reactor after allowing them to be set properly in an anaerobic environment (without air). Furthermore, the wastewater overlay was cleaned at the same time. By securing the reactor inlet, anaerobic settings were kept constant inside the reactor. Anaerobic sludge was allowed to settle for 13 days during the settling stage.

Initial HRT = 24 h, initial SRT = 2.5 days, initial MLVSS = 11 656 mg L^{-1}, and initial COD = 80 000 mg L^{-1} were the ideal parameters found for the anaerobic sludge treatment. A growth yield of 0.46 mg VSS mg^{-1} COD, a saturation coefficient of 3500 (mg L^{-1} of COD), a maximum substrate utilization of 0.00466 (mg L^{-1} COD), a growth rate of microbes of 0.03833 h^{-1}, and a maximum growth rate of microbes of 0.3672 h^{-1} were also obtained from the computation of kinetic coefficients based on COD values. Additionally, it was found that the development of the bacteria was necessary for the decomposition of organic matter. Additionally, a COD reduction degree analysis over 13 days revealed a 99.31% efficacy for their suggested approach. As a result, this research is important and helpful in developing a novel anaerobic sludge-based technique for treating beverage sector wastewater as well as working for current anaerobic sludge-based techniques.

Prakash *et al.*[72] concentrated on developing graphene-based thiourea (N and S atom)-doped electrochemical sensors for MO sensing. The addition of nitrogen and sulfur (N and S) atoms to graphene improved its electrochemically active surface area and active sites, which sped up the electron transfer process. The successful formation of N- and S-doped graphene was supported by a variety of microscopic and spectroscopic characterization techniques, including XRD, SEM, and XPS. Cyclic voltammetry (CV) and electrochemical impedance spectroscopy (EIS) were used to examine the electrochemical characteristics of an NSG modified glassy carbon electrode (NSG 4/GCE). Further to document the dosage-related variation in the redox current reactions, differential pulse voltammetry (DPV) measurements of NSG 4/GCE in the presence of MO were performed. With a dynamic concentration range of 0.57 $g\,mL^{-1}$ to 294 $g\,mL^{-1}$ (at a potential of 0.37 V) and a LOD of 0.26 $g\,mL^{-1}$, the built sensor showed a dose-dependent rise in DPV anodic peak current. Furthermore, real samples like hard and soft drinks were used to assess the viability of the suggested electrochemical sensor for the detection of tiny amounts of morphine. As a result, this electrochemical sensor, which employs NSG substances as electrochemical interfaces, may be a candidate for morphine identification in clinical contexts as well as for forensic purposes.

Utilizing the photoluminescence (PL) of carbon nanoparticles (NPs), Ivanov *et al.*[73] developed a straightforward, precise approach for the general identification of whiskey, armagnac, cognac, and an ethanol/water mixture. The anhydrous citric acid and urea mixture was used to synthesize carbon NPs, which were then subjected to a series of annealing processes to fully realize the benefits of solvothermal carbonization. The carbon NPs' PL characteristics were influenced by the alcohol environment in which they were dispersed. After computationally processing the PL/PL excitation maps of the ABs, the resultant principal components analysis diagram permits viewing of the various clusters related to each beverage. To guarantee a trustworthy recognition grade, the ideal measurement conditions (NP content in colloidal solution and excitation wavelength) were established.

Three up-flow anaerobic sludge blanket (UASB) reactors functioning isothermally at 45, 55, and 65 °C were used by Contreras *et al.*[74] to remove polyphenols from sugarcane vinasses commonly utilized for the production of liquor. The study began by quantifying four crucial procedure variables: alkalinity component, temperature, pH, and acetic acid generation. This helped determine the durability of the reactors. To assess the efficacy of the method, the vinasses were subjected to physicochemical treatment. Utilizing the Folin–Ciocalteu technique, residual polyphenols were measured both before and after the biochemical treatment. The obtained findings demonstrated that the phenolic chemicals included in the vinasse feed could not be completely removed by this biochemical procedure on its own. Additionally, the supplementary treatment was applied to remove polyphenols using AC. Utilizing the gathered experimental results, the efficiency of the adsorption mechanism was examined and contrasted utilizing the Freundlich, Langmuir, and Dubinin–Radushkevich models. The Langmuir model, with a correlation coefficient of 0.991, offered the greatest fit for the experimental results. By using a single phase of adsorption, it was possible to lower the number of polyphenols in the vinasses from 3600 mg of gallic acid equivalent $(GAE) L^{-1}$ to 1072 $mg\, GAE\, L^{-1}$. This amounts to a general decrease of 70.2%, which contributed to the high yield of alcohol.

8.4 Conclusion

Several contaminating compounds such as purines can be removed from aqueous phases and alcohols using a variety of techniques, notably biological, enzymatic, and adsorption techniques. The majority of biological and enzymatic approaches have limitations that might prevent their use in removing purine chemicals from food items, including the frequency of enzyme breakdown and the compliance of the enzyme to food storage environments. Adsorption techniques might be thought of as distinct, potential technologies in this regard. Notwithstanding their continued importance, only a small number of studies have applied adsorption technologies to extract such molecules from wort or beer samples, to produce and purify a final product containing a reduced harmful chemical content in the AB. For instance, the analysis of a biologically processed agro product utilized involved sand filtration and AC processing, which conclusively demonstrated the higher potential of the former procedure in the elimination of macronutrients, suppressing molecules and biosolids from ABs. The cumulative COD level $(50 \quad mg\, L^{-1})$, EC, cumulative nitrogen (from $7.94 \pm 0.02 \quad mg\, L^{-1}$ to $2.44 \pm 0.02\, mg\, L^{-1})$, and NH^{4+}–N $(1\, mg\, L^{-1})$ were all significantly reduced as a consequence of this treatment process using high-scale ACs. Similar such potent studies utilized substances including AC which play a crucial role as discussed in the purification of the end product of ABs. These materials should also be looked into for the elimination of other inhibitory compounds from beverages, especially beer, depending on their excellent adsorption capability. Additionally, in contrast to demonstrating their strong ability to adsorb such molecules and, as a result, their significant possibility to be utilized in the elimination of such compounds, these investigations were primarily centred on the examination of the interactions implicated.

References

1. J. Bliss, B. Aakriti, F. Thomas, K. Smita, R. Kashane and W. Zhixin, Worchester Polytech. Inst., 2012.
2. A. S. Milliporesigma, V. P. View and A. Saksule, *Int. J. Chem. Eng. Appl. Sci.*, 2015, 13–23.

3. B. Hansdah, *J. Natl. Inst. Technol.*, 2016, 1–7.
4. V. K. Gupta, A. Nayak, B. Bhushan and S. Agarwal, *Crit. Rev. Environ. Sci. Technol.*, 2015, **45**, 613–668.
5. V. N. Alves, S. S. O. Borges and N. M. M. Coelho, *Int. J. Anal. Chem.*, 2011, **2011**, 1–8.
6. M. Pauley and D. Maskell, *Beverages*, 2017, **3**, 1–11.
7. X. Xing, W. Jiang, S. Li, X. Zhang and W. Wang, *Waste Manage.*, 2019, **89**, 64–72.
8. A. Thakur and A. Kumar, *Sci. Total Environ.*, 2022, **834**, 155219.
9. A. Thakur and A. Kumar, *J. Bio- Tribo-Corrosion*, 2021, **7**, 1–48.
10. A. Thakur, S. Kaya, A. S. Abousalem and A. Kumar, *Sustainable Chem. Pharm.*, 2022, **29**, 100785.
11. A. Thakur, S. Kaya, A. S. Abousalem, S. Sharma, R. Ganjoo, H. Assad and A. Kumar, *Process Saf. Environ. Prot.*, 2022, **161**, 801–818.
12. G. Parveen, S. Bashir, A. Thakur, S. K. Saha, P. Banerjee and A. Kumar, *Mater. Res. Express*, 2020, **7**, 016510.
13. S. Bashir, A. Thakur, H. Lgaz, I.-M. Chung and A. Kumar, *Arab. J. Sci. Eng.*, 2020, **45**, 4773–4783.
14. S. Bashir, A. Thakur, H. Lgaz, I. M. Chung and A. Kumar, *Surf. Interfaces*, 2020, **20**, 100542.
15. A. Kumar and A. Thakur, *Encapsulated nanoparticles in organic polymers for corrosion inhibition*, Elsevier Inc., 2020.
16. R. A. Rashid, A. H. Jawad, M. A. B. M. Ishak and N. N. Kasim, *Sains Malays.*, 2018, **47**, 603–610.
17. H. K. Atiyeh and Z. Duvnjak, *Sep. Sci. Technol.*, 2005, **39**, 341–362.
18. J. Shibata, N. Murayama and M. Tateyama, *Resour. Process.*, 2009, **56**, 120–126.
19. P. R. Oliveira, A. C. Lamy-Mendes, E. I. P. Rezende, A. S. Mangrich, L. H. Marcolino Junior and M. F. Bergamini, *Food Chem.*, 2015, **171**, 426–431.
20. L. M. Leonardo, C. de, O. S. Anna, S. Murilo, F. de, P. Nádia, M. S. Luciana and C. F. Mariana, *African J. Microbiol. Res.*, 2017, **11**, 287–295.
21. G. S. Murthy, S. Sridhar, M. Shyam Sunder, B. Shankaraiah and M. Ramakrishna, *Sep. Purif. Technol.*, 2005, **44**, 221–228.
22. F. C. Duarte, M. das G. Cardoso, A. C. M. Pinheiro, W. D. Santiago and L. L. de Carvalho, *Food Sci. Technol.*, 2012, **32**, 471–477.
23. L. Marynchenko, V. Marynchenko and M. Hyvel, *East.-Eur. J. Enterp. Technol.*, 2017, **4**, 10–15.
24. A. Thakur, S. Kaya and A. Kumar, *Curr. Nanosci.*, 2021, **18**, 203–216.
25. C. Almeida, M. C. Neves and M. G. Freire, *Molecules*, 2021, **26**, 6460.
26. R. Zhao, F. Duan, J. Yang, M. Xiao and L. Lu, *Catalysts*, 2021, **11**, 1–12.
27. L. N. Dos Santos, A. S. Santos, K. D. G. F. Dantas and N. R. Ferreira, *Polymers*, 2022, **14**, 573.

28. M. A. Gantumur, N. Sukhbaatar, A. Qayum, A. Bilawal, B. Tsembeltsogt, K. C. Oh, Z. Jiang and J. Hou, *J. Dairy Sci.*, 2022, **105**, 83–96.
29. R. Liu, L. Mao, Z. Guan, C. Wang, J. Xu, L. Huang, P. Wang, G. Xin, R. Hu, C. Zhang, Z. Zhao, Y. Lin and X. Hu, *Anal. Bioanal. Chem.*, 2022, **414**, 5009–5022.
30. M. S. Mauter and M. Elimelech, *Am. Chem. Soc.*, 2008, **42**, 5843–5859.
31. J. Guo, Y. Song, X. Ji, L. Ji, L. Cai, Y. Wang, H. Zhang and W. Song, *J Sci.: Adv. Mater. Dev.*, 2019, **4**, 544–553.
32. A. Macías-García, J. García-Sanz-Calcedo, J. P. Carrasco-Amador and R. Segura-Cruz, *Sustainability*, 2019, **11**, 1–11.
33. S. He, G. Chen, H. Xiao, G. Shi, C. Ruan, Y. Ma, H. Dai, B. Yuan, X. Chen and X. Yang, *J. Colloid Interface Sci.*, 2021, **582**, 90–101.
34. S. García, J. J. Pis, F. Rubiera and C. Pevida, *Langmuir*, 2013, **29**, 6042–6052.
35. D. Peredo-Mancilla, I. Ghouma, C. Hort, C. M. Ghimbeu, M. Jeguirim and D. Bessieres, *Energies*, 2018, **11**, 1–13.
36. T. C. Drage, J. M. Blackman, C. Pevida and C. E. Snape, *Energy Fuels*, 2009, **23**, 2790–2796.
37. T. C. Drage, O. Kozynchenko, C. Pevida, M. G. Plaza, F. Rubiera, J. J. Pis, C. E. Snape and S. Tennison, *Energy Proc.*, 2009, **1**, 599–605.
38. K. S. Ukanwa, K. Patchigolla, R. Sakrabani, E. Anthony and S. Mandavgane, *Sustainability*, 2019, **11**, 1–35.
39. X. He, J. Zhu, H. Wang, M. Zhou and S. Zhang, *Coatings*, 2019, **9**, 590.
40. Z. Sabara, A. Anwar, S. Yani, K. Prianto, R. Junaidi, R. Umam and R. Prastowo, *Sustainability*, 2022, **14**, 1026.
41. Í. N. Raupp, A. V. Filho, A. L. Arim, A. R. C. Muniz and G. S. da Rosa, *Materials*, 2021, **14**, 6820.
42. A. A. Pam, *C*, 2019, **5**, 43.
43. F. C. Chang, S. H. Yen and S. H. Wang, *Materials*, 2018, **11**, 1877.
44. S. Suhdi and S. C. Wang, *Nanomaterials*, 2021, **11**, 2038.
45. Ç. Şentorun-Shalaby, M. G. Uçak-Astarlioğlu, L. Artok and Ç. Sarici, *Microporous Mesoporous Mater.*, 2006, **88**, 126–134.
46. S. S. Lam, M. H. Su, W. L. Nam, D. S. Thoo, C. M. Ng, R. K. Liew, P. N. Yuh Yek, N. L. Ma and D. V. Nguyen Vo, *Ind. Eng. Chem. Res.*, 2019, **58**, 695–703.
47. K. Krochmalny, H. Pawlak-Kruczek, N. Skoczylas, M. Kudasik, A. Gajda, R. Gnatowska, M. Serafin-Tkaczuk, T. Czapka, A. K. Jaiswal, Vishwajeet, A. Arora, T. Hardy, M. Jackowski, M. Ostrycharczyk and Ł. Niedźwiecki, *Energies*, 2022, **15**, 4396.
48. A. A. Bhusari, B. Mazumdar and A. P. Rathod, *Waste Biomass Valorization*, 2021, **12**, 1303–1312.
49. Z. Yang, R. Gleisner, D. H. Mann, J. Xu, J. Jiang and J. Y. Zhu, *Polymers*, 2020, **12**, 1–16.

50. R. El Khoury, E. Choque, A. El Khoury, S. P. Snini, R. Cairns, C. Andriantsiferana and F. Mathieu, *Toxins*, 2018, **10**, 1–16.
51. D. Figueira, J. Cavalheiro and B. Sommer Ferreira, *Fermentation*, 2017, **3**, 13.
52. D. Valero, C. Rico, B. Canto-Canché, J. A. Domínguez-Maldonado, R. Tapia-Tussell, A. Cortes-Velazquez and L. Alzate-Gaviria, *Energies*, 2018, **11**, 2101.
53. F. Xu, J. Chen, G. Yang, X. Ji, Q. Wang, S. Liu and Y. Ni, *Polymers*, 2019, **11**, 1558.
54. A. P. M. Tavares, M. J. A. Gonçalves, T. Brás, G. R. Pesce, A. M. R. B. Xavier and M. C. Fernandes, *Energies*, 2022, **15**, 1993.
55. A. Boondaeng, N. Khanoonkon, P. Vaithanomsat, W. Apiwatanapiwat, C. Trakunjae, P. Janchai and N. Niyomvong, *Fermentation*, 2022, **8**, 121.
56. Y. Zhou, B. Zhao, L. Wang, T. Li, H. Ye, S. Li, M. Huang and X. Zhang, *Foods*, 2022, **11**, 2114.
57. L. Alcaraz, D. N. Saquinga, F. J. Alguacil, E. Escudero and F. A. López, *Metals*, 2021, **11**, 630.
58. K. Azis, Z. Mavriou, D. G. Karpouzas and P. Melidis, *Processes*, 2021, **9**, 1223.
59. S. Onuki, J. A. Koziel, W. S. Jenks, L. Cai, S. Rice and J. Van Leeuwen, *Sep. Purif. Technol.*, 2015, **151**, 165–171.
60. M. Balcerek, K. Pielech-przybylska, P. Patelski and T. Jusel, *Food Addit. Contam., Part A*, 2017, **34**, 714–727.
61. P. Raspor and D. Goranovič, *Crit. Rev. Biotechnol.*, 2008, **28**, 101–124.
62. L. Marynchenko, V. Marynchenko and M. Hyvel, *EUREKA Phys. Eng.*, 2017, **4**, 3–10.
63. A. Alham, A. Ibraimov, M. Alimzhanova and M. Mamedova, *Food Anal. Methods*, 2022, **15**, 707–716.
64. A. Aït-Aissa, N. Gerliani, T. Orlova, B. Sadeghi-Tabatabai and M. Aïder, *ACS Omega*, 2021, **6**, 17748.
65. G. K. P. Lopes, H. G. Zanella, L. Spessato, A. Ronix, P. Viero, J. M. Fonseca, J. T. C. Yokoyama, A. L. Cazetta and V. C. Almeida, *Arab. J. Chem.* 2021, **14**, 103001.
66. P. I. Omwene, M. Yagcioglu, Z. B. Ocal Sarihan, A. Karagunduz and B. Keskinler, *J. Environ. Chem. Eng.*, 2020, **8**, 104216.
67. X. Zhang, C. Wang, L. Wang, S. Chen and Y. Xu, *J. Chromatogr. A*, 2020, **1610**, 460584.
68. R. J. Ferreira, T. R. Rosa, J. Ribeiro and R. C. Barthus, *Food Chem.*, 2020, **314**, 126126.
69. G. Jashari, I. Švancara and M. Sýs, *Electroanalysis*, 2020, **32**, 1949–1956.
70. G. Dattatraya Saratale, R. Bhosale, S. Shobana, J. R. Banu, A. Pugazhendhi, E. Mahmoud, R. Sirohi, S. Kant Bhatia, A. E. Atabani, V. Mulone, J. J. Yoon, H. Seung Shin and G. Kumar, *Bioresour. Technol.*, 2020, **314**, 123800.
71. Um-E. Habiba, M. S. Khan, W. Raza, H. Gul, M. Hussain, B. Malik, M. Azam and F. Winter, *Separations*, 2021, **8**, 43.

72. V. Prakash, Garima, N. Prabhakar, G. Kaur, A. Diwan, S. K. Mehta and S. Sharma, *Curr. Res. Green Sustainable Chem.*, 2022, **5**, 100267.
73. I. I. Ivanov, A. N. Zaderko, V. Lysenko, T. Clopeau, V. V. Lisnyak and V. A. Skryshevsky, *ACS Omega*, 2021, **6**, 18802–18810.
74. J. A. Contreras-Contreras, M. Bernal-González, J. A. Solís-Fuentes and M. del C. Durán-Domínguez-de-Bazúa, *Water, Air, Soil Pollut.*, 2020, **231**, 401.

CHAPTER 9

Fuel Storage Application of Activated Carbon

PATRICK U. OKOYE,*[a] CESAER GIOVANNI MONDRAGON[b] AND JUDE A. OKOLIE[c]

[a] Instituto de Energías Renovables, Universidad Nacional Autónoma de México, Priv. Xochicalco s/n, Col. Centro, Temixco CP 62580 Morelos, Mexico; [b] School of Engineering, Av. Universidad 3000, Ciudad Universitaria, Coyoacán, Cd. Mx., CP 04510, Mexico; [c] Gallogly College of Engineering, University of Oklahoma, 73019, Norman, USA
*Email: ugopaok@ier.unam.mx

9.1   Introduction

The increase in global population has resulted in a rise in the demand for energy mainly derived from fossil sources. The awareness of the harmful effects of greenhouse gas emissions from fossil fuels has resulted in the search for alternative energy sources that are green and sustainable. Transitioning from a linear economy to a circular economy and achieving decarbonization require social transitioning coupled with institutional innovation and scientific and technological changes. This ambition has already been incorporated in many policy documents; however, given the interests at play, it is a very complex process, but achievable. Biomass is an indispensable alternative energy source that can be used for the production of high energy-density materials, food, animal feed, chemicals, *etc.* Active materials derived from biomass could play a vital role in energy storage.

Microporous and mesoporous materials from biomass have been the focus due to their physicochemical properties that allow energy storage and ion

Activated Carbon: Progress and Applications
Edited by Chandrabhan Verma and Mumtaz A. Quraishi
© The Royal Society of Chemistry 2023
Published by the Royal Society of Chemistry, www.rsc.org

transport. Other materials that have been proposed for fuel storage purposes include zeolites, metal–organic frameworks, complex hydrides, porous organic membranes, and porous membranes. While these materials present possible solutions to fuel storage, they have limitations such as appreciably high synthesis cost and finite lifecycles because they are easily deactivated by chemical attack and thermally unstable.[1] Activated carbon has been one of the materials widely investigated for the adsorption of gas molecules.[2] This is because it can be derived from abundant and low-cost biomass *via* either physical or chemical activation or both processes. It is characterized by high specific surface area, well-developed pore structure, good mechanical properties, cost-effectiveness, resistance to chemical attack, and an excellent adsorbing ability.[3] These properties make activated carbon appealing for wide application prospects in wastewater treatment,[4] catalysis,[5] gas purification, *etc.* Activated carbon with hierarchical pores exhibits interesting properties for gas storage because the micropores (<2 nm) act mainly as storage sites, whereas the mesopores (2–50 nm) and macropores (>50 nm) are channels to allow mass transport of gas molecules.[6]

Gas storage involves either physisorption or chemisorption onto a solid material. Gas storage with activated carbon has been achieved under cryogenic conditions, low temperatures, and atmospheric pressure. Largely weak carbon–hydrogen interactions *via* physisorption have been exploited.[7] Physisorption is generally preferred because adsorbed gas molecules could easily be desorbed with minimal energy and facile regeneration of the adsorbent. Gases such as butane, propane, methane, and hydrogen have been stored with activated carbon. Two cases of hydrogen and methane storage are presented to provide an idea of the challenges and way forward. For hydrogen storage *via* activated carbons, the challenges are well known because the carbon–hydrogen interaction is not appropriate to retain and release hydrogen under ideal fuel cell working conditions. For instance, most activated carbons have carbon atoms with the sp^2 hybridization configuration and bind to the nearest three neighbor atoms through π or σ bonds. Under such conditions, the carbon atoms interact with the hydrogen molecule *via* weak van der Waals forces with about 0.1 eV binding energies, which is insufficient to retain the hydrogen molecule under ambient conditions (temperature and pressure). In contrast, atomic hydrogen interaction with carbon atoms could cause an electronic shift in the hybridization from sp^2 to sp^3 with relatively strong C–H bonding and an associated binding energy of several electronvolts. In this scenario, the release of hydrogen would require heating at a high temperature, which makes it impractical. Hence, the idea would be to tune C–H interactions such that their magnitudes lie mid-way to physisorption and chemisorption with associated binding energies of 0.2 to 0.8 eV.

Similar conditions exist for methane storage, which requires high compression energy at 15–27 °C and 18–20 MPa or in the liquefied form at −160 °C and 2–6 bar. The volumetric density is usually 470 to 570 m^3 for liquefied natural gas (LNG) and 220 to 260 m^3 for compressed natural gas (CNG) per 1 m^3 of the tank under STP conditions.[8] The properties of LNG and its storage conditions suggest that specialized equipment, which is usually very costly, and careful safety measures are needed. In fact, at -80 °C, natural gas transitions to supercritical states and cannot be liquefied by compression.[9] Also, the compression process is energy intensive. CNG is characterized by a low volumetric energy density of about 9.2 MJ L^{-1} at 25 MPa compared to gasoline with 34.2 MJ L^{-1}.[10] These drawbacks require alternative methods or technologies to store natural gases, and activated carbon could provide the solution. Studies have shown that the adsorption system of natural gas could accumulate a 10 fold higher volume of methane than the compression system at 20 bar and room temperature.[11] The adsorption natural system is presumed to operate at lower pressures from 30 to 100 bar and room temperature, which significantly reduces the safety concern.[12]

However, the adsorption system presents some disadvantages which require technological improvement. The system is designed to provide fuel injection at a slow rate under turbulent conditions because of the resistance of the adsorbent bed, packing style, and pore structure.[13] In a practical scenario, adsorption kinetics requires fast fill-up and slow discharge of the tank. However, due to the exothermic nature of adsorption, the heating of the bed during storage (adsorption) impacts the volume of the gas stored and the efficiency of adsorption.[14–16] Likewise, during desorption, which is an endothermic process, temperature fluctuations adversely impact the adsorption efficiency. Hence, to reduce the temperature fluctuations during fill and discharge cycles, a process design based on low heat adsorption with high adsorbent heat capacity is required.[17–20] Also, heavier gases such as CO_2, butane, and propane in natural gas present higher polarization, which is responsible for higher binding energies between the gas molecules and the adsorbent, thereby reducing the natural gas discharge efficiency.

A solution to the enumerated problems would be to produce activated carbon with a large surface area and high micropore volume. The problem with very high micropore volume and high surface area activated carbon is that it often has low bulk density and consequently has a low volumetric capacity, which increases material costs. Hence, a balance has to be struck to maintain a high surface area and high bulk density. Also, the presence of heteroatoms could improve the binding energies, resulting in electron shift to accumulate more gases. Hence, this chapter elucidates the fuel storage capacity of activated carbon. Particular interest is in gas storage especially methane, hydrogen, butane, and propane storage. The characteristics and isotherm studies of activated carbon are elucidated. However, the chapter does not consider the storage of carbon dioxide or other gases or liquid fuel or the simulation aspect of gas storage.

9.1.1 Adsorption Phenomena and Activated Carbon Production for Fuel Storage

Adsorption can be defined as a process that leads to an increase in gas density or solute in the surroundings of the substrate, called adsorbent, because of the interaction between the adsorbent and the adsorbate.[21] The adsorbate molecules far from the vicinity of the adsorbent are referred to as the bulk and they correspond to the density of free adsorbate molecules under the same thermodynamic conditions. Adsorption can be categorized into two regimes for gases, namely, subcritical and supercritical adsorption. Subcritical adsorption is observed when adsorption occurs below the critical temperature of the adsorbent.[22–25] The adsorption isotherm is usually characterized by a steep increase of the adsorbed gas density near the saturation pressure, which normally results in condensation and liquid formation on the surface of the adsorbate.[26–28] On the other hand, supercritical adsorption occurs above the critical temperature of the adsorbent and without any visible gas–liquid transition. In hydrogen adsorption with activated carbon, a supercritical regime is normally considered because of the very low critical temperature of hydrogen and to achieve the operation of this system as close to room temperature as possible, taking safety measures into account. For methane adsorption, studies have shown that the isotherm of adsorption on activated carbon shows no step-wise changes when transitioning from subcritical to supercritical temperature.[22,24]

Adsorption measuring methods could be classified into gravimetric, gas flow, and volumetric techniques. The gas flow technique is similar to gas chromatography and was first proposed by Nelson and Eggertson. The adsorbed volume can be determined from the peak area of the adsorption/desorption profile, which is obtained using a potentiometer over a period of time. Helium is used as a carrier gas and the partial pressure of the adsorbate (gas) is determined using a gas flow meter. The apparatus is simple, can be easily operated, and requires no vacuum. However, it is used for single-point isotherm measurement, and multipoint adsorption measurements could be problematic. For volumetric adsorption, which is a favorable choice in most experiments and industrial applications, the gas is expanded into a vessel containing an adsorbent that has been degassed. Part of the expanded gas is adsorbed on the surface of the adsorbent, whereas the excess gas remains in the bulk (surrounding the adsorbent). Using a mass balance, the amount of adsorbed gas could be determined if the void volume, *i.e.*, the volume that is impenetrable by the gas, is known. Gravimetric methods deal with weight differences. In this case, the weight changes of the adsorbent due to the adsorbed gas from the gas phase are determined using a sensitive microbalance.

Several models have been deployed to describe the isotherm of microporous activated carbons for the adsorption of gases. Models are very vital because they could be used to determine the total amount of hydrogen in the system at any temperature and pressure (which cannot be determined experimentally) to be able to compare the behavior of different adsorbents. Also, the principles of adsorption, kinetics, and thermodynamics could be learned from the model to guide the design of materials and the selection of the optimum pore size necessary to obtain maximum adsorption. The models must be framed to consider excess hydrogen mass to fit the experimental data obtained by volumetric and gravimetric techniques. Different types of Type I isotherm (based on IUPAC) equations have been used to fit the experimental data. These models are derived based on assumptions that are only practicable in an ideal scenario. Hence, the preferred equation for comparison of different adsorption materials should adequately fit the data with minimal error and should have fewer parameters, which agrees with Occam's razor rule of model selection based on fewer assumptions. Some of the reported models and the number of parameters to be determined are shown in Table 9.1.

Despite the intensified research efforts in the use of biomass to produce activated carbon for gas storage, the methods of characterization vary widely and make it difficult to compare. Also, the characteristics of different activated carbons vary depending on the parent material, conditions of activation, and post-treatment. Over the years, activated carbon production has been largely categorized into two types, namely, physical and chemical activation[35,36] (Figure 9.1). Physical activation is generally a two-step technique that involves carbonization at a temperature between 500 and 800 °C and activation at above 800 °C using gaseous agents like steam and CO_2. This process is widely used in industry; however, it involves higher energy demand and the tuning of pore structure could be problematic. In contrast, chemical activation involves the addition of alkaline salts or acids *via* impregnation and subsequent activation around 800 °C.[37,38] This process generates activated carbons with well-developed pores and generally higher surface area compared to physical activation.[35] The ratio of the activating agent to precursor, type of activating agent, and temperature of activation are vital factors that must be manipulated, and they affect the properties of activated carbon. Table 9.2 shows reported activated carbons derived from different biomasses and their activation conditions, techniques (including one-step or two-step, chemical, and physical activation), and textural characteristics. Another aspect is that these chemical agents in some cases could introduce foreign functional groups that are heteroatoms and high oxygen sites, which could facilitate gas adsorption.

Engineering the production of activated carbons to suit specific applications is necessary considering that different gas compounds have different characteristics. However, there is some kind of consensus that all activated carbons for gas storage purposes must have high porosity and density, especially for high-pressure adsorption and the gravimetric and volumetric capacity must be increased simultaneously. The next section will discuss in detail the trends in methane, hydrogen, butane, and propane storage in activated carbons.

9.2 Fuel Storage Applications of Activated Carbon

9.2.1 Methane Storage on Activated Carbon

Methane (CH_4) has been described as an excellent fuel for motor vehicle engines with several advantages such as availability and low price, while

Table 9.1 Type I isotherm equations used in gas adsorption analysis.

Model	Equation	Description of model terms	Parameters
Langmuir[29]	$C = C_0 \dfrac{k_0 e^{\frac{q_{st}}{RT}} P}{1 + k_0 e^{\frac{q_{st}}{RT}} P}$	$C/\text{kg kg}^{-1}$ is the adsorption uptake, $C_0/\text{kg kg}^{-1}$ is the saturated amount adsorbed, $q_{st}/\text{J mol}^{-1}$ represents the isosteric heat of adsorption, $R/\text{J mol}^{-1}\,\text{K}^{-1}$ is the universal gas constant, and P/kPa is the equilibrium pressure. The parameter t characterizes the heterogeneity of the adsorbent–adsorbate pair, T/k stands for the adsorption temperature, and k_0 is an equilibrium constant	1
Tóth[30]	$C = C_0 \dfrac{k_0 e^{\frac{q_{st}}{RT}} P}{\left(1 + \left(k_0 e^{\frac{q_{st}}{RT}}\right)^t\right)^{1/t}}$		2
Dubinin–Astakhov (D–A)[31] without volume correction	$C = C_0 \exp\left[-\left(\dfrac{RT}{E}\ln\left(\dfrac{P_s}{P}\right)\right)^n\right]$	$E/\text{J mol}^{-1}$ is an adsorption characteristic parameter and P_s is the saturation pressure. $W/\text{cm}^3\,\text{g}^{-1}$ is the volumetric adsorption uptake and $W_0/\text{cm}^3\,\text{g}^{-1}$ is the maximum volumetric adsorption capacity	4
Dubinin–Astakhov (D–A)[31] with volume correction	$W = W_0 \exp\left[-\left(\dfrac{RT}{E}\ln\left(\dfrac{P_s}{P}\right)\right)^n\right]$		4
Sips[32]	$C = \dfrac{(b_s P)^{\frac{1}{m_s}}}{1 + (b_s P)^{\frac{1}{m_s}}}$	b_s is the Sips affinity parameter and m_s is the Sips heterogeneity parameter	2
Generalized Freundlich[33]	$C = \left(\dfrac{b_F P}{1 + b_F P}\right)^q$	b_F is the generalized Freundlich affinity parameter and q is the generalized Freundlich heterogeneity parameter	2
Unilan-Q[34]	$C = \dfrac{RT}{Q_2 - Q_1}\ln\left(\dfrac{1 + b_0 \exp\left(\frac{Q_2}{RT}\right)P}{1 + b_0 \exp\left(\frac{Q_1}{RT}\right)P}\right)$	Q_1 and Q_2 are the minimum and maximum values of the energy parameter and b is the affinity parameter	3

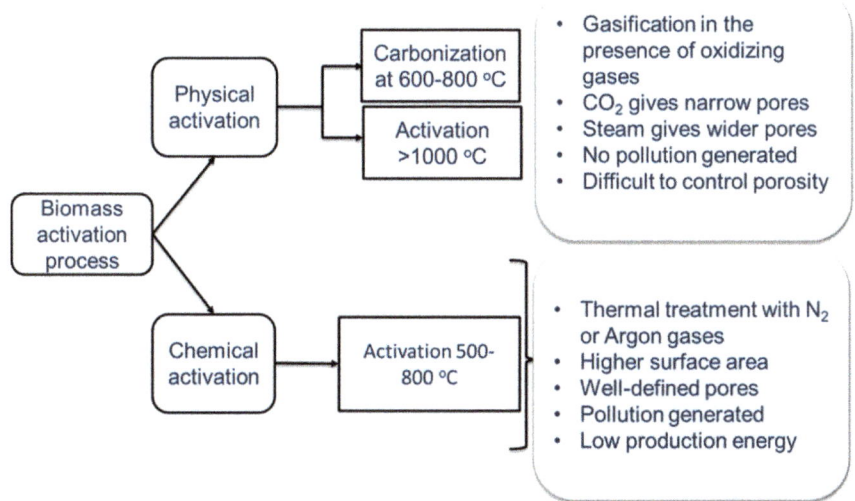

Figure 9.1 Preparation methods of activated carbons with their merits and demerits.

decreasing harmful combustion compared to fuels obtained from fossil fuels. Transportation of natural gas is one of the obstacles to utilizing methane gas in transportation fuel. Methane has a high heating value of 50.1 MJ kg^{-1} unit of mass of hydrocarbon fuels. Methane in a liquid state has an energy density of 21.2 MJ L^{-1}, a specific energy of 50 MJ kg^{-1}, and a density of 0.424 g m^{-3} (at 15 °C).[61] It is relatively easy to compress this gas; for instance, the energy required to reach 600 bar is just 1.25 MJ kg^{-1}.[62] As a transportation fuel, methane can be stored or transported in three forms: namely, CNG, Adsorbed Natural Gas (ANG), and LNG.[63] CNG is obtained when the gas is compressed to <1% of its volume that it occupies at STP. The compressed gas is distributed in a pressurized container at a pressure in the range of 20–25 MPa. CNG can be used in the conventional internal combustion diesel engine with modification. Some of the main advantages are low maintenance cost compared to fuel-powered vehicles, and CNG is less likely to be flammable because its autoignition temperature is high (about 540 °C) with a narrow range of flammability.[64] On the other hand, LNG is a liquefied gas obtained at −161.5 °C under 1–2 MPa pressure.[16] This process is very expensive and consumes a lot of energy. An alternative to CNG is ANG whereby natural gas is stored in a porous material at a relatively lower pressure of 0.7–4 MPa and room temperature.[10,19] This process has several advantages including the potential for removal of fuel tanks for filling up, lower hazard potential, high volumetric storage capacity, *etc.* The target set by the US Department of Energy for any porous material used in methane storage is to achieve 80% v/v, and activated carbon depending on the raw materials and activation conditions could store 100–150% v/v methane, which makes it appealing. This section is focused is on adsorbed methane on porous activated carbon.

Table 9.2 Some reported activated carbons, preparation conditions, techniques, and textural characteristics.

Feedstock (F)	Activation agent (AA)	Technique	Activation conditions			Pore size (nm)	Pore volume (cm³ g⁻¹)	Specific surface area (m² g⁻¹)
			Impregnation ratio (AA:F)	Temp. (°C)	Time (min)			
Pine sawdust[39]	Steam (300 psi)	One-step physical	—	900	60	—	0.526	962
Wine stone waste[40]	N₂	One-step physical	—	900	120	1.82	0.7767	1814
Date seeds[41]	CO₂	Two-step physical	—	900	60	—	0.28	798.38
Coconut shells[42]	Steam	Two-step physical	—	800	60	1.87	0.45	1011
Sargassum fusiforme[43]	CO₂	Two-step physical	—	900	29.05	4.00	—	1329
Coconut shells[44]	CO₂	Two-step physical	—	700	60	—	1.22	2276
Coal fly ash[45]	KOH	One-step chemical	4:1	800	60	—	0.477	946.77
Kanlow switchgrass[46]	KOH	One-step chemical	1:1	900	60	1.69	0.59	1271.66
	H₃PO₄			900	60	3.24	1.44	1372.93
Hemp residue[47]	H₃PO₄	One-step chemical	1:1	900	15	—	0.86	1630
Oleaster fruit[48]	ZnCl₂	One-step chemical	1.5:1	550	60	—	0.735	2021
	KOH	One-step chemical	3:1	800	60	—	0.39	1816
Almond shells[49]	KOH	One-step chemical	5:1	800	180	0.5	1.19	2054
Arundo donax[50]	KOH	One-step chemical	3:1	600	120	0.82	1.01	2232
Rattan stalks[51]	NaOH	Two-step chemical	3:1	600	60	3.55	0.60	1135
Garlic peels[52]	KOH	Two-step chemical	2:1	800	60	2.20	0.54	1309
Cotton stalks[37]	H₃PO₄	Two-step chemical	1.5:1	600	120	3.43	1.476	1424
Olive stones[53]	KOH	Two-step chemical	2:1	900	120	—	0.50	1478
Sorghum stem[54]	KOH	Two-step chemical	1:1	900	180	—	1.076	1674
Jute plant[55]	KOH	Two-step chemical	4:1	800	60	2.12	1.01	1903
Cupuassu shells[56]	KOH	Two-step chemical	2:1	500	120	<2	0.79	2004
Black locust[57]	KOH	Two-step chemical	6:1	830	90	1.89	1.16	2064
Cow dung[58]	H₃PO₄	Two-step chemical	4:1	900	120	—	—	2450
Date sheets[59]	KOH	Two-step chemical	6:1	800	120	0.818	1.632	3337
Coconut shells[60]	ZnCl₂/CO₂	Physicochemical	0.15:1	500	120	—	0.77	1599
	H₃PO₄/CO₂	Physicochemical	0.21:1	500	120	—	1.12	2191

Activated carbon used in methane storage must have high micropore volume, optimal micropore half-width or radius, and energy of adsorption. Also, it must have a high packing density to ensure high volumetric and gravimetric capacity.[24] The packing density is normally measured as the ratio of the mass of the adsorbent to the occupied volume in a cylinder according to ISO standards 60 and 697.[65,66] Studies on methane ANG adsorption could be grouped into materials development and kinetic studies on adsorption parameters,[67–69] implementation of effective cooling/heating systems for temperature control of the adsorbent layer during charge/discharge cycles,[20,70] study on the vessel performance,[71] and control of thermal energy in the storage tank.[18,19] Most of the researchers conduct theoretical and experimental studies to validate their results.

Theoretical and experimental studies of methane adsorption over AU-4 and AU-5 porous carbon adsorbents in the subcritical and supercritical temperature regions (178–360 K) and in the pressure range of 20–25 MPa revealed no step-wise change in the first-order phase transition.[72] The maximum adsorbed methane at 178 K was 11.5 $mmol\,g^{-1}$ for AU-4 and between 0.2 and 11 $mmol\,g^{-1}$ for AU-5.[22,72] The adsorption isosteres were well approximated with a straight line. At room temperature, the absolute effectiveness of methane accumulation shifted toward higher pressure for both evaluated adsorbents, corresponding to 3–7 MPa for AU-4 and 8–12 MPa for AU-5.

The thermal effect resulting from the heat of adsorption on the charge/discharge performance of the ANG storage system was evaluated from the viewpoints of storage tank configurations and the possible addition of a heat exchanger to increase the discharge velocity. Two ANG storage systems of both horizontal and vertical tanks were elucidated to investigate the thermal effects resulting from adsorption heat charge/discharge at 4 MPa (Figure 9.2).

The results revealed that the middle region suffers the most temperature fluctuation during charge/discharge cycles and the adsorption/desorption

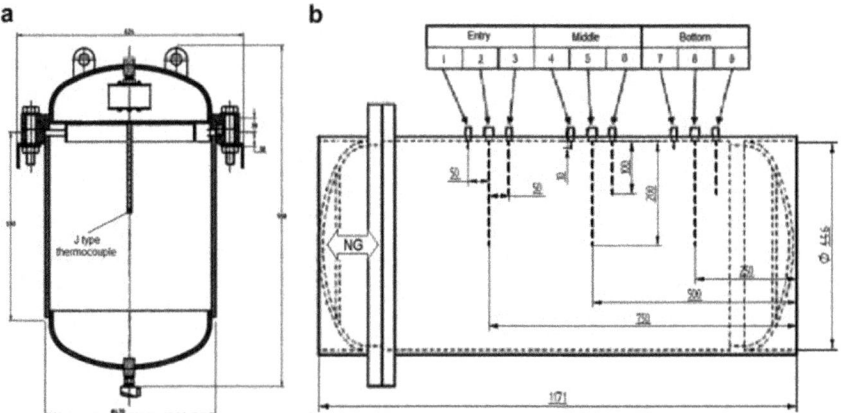

Figure 9.2 ANG storage vessels of (a) vertical and (b) horizontal configurations.[18] Reproduced from ref. 18 with permission from Elsevier, Copyright 2009.

temperature profile strongly depends on the fidelity of the flow control in the desorption line, the type of activated carbon used, and the adsorption pressure. In the horizontal tank, the temperature initially increased to 65 K and stabilized at 63 K, whereas in the vertical tank the temperature increased to 34 K and then stabilized at 30 K. The discharge amount was higher for the horizontal tank (13 547 L) with a longer time (33 780 s) compared to the vertical tank discharge amount (8019 L) with a shorter time (1400 s).[18] This study recommended that the efficiency of desorption could be enhanced *via* an efficient heating system to increase the desorption velocity.

Another study on the thermal effects of adsorption heat considered the incorporation of a heat-exchanging device in their 2D model of the adsorbent layer. Also, 3D modeling of the heat transfer was provided to understand the heat transfer mechanisms. Their study demonstrated that the amount of filled gas doubled when the heat-exchanging device was incorporated compared to the adiabatic process. Also, the high temperature was located in the middle part of the initial charging process, which was dispersed after some time. Optimization of the fin and tube heat exchanger resulted in a 10% increase in the ANG storage capacity. The overall mass-to-volume ratio decreased from 796 $kg\,m^{-3}$ to 518.01 $kg\,m^{-3}$, which suggests a significant reduction in the weight of the ANG vessel.[19]

Tailoring the characteristics (surface area, density, pore size, and micropore volume) of activated carbon to increase methane storage capacity at ambient temperature is a noteworthy research effort.[73,74] A recent study showed that alkaline (KOH, K_2CO_3) modification of commercial activated carbons (FPV, CWZ-22, and WG-12) in the temperature range of 600 to 850 °C and with an activating agent to precursor ratio of 1 : 4 improved the textural properties (BET surface area of up to 2406 $m^2\,g^{-1}$ for KOH activation).[75] The study revealed a correlation between the amount of CH_4 adsorbed by the activated carbon and microporosity and surface area. The modification temperature and agent slightly influenced the bulk density no matter what the type of carbon material is, whereas the skeleton density increased significantly. At 35 bar and 25 °C, a maximum of 9.7 $mmol\,g^{-1}$ of methane could be stored using the FPV activated carbon. The experimental data were best fitted with the Sips isotherm model. Another study using commercial activated carbon with a surface area of 1007 $m^2\,g^{-1}$ and a total pore volume of 0.366 $cm^3\,g^{-1}$ revealed that about 8.52 wt% methane could be stored.[76]

Conclusively, activated carbon has suitable characteristics of large surface area and high micropore volume to store high amounts of methane. The amount of adsorbed methane is strongly correlated with the activated carbon's surface area and microporosity. The main challenge is to increase the bulk density of the activated carbons which can be achieved by producing different types of packing and shapes (monolithic, granular, and spherical shapes). Also, the thermal effects resulting from the adsorption/desorption of methane could be better managed by providing heat-exchanging devices to increase the discharge velocity and a high-fidelity control system to achieve uniform bed temperature and increase the capacity of adsorption.

9.2.2 Butane Storage on Activated Carbon

Butane is a colorless gas with a petroleum-like odor. It has two isomers, namely, *n*-butane and iso-butane, which exist as gases at 25 °C and 1 atm. It could be easily liquefied and shipped under its vapor pressure. The raw material for its production is natural gas and petroleum and it's used as a fuel and an aerosol propellant, in cigarette lighters, and for making chemicals. To regulate the emissions from gasoline vehicles, today, all gasoline vehicles have an evaporative emission control system, called the EVAP system.[77] The main component of the EVAP system is a carbon filter canister, which is a plastic container that is packed with activated carbon to help adsorb fuel vapors and prevent the same from being released into the environment. The canister could be defined by the carbon density and the butane working capacity (BWC).[78,79] The BWC is the amount of *n*-butane that can be adsorbed by the carbon in the canister before it gets saturated. After saturation, the canister is regenerated *via* desorption while the engine is running (endothermic process) by activation of the purge control valves which creates a communication with the engine's intake manifold. In this process, a vacuum is generated, which allows air from the environment into the canister for purging of the activated carbon, releasing fuel vapors for combustion in the engine (Figure 9.3).[77,80]

When the engine is running but not in motion, the air–fuel ratio can be significantly changed by the fuel vapor coming from the canister filter; hence, a closed loop is normally required for the purging phase to maintain the air–fuel ratio mixture at a stoichiometric value. Another problem is that this canister could be saturated after long parking, which is dependent on the previous use (milage), how long it was parked, daily temperature, and uncontrollable parameters like the car designer. In this condition, the canister could easily become unfunctional. Hence, EVAP system regulations are needed to monitor vehicle emissions for a cleaner environment. In the case of hybrid vehicles, it is even more challenging since it has less time to purge and

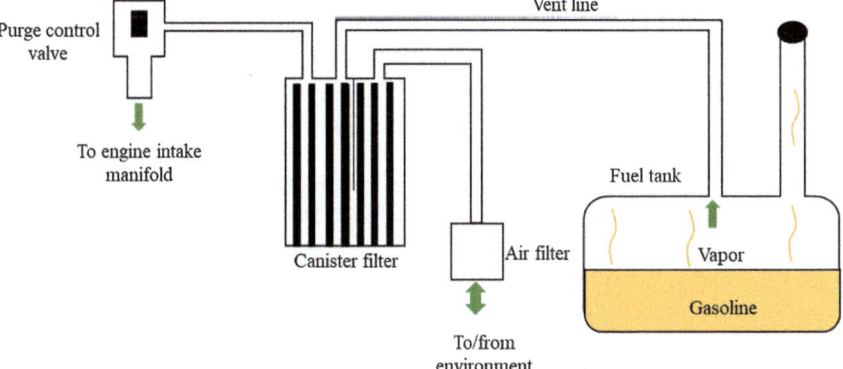

Figure 9.3 Evaporation loss control system for *n*-butane adsorption.[77] Reproduced from ref. 77 with permission from Elsevier, Copyright 2022.

release the fuel vapor. Hence, the preparation of activated carbon for n-butane adsorption and the design of the canisters become imperative.

According to the United Nations Global Technical Regulations, to determine the BWC, about 40 gh^{-1} of the n-butane with a 50% nitrogen mixture shall be charged to the canister until it reaches a 2.0 g breakthrough. Then the canister should be purged with nitrogen (300-bed volume, 25 $Lmin^{-1}$).[78] A study demonstrated that a canister containing BAX 1100 activated carbon used to adsorb gasoline vapor could reach the breakthrough and the bed temperatures reach as high as 70 °C. The temperature profile in the adsorbent beds showed linear adsorption behavior, and the internal temperature is vital to model the ad/desorption of the bed. The experiment lasted 6 h, and after several cycles of purging and loading, the canister adsorption capacity decreased by 20%, due to the aging trend of the activated carbon.[81]

Some very important factors in the adsorption/desorption of n-butane in canister filters are the porosity and textural characteristics of the activated carbon used. Butane has a kinetic diameter of 0.42 nm,[79] and most studies have demonstrated that micropores and ultramicropores normally used in CO_2, methane, and hydrogen adsorption suffer pore blocking and eventual deactivation. Studies have shown that the n-butane adsorption capacity could be determined in the pore size range of 1.5 to 4.5 nm, and the adsorption was strongly dependent on the pore size range of 2 to 2.5 nm.[82] Evaluation of the activated carbons from beech wood obtained at different activation times at the same activation temperature revealed that when the activation time was increased from 20 min to 60 min, micropores were converted to mesopores, and the obtained activated carbon had about 54% mesopore volume and 46% micropore volume. This carbon was suitable for the adsorption of other gases like CO_2 and CH_4 because of its meso-microporosity. Consequently, the n-butane adsorption increased from 3.8 $mmol\,g^{-1}$ to 5.2 $mmol\,g^{-1}$ when the time was increased from 30 min to 60 min. Another study showed that the presence of water vapor could not hinder n-butane adsorption in mesoporous activated carbon. In fact, n-butane was selectively adsorbed, hence inhibiting the adsorption of water. Wide temperature variations were observed at the initial adsorption and desorption stage.[80] These temperature variations in the adsorption–desorption regime have been modeled using computational fluid dynamics simulations to present real-life scenarios.[77,83,84]

In summary, fuel vapor consisting of mainly n-butane could be stored in activated carbon produced from biomass in a canister filter. Several factors influence the adsorption capacity, including the textural and pore characteristics. Mainly mesopores in the range of 2–2.5 nm strongly influence the adsorption. Also, due to the desorption phenomenon facilitated by air purging, a temperature drop during desorption is normally encountered. Further studies on the thermal effects of the adsorption bed phenomenon are required, especially to have a deep understanding of the canister's performance under different temperature conditions. Deep characterization of the EVAP system for hybrid vehicles is also needed to provide policy documents for regulatory purposes.

9.2.3 Hydrogen Storage on Activated Carbon

Another important activated carbon application is in hydrogen gas storage. Although hydrogen in itself is not a fuel, it's an energy carrier with a calorific value of 126 MJ kg^{-1}. Hydrogen is the lightest element and is the most abundant element in the universe. It can be produced *via* electrochemically, thermally, biologically, and solar-driven processes. In the year 2020, the global market of hydrogen was about \$3.9 billion with a projected annual growth rate of 8% until 2026 and may represent 18% of the global energy consumption in the year 2050.[85-87] Hydrogen combustion does not produce CO_2 (water and heat are the only byproducts); hence, it is viewed as a clean fuel. Hydrogen gas is very light and occupies a large volume at room temperature and its storage requires a very high pressure (350–700 bar). Compressed hydrogen has been applied in space shuttle propulsion; however, the boil-off temperature, low condensation, and high-fidelity and sophisticated tanks used for isolation make its application in vehicles unattractive. Moreover, it takes about 30% of the energy content of hydrogen to achieve its liquefaction.[88]

The drawbacks of safe hydrogen transportation and use require the development of safe storage facilities as well as increasing the volumetric capacity to optimize cost. The United States Department of Energy proposed that for hydrogen to be used in vehicles and other applications,[89] a target storage capacity of 5.5 wt% and volumetric capacity of 40 g l^{-1} in porous materials[90] should be attained. Hydrogen has been stored in different adsorbents including metal–organic frameworks, metal hydrides, complex metal hydrides, and ammonia borane (Figure 9.4).

The use of carbon adsorbents offers several advantages such as fast adsorption/desorption kinetics, facile synthesis routes, and avoiding the need for exothermic processes that characterize most metal hydrides.[92] The interaction of carbon atoms and hydrogen could occur *via* electrostatic

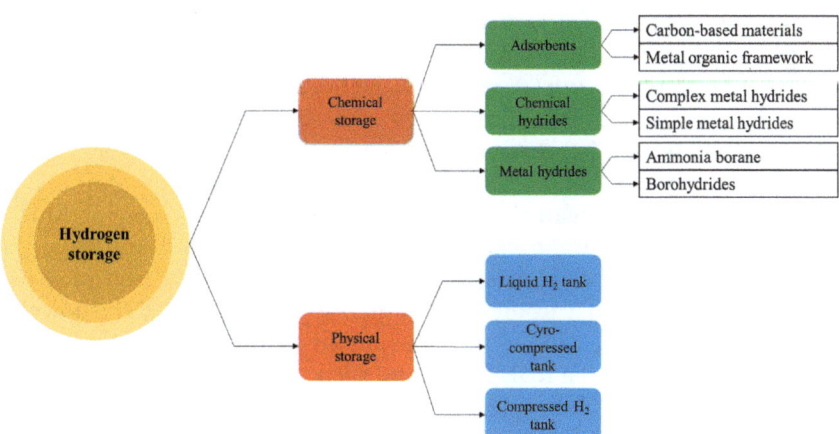

Figure 9.4 Mechanisms of hydrogen storage and explored materials.[91] Reproduced from ref. 91 with permission from Elsevier, Copyright 2014.

interaction between the oppositely charged surface and the hydrogen atom or through hydrogen atom or ion intercalation into the carbon lattice.[93] Also, hydrogen could be stored *via* electrosorption or reverse redox reaction where functional groups (heteroatoms) could react to adsorb or release hydrogen.

For electrochemical hydrogen storage, electrolytes such as 1 and 4 M H_2SO_4 and 6 M KOH have been used extensively, with 6 M KOH being the preferred electrolyte. The conducted study revealed that about 1 wt% of hydrogen can be stored in acidic media, whereas in alkaline media, 2 wt% was achieved.[94] This performance in an electrochemical cell can be enhanced by the functionalization of the activated carbon with metal oxide such as CuO. The presence of CuO improved the carbon conductivity and interaction with hydrogen to realize about 11.2 wt% hydrogen storage capacity.[95] Several adsorbents derived from biomass have been studied for hydrogen sorption. The main challenge is to tailor the carbon binding energy mid-way between physisorption and chemisorption to allow higher storage capacity and facile desorption.[90] To achieve this, heteroatoms, noble metals, transition metals, and other functional groups have been used. Also, textural characteristics are very important because it has been demonstrated that high surface area and micropores facilitate the insertion of hydrogen ions or atoms in the carbon lattice. For instance, empty fruit bunch-derived activated carbon with KOH impregnation (0.5 to 2 M) revealed a microporosity of 94% and a surface area in the range of 305 to 687 $m^2 g^{-1}$. When this material was tested for hydrogen adsorption at 19 bar and 77 K, about 2.1 wt% hydrogen storage capacity was achieved.[96] A similar study with rice husk activated carbon produced by KOH activation achieved a storage capacity of about 2.85 wt% at 77 K and 1 atmosphere. The nanometer size of the carbon and microporosity (0.6–0.8 nm pore size distribution) had a greater influence than the surface area and total pore volume.[97]

However, the physisorption of supercritical gases proceeds *via* monolayer adsorption on the carbon surface and decreases as the temperature increases, indicating that surface area and pore distribution are very vital for hydrogen storage. The surface and the microporosity could be influenced by the type of activating agent used and the activation temperature. In hydrogen storage using activated carbons, it has been demonstrated that increasing KOH and $ZnCl_2$, which are popularly used activating agents, could tailor the carbons so that they have high surface area and predominantly micropores.[86] However, using $ZnCl_2$ presents pollution problems, and due to its highly corrosive nature, it can damage reactors and pipelines. Also, it is possible to obtain higher hydrogen storage capacity at increased pressure. A study using activated carbon from tamarind seeds and KOH showed a remarkable surface area of 1785 $m^2 g^{-1}$ and pore distribution in the range of 0.8–1.1 nm at 700 °C and a KOH : tamarind seed impregnation ratio of 3. The storage capacity of the carbon reached 4.85 wt% at a pressure of 40 bar and room temperature.[98]

Oxygen, noble metals, and transition metals have been doped on activated carbon to increase its hydrogen storage capacity. These metals include Pd, Ag,

Co, Cu, Ni, and Pt and were functionalized on olive pomace-derived activated carbon (phosphoric acid was used as an activating agent). Oxygen functional groups developed on the active carbon promoted the steric hindrance effect and saturation of active sites to suppress hydrogen uptake. Also, the Pt and Pd metals on the activated carbon propagated the spillover mechanism, which accelerated the adsorption of hydrogen at room temperature and 25 bar.[88] Another study countered the results, which posits that the oxygen functional group hinders hydrogen sorption. In this study, cellulose acetate with a high oxygen/carbon ratio of 0.93 was employed as a precursor to producing an oxygen-rich activated carbon *via* KOH activation. The obtained carbon revealed a remarkably high surface area of 3800 $m^2 g^{-1}$ and a distinctive micropore volume of 1.8 $cm^3 g^{-1}$.[99] The material could store up to 8 wt% at 77 K and 20 bar. The hydrogen uptake by the activated carbon increased when the pressure was increased from 20 to 30 bar, reaching a value of 8.9 wt%. Table 9.3 shows the preparation conditions, textural properties, and hydrogen storage capacity of lignin-derived activated carbon.

It is noteworthy to mention that other non-biomass sources have been explored to produce hydrogen storage materials. For instance, polymeric materials, containing mainly carbon chains, have been used to produce activated carbon. In this case, a polythiophene polymer, which has some sulfur elements, was activated using KOH at a 2:1 molar ratio, and in the temperature range of 600–850 for 1 h. At 700 °C the surface area could reach 1700 $m^2 g^{-1}$ and its performance for hydrogen adsorption showed that about 5.71 wt% storage capacity could be reached at 77 K and 20 bar.[111] The high sulfur content modified the carbon binding energy and promoted hydrogen interaction with the carbon structure. However, the use of polymeric materials could be expensive considering the complex production systems.

Fullerenes are very important carbon-based materials that have been tested for hydrogen storage. These materials are obtained by very slow condensation of vaporized carbon. They were discovered when experiments were conducted to understand how long-chain molecules could be formed in the circumstellar and interstellar spaces in the presence of a laser beam.[112] This carbon structure is very similar to graphite but with rolled-up layers and can take tubular, spherical, and ring-like geometric shapes. The difference between graphite and fullerenes is that while graphite has a hexagonal structure of carbon atoms, the latter have pentagonal carbon rings.[90] Normally, the slow condensation of carbon vapor results in spherical fullerenes; however, the use of a catalyst during synthesis could yield tubular or ring-like structures. The hydrogen storage capacity of raw fullerenes and sodium impregnated fullerenes (Na:C_{60} of 10) was compared and the results showed that a higher hydrogen capacity of about 3.5 wt% could be realized with sodium functionalized fullerenes than the pristine materials.[113] The problem with sodium functionalization is that there is a tendency for ionic bonding with the sodium atom, which will require higher energy to desorb. However, the majority of the carbon–hydrogen interactions are covalent bonding in $C_{60}H_x$ requiring mild temperature (250 °C) to reversibly desorb

Table 9.3 Hydrogen storage on activated carbons and conditions.

Precursor	Activating agent	Activation conditions			S_{BET}	Pore volume ($cm^3\ g^{-1}$)	Hydrogen storage conditions		Storage capacity (wt%)
		Molar ratio (agent:biomass)	Temp. (°C)	Time (min)			Temp. (K)	Pressure (bar)	
Palmyra sprouts[100]	KOH	2	900	120	2090	1.441	298.15	15	1.06
Walnut shells[101]	H_3PO_4	—	450	90	420.5	0.24	298.15	1	5.66
Tangerine peels[86]	$ZnCl_2$	5	700	120	1230	—	77	30	1.67
Olive pomace[88]	H_3PO_4	40%	550	—	1074	0.45	77	40	2.46
Coffee shells[102]	KOH	5	800 W	9	3149	1.67	298.15	140	0.91
Oil palm ash[96]	KOH	2	900	15	687	0.297	77	20	2.14
Empty fruit bunch[96]	KOH	2	900	15	687	0.297	77	1	1.97
Pinecone[103]	KOH	1	900	120	1173	0.451	77	80	5.25
Tamarind seeds[98]	KOH	3	700	—	1785	0.93	298.15	40	4.73
Coconut shells[104]	KOH	4	900	120	2800	1.89	298.15	100	0.85
Onion peels[105]	KOH	5	750	60	3150	1.64	77	1	3.67
Corncob[106]	KOH	4	850	180	3708	2	77	40	5.80
Olive pomace[107]	KOH	4	800	120	1269	0.48	77	209	6.11
Sucrose[108]	KOH/urea based	3	900	120	2842	1.35	77	1	2.21
Loblolly pine[109]	KOH	4	800	120	3666	1.56	77	55	10.2
Rice husk[97]	KOH	1	900	60	2682	0.792	77	1	2.85
Melaleuca bark[110]	KOH	5	750	60	3170	1.07	77	10	4.08

the adsorbed hydrogen. Some studies have investigated the thermodynamic properties of the hydrogenation of fullerenes.[114]

Other carbon sources are heavy oil and coal, which are hydrocarbon fuels. Coal, for instance, has a high calorific value, and its conversion into activated carbon yields a relatively high surface area with appreciably high hydrogen sorption capacity. Four types of coal, namely, bituminous, low-bituminous, anthracite, and sub-bituminous coal, have been used to produce activated carbon with KOH as an activating agent. The activated carbon could reach $3000 \text{ m}^2\text{g}^{-1}$ specific surface area with high micropore volume. The result of hydrogen sorption measurement demonstrated that these materials could reach 6.8 wt% at 77 K and 4 MPa over activated carbon derived from bituminous coal.[115] Also, other works reported the use of anthracite as a precursor to developing activated carbon at 725 °C to 775 °C and with the KOH impregnation ratio in the range of 2–4, which could store about 6.6 wt% of hydrogen.[116,117] In all these studies, the main factors that determine the sorption capacity of coal were the surface area and the microporosity.

9.2.4 Propane Storage on Activated Carbon

Propane, also known as propane autogas or liquefied petroleum gas, is a clean burning fuel used in light, medium, and heavy-duty vehicles powered with propane. It is a three-carbon atom alkane gas usually stored in pressurized tanks. It is colorless and odorless, and it is vaporized from its liquid form when the pressure of the tank is released. Propane has a high octane number, which makes it a candidate fuel for the internal combustion engine. Also, its spill presents no hazard to the environment and water bodies. As in the production of butane, propane gas is produced by the refining of crude oil and processing of natural gas. The purification of propane from different hydrocarbon mixtures could be achieved by membrane separation,[118] adsorption,[119] absorption,[120] and cryogenic distillation.[121] In the United States, propane accounts for 2% of the energy used. Its main applications are in refrigerating gas, cooking, water heating, powering industrial equipment, and farm tools. Also, propane can be used in chemical industries as a raw material for plastic making.

Propane is a commonly used fuel behind diesel and gasoline, and propane used in vehicles is designated as HD-5, which contains about 90% propane, 5% propylene, and 5% of other gases.[122] In vehicles, propane can be stored in pressurized tanks of about 10 bar as a liquid, which has about 270 times the energy density of its gaseous form. Adsorption of ethane, methane, and propane by a pressure swing adsorption process on different commercial activated carbons (granules and pellets) at 303 K and in the pressure range of 0–3 MPa revealed that preferential adsorption was observed for greater carbon chain lengths at low pressure.[123] The carbon materials displayed Type I isotherms, and the adsorption data could be fitted with the Langmuir model. As the pressure increased from 0.5 to around 3 MPa, the monolayer coverage of the micropores was achieved and the mesopore saturation became slower.

Further studies on the adsorption characteristics of mixed gases including nitrogen, propane, ethane, and methane using commercial activated carbon (Norit-RB2) noted its interesting adsorption properties. Other gases with low carbon chains showed homogeneous adsorption and the adsorption decreases as the temperature increases. On the other hand, propane gas with greater chain length revealed an increase in adsorption as the temperature increased, which indicates the heterogeneity of the adsorption process. Again, the adsorption isotherm of propane showed an increase in adsorption at low partial pressure, followed by moderate adsorption as the pressure increased. Since the adsorption process of propane was different from the conventional adsorption process, a modified Sips model which includes the temperature dependence is suggested to clearly describe the propane isotherm.[124]

The adsorption process below 0 °C has great potential for higher storage capacity and environmental regulation. Hence, understanding the adsorption thermodynamics at such temperatures will provide insights into the isosteric heats and isotherms. Adsorption of hydrocarbon gas mixtures, namely, ethane, propane, butane, and methane, has been evaluated from −40 °C to 60 °C and a pressure range of 5 Pa to 1 kPa. The activated carbon used is commercial activated carbon (Norit RX 1.5) with a surface area of 1600 $m^2 g^{-1}$ and a micropore volume of 0.605.[125] The propane molecule has a critical diameter (nm) and polarizability ($10^{-24} cm^3$) of 0.5230 and 6.29, respectively.[126] The results show that the chain length of the hydrocarbon gases has a profound effect on the adsorbed amount. This is because the polarizability and number of molecules that bind with the carbon atom increase. Similarly, the heat of adsorption increases with chain length; however, the isosteric heat decreases due to the heterogeneous energy distribution of the adsorption sites. Other applications of propane storage on activated carbon for the cooling system have been reported; however, the performance of adsorptive carbon was poor compared to the traditional process because work is required to extract the propane (refrigerant) inserted in the activated carbon adsorbent.[127]

9.3   Conclusions

In conclusion, activated carbon derived from biomass presents tunable structural and surface functionalities required for fuel storage. The activation with chemical agents promotes well-defined pores and higher surface area. Especially, activation with $ZnCl_2$ and KOH in the temperature range from 500 °C to 800 °C, depending on the activating agent, results in high micropore volume and specific surface area, but with low density. The low-density activated carbon will result in low volumetric and gravimetric storage capacity; hence, granular and monolith activated carbon are normally used to achieve high packing density. Overall, activated carbon from biomass is a suitable material to store hydrogen, methane, propane, and n-butane. Preferential adsorption of heavier hydrocarbons occurs as the carbon chain length increases because of the polarizability and molecular geometry of the fuels. Also, the thermal effects resulting from the adsorption/desorption of methane could be better managed by providing heat-exchanging devices to increase the discharge velocity and a high-fidelity control system to achieve uniform bed temperature and increase the capacity of adsorption. The presence of heteroatoms such as oxygen and sulfur could tailor the binding energy of the activated carbon mid-way physisorption and chemisorption to promote higher hydrogen storage capacity. The adsorption isotherms of fuels are mainly described by the Sips, Langmuir, Tóth, and Dubinin–Astakhov models.

It is recommended that new waste streams should be explored, and integrated wastes should be optimized for the production of activated carbons for fuel storage. Also, the development of hierarchically porous (mesoporous–microporous) carbons with heteroatoms will allow beneficial mass transport and versatile applications, especially for longer carbon chain hydrocarbons that require mesopores. Finally, detailed thermodynamic studies on different adsorption systems are further needed.

Acknowledgements

The authors acknowledge the financial support provided by the Dirrecion General de Asuntos del Personal Academico under PAPIIT Project No. IA102522 and IG100720.

References

1. D. R. Lobato-Peralta, *et al.*, A review on trends in lignin extraction and valorization of lignocellulosic biomass for energy applications, *J. Clean. Prod.*, 2021, **293**, 126123.
2. A. Pal, *et al.*, A benchmark for CO_2 uptake onto newly synthesized biomass-derived activated carbons, *Appl. Energy*, 2020, **264**, 114720.
3. M. Gayathiri, T. Pulingam, K. T. Lee and K. Sudesh, Activated carbon from biomass waste precursors: Factors affecting production and adsorption mechanism, *Chemosphere*, 2022, **294**, 133764.
4. S. Gidstedt, *et al.*, A comparison of adsorption of organic micropollutants onto activated carbon following chemically enhanced primary treatment with microsieving, direct membrane filtration and tertiary treatment of municipal wastewater, *Sci. Total Environ.*, 2022, **811**, 152225.
5. S. A. Sadeek, E. A. Mohammed, M. Shaban, M. T. H. Abou Kana and N. A. Negm, Synthesis, characterization and catalytic performances of activated carbon-doped transition metals during biofuel production from waste cooking oils, *J. Mol. Liq.*, 2020, **306**, 112749.
6. D. R. Lobato-Peralta, D. M. Arias and P. U. Okoye, Polymer superabsorbent from disposable diaper as a sustainable precursor for the development of stable supercapacitor electrode, *J. Energy Storage*, 2021, **40**, 102760.

7. R. Mustapha, M. H. C. Harun, A. Manas, A. Ali and S. Hamzah, Preparation and characterization of bio-adsorbent from coconut husk for remazol red dye removal, *Biointerface Res. Appl. Chem.*, 2021, **11**, 10006–10015.

8. E. I. Knerelman, Y. A. Karozina, I. G. Shunina and I. V. Sedov, Highly Porous Materials as Potential Components of Natural Gas Storage Systems: Part 1 (A Review), *Pet. Chem.*, 2022, **62**, 561–582.

9. M. Beckner and A. Dailly, A pilot study of activated carbon and metal–organic frameworks for methane storage, *Appl. Energy*, 2016, **162**, 506–514.

10. K. V. Kumar, K. Preuss, M. M. Titirici and F. Rodríguez-Reinoso, Nanoporous Materials for the Onboard Storage of Natural Gas, *Chem. Rev.*, 2017, **117**, 1796–1825.

11. J. P. Marco-Lozar, M. Kunowsky, J. D. Carruthers and Á. Linares-Solano, Gas storage scale-up at room temperature on high density carbon materials, *Carbon*, 2014, **76**, 123–132.

12. I. E. Men'shchikov, *et al.*, Adsorption accumulation of natural gas based on microporous carbon adsorbents of different origin, *Adsorption*, 2017, **23**, 327–339.

13. Z. Nie, Y. Lin and X. Jin, Research on the theory and application of adsorbed natural gas used in new energy vehicles: A review, *Front. Mech. Eng.*, 2016, **11**, 258–274.

14. N. Bimbo, *et al.*, Kinetics and enthalpies of methane adsorption in microporous materials AX-21, MIL-101 (Cr) and TE7, *Chem. Eng. Res. Des.*, 2021, **169**, 153–164.

15. L. E. Kanonchik and L. L. Vasiliev, Charge dynamics of a low-pressure natural gas accumulator with solid adsorbent, novel thermosyphon and recirculation loop, *Int. J. Heat Mass Transfer*, 2019, **143**, 118374.

16. M. Prosniewski, *et al.*, Effect of cycling and thermal control on the storage and dynamics of a 40-L monolithic adsorbed natural gas tank, *Fuel*, 2019, **244**, 447–453.

17. J. Wieme, *et al.*, Thermal Engineering of Metal-Organic Frameworks for Adsorption Applications: A Molecular Simulation Perspective, *ACS Appl. Mater. Interfaces*, 2019, **11**, 38697–38707.

18. A. Sáez and M. Toledo, Thermal effect of the adsorption heat on an adsorbed natural gas storage and transportation systems, *Appl. Therm. Eng.*, 2009, **29**, 2617–2623.

19. D. Ybyraiymkul, K. C. Ng and A. Kaltayev, Experimental and numerical study of effect of thermal management on storage capacity of the adsorbed natural gas vessel, *Appl. Therm. Eng.*, 2017, **125**, 523–531.

20. P. K. Sahoo, *et al.*, Influence of exhaust gas heating and L/D ratios on the discharge efficiencies for an activated carbon natural gas storage system, *Appl. Energy*, 2014, **119**, 190–203.

21. P. Bénard and R. Chahine, Carbon nanostructures for hydrogen storage, *Solid-State Hydrogen Storage Materials and Chemistry*, 2008, pp. 261–287.

22. A. V. Shkolin, *et al.*, Experimental study and numerical modeling: Methane adsorption in microporous carbon adsorbent over the subcritical and supercritical temperature regions, *Prot. Met. Phys. Chem. Surf.*, 2016, **52**, 955–963.

23. I. I. El-Sharkawy, M. H. Mansour, M. M. Awad and R. El-Ashry, Investigation of Natural Gas Storage through Activated Carbon, *J. Chem. Eng. Data*, 2015, **60**, 3215–3223.

24. I. E. Men'shchikov, *et al.*, Thermodynamic behaviors of adsorbed methane storage systems based on nanoporous carbon adsorbents prepared from coconut shells, *Nanomaterials*, 2020, **10**, 1–26.

25. J. P. Marco-Lozar, M. Kunowsky, F. Suárez-García, J. D. Carruthers and A. Linares-Solano, Activated carbon monoliths for gas storage at room temperature, *Energy Environ. Sci.*, 2012, **5**, 9833–9842.

26. Q. R. Zheng, A. Z. Gu, X. S. Lu and W. S. Lin, Adsorption equilibrium of supercritical hydrogen on multi-walled carbon nanotubes, *J. Supercrit. Fluids*, 2005, **34**, 71–79.

27. L. Zhou and Y. Zhou, A comprehensive model for the adsorption of supercritical hydrogen on activated carbon, *Ind. Eng. Chem. Res.*, 1996, **35**, 4166–4168.

28. J. E. Sharpe, *et al.*, Supercritical hydrogen adsorption in nanostructured solids with hydrogen density variation in pores, *Adsorption*, 2013, **19**, 643–652.

29. B. B. Saha, S. Jribi, S. Koyama and I. I. El-Sharkawy, Carbon dioxide adsorption isotherms on activated carbons, *J. Chem. Eng. Data*, 2011, **56**, 1974–1981.

30. J. Tóth, State Equation of the Solid-Gas Interface Layers, 1971.

31. M. M. Dubinin, The Potential Theory of Adsorption of Gases and Vapors for Adsorbents with Energetically Nonuniform Surfaces, *Chem. Rev.*, 1960, **60**, 235–241.

32. R. Sips, On the structure of a catalyst surface, *J. Chem. Phys.*, 1948, **16**, 490–495.

33. R. Sips, On the structure of a catalyst surface II, *J. Chem. Phys.*, 1950, **18**, 1024–1026.

34. J. M. Honig and L. H. Reyerson, Adsorption of nitrogen, oxygen, and argon on rutile at low temperatures; applicability of the concept of surface heterogeneity, *J. Phys. Chem.*, 1952, **56**, 140–144.

35. S. Z. Naji and C. T. Tye, A review of the synthesis of activated carbon for biodiesel production: Precursor, preparation, and modification, *Energy Convers. Manage.: X*, 2022, **13**, 100152.

36. R. T. Ayinla, *et al.*, A review of technical advances of recent palm bio-waste conversion to activated carbon for energy storage, *J. Clean. Prod.*, 2019, **229**, 1427–1442.

37. J. Cheng, *et al.*, Comparison of activated carbons prepared by one-step and two-step chemical activation process based on cotton stalk for supercapacitors application, *Energy*, 2021, **215**, 119144.

38. M. Om Prakash, G. Raghavendra, S. Ojha and M. Panchal, Characterization of porous activated carbon prepared from arhar stalks by single step chemical activation method, *Mater. Today Proc.*, 2021, **39**, 1476–1481.

39. Y. Yang and F. S. Cannon, Biomass activated carbon derived from pine sawdust with steam bursting pretreatment; perfluorooctanoic acid and methylene blue adsorption, *Bioresour. Technol.*, 2022, **344**, 126161.

40. H. Arslanoğlu, Direct and facile synthesis of highly porous low cost carbon from potassium-rich wine stone and their application for high-performance removal, *J. Hazard. Mater.*, 2019, **374**, 238–247.

41. A. E. Ogungbenro, D. V. Quang, K. A. Al-Ali, L. F. Vega and M. R. M. Abu-Zahra, Physical synthesis and characterization of activated carbon from date seeds for CO_2 capture, *J. Environ. Chem. Eng.*, 2018, **6**, 4245–4252.

42. A. R. Hidayu and N. Muda, Preparation and Characterization of Impregnated Activated Carbon from Palm Kernel Shell and Coconut Shell for CO_2 Capture, *Procedia Eng.*, 2016, **148**, 106–113.

43. M. Ma, H. Ying, F. Cao, Q. Wang and N. Ai, Adsorption of congo red on mesoporous activated carbon prepared by CO_2 physical activation, *Chin. J. Chem. Eng.*, 2020, **28**, 1069–1076.

44. M. J. Prauchner, K. Sapag and F. Rodríguez-Reinoso, Tailoring biomass-based activated carbon for CH_4 storage by combining chemical activation with H_3PO_4 or $ZnCl_2$ and physical activation with CO_2, *Carbon*, 2016, **110**, 138–147.

45. N. M. Musyoka, *et al.*, Synthesis of activated carbon from high-carbon coal fly ash and its hydrogen storage application, *Renewable Energy*, 2020, **155**, 1264–1271.

46. T. Yumak, *et al.*, Comparison of the electrochemical properties of engineered switchgrass biomass-derived activated carbon-based EDLCs, *Colloids Surf., A*, 2020, **586**, 124150.

47. J. Chaparro-Garnica, D. Salinas-Torres, M. J. Mostazo-López, E. Morallón and D. Cazorla-Amorós, Biomass waste conversion into low-cost carbon-based materials for supercapacitors: A sustainable approach for the energy scenario, *J. Electroanal. Chem.*, 2021, **880**, 114899.

48. E. Yagmur, Y. Gokce, S. Tekin, N. I. Semerci and Z. Aktas, Characteristics and comparison of activated carbons prepared from oleaster (Elaeagnus angustifolia L.) fruit using KOH and $ZnCl_2$, *Fuel*, 2020, **267**, 117232.

49. H. Ait Ahsaine, M. Zbair, Z. Anfar, Y. Naciri, R. El Haouti, N. El Alem and M. Ezahri, Cationic dyes adsorption onto high surface area 'almond shell' activated carbon: Kinetics, equilibrium isotherms and surface statistical modeling, *Mater. Today Chem.*, 2018, **8**, 121–132.

50. G. Singh, *et al.*, A combined strategy of acid-assisted polymerization and solid state activation to synthesize functionalized nanoporous activated biocarbons from biomass for CO_2 capture, *Microporous Mesoporous Mater.*, 2018, **271**, 23–32.

51. M. A. Islam, M. J. Ahmed, W. A. Khanday, M. Asif and B. H. Hameed, Mesoporous activated carbon prepared from NaOH activation of rattan (Lacosperma secundiflorum) hydrochar for methylene blue removal, *Ecotoxicol. Environ. Saf.*, 2017, **138**, 279–285.

52. Y. Luo, F. Zhang, C. Li and J. Cai, Biomass-based shape-stable phase change materials supported by garlic peel-derived porous carbon for thermal energy storage, *J. Energy Storage*, 2022, **46**, 103929.

53. S. Schaefer, *et al.*, Oxygen-promoted hydrogen adsorption on activated and hybrid carbon materials, *Int. J. Hydrogen Energy*, 2020, **45**, 30767–30782.

54. M. Kim, *et al.*, Sorghum biomass-derived porous carbon electrodes for capacitive deionization and energy storage, *Microporous Mesoporous Mater.*, 2021, **312**, 110757.

55. P. Manasa, Z. J. Lei and F. Ran, Biomass Waste Derived Low Cost Activated Carbon from Carchorus Olitorius (Jute Fiber) as Sustainable and Novel Electrode Material, *J. Energy Storage*, 2020, **30**, 101494.

56. J. Serafin, *et al.*, Conversion of fruit waste-derived biomass to highly microporous activated carbon for enhanced CO_2 capture, *Waste Manage.*, 2021, **136**, 273–282.

57. C. Zhang, *et al.*, Enhancement of CO_2 Capture on Biomass-Based Carbon from Black Locust by KOH Activation and Ammonia Modification, *Energy Fuels*, 2016, **30**, 4181–4190.

58. R. Jothi Ramalingam, M. Sivachidambaram, J. J. Vijaya, H. A. Al-Lohedan and M. R. Muthumareeswaran, Synthesis of porous activated carbon powder formation from fruit peel and cow dung waste for modified electrode fabrication and application, *Biomass Bioenergy*, 2020, **142**, 105800.

59. J. Li, *et al.*, Selective preparation of biomass-derived porous carbon with controllable pore sizes toward highly efficient CO_2 capture, *Chem. Eng. J.*, 2019, **360**, 250–259.

60. M. J. Prauchner, K. Sapag and F. Rodríguez-Reinoso, Tailoring biomass-based activated carbon for CH_4 storage by combining chemical activation with H_3PO_4 or $ZnCl_2$ and physical activation with CO_2, *Carbon*, 2016, **110**, 138–147.

61. J. A. Okolie, *et al.*, Futuristic applications of hydrogen in energy, biorefining, aerospace, pharmaceuticals and metallurgy, *Int. J. Hydrogen Energy*, 2021, **46**, 8885–8905.

62. E. I. Epelle, *et al.*, A comprehensive review of hydrogen production and storage: A focus on the role of nanomaterials, *Int. J. Hydrogen Energy*, 2022, **47**, 20398–20431.

63. S. Askari and A. Jafari, A novel model for natural gas storage on carbon nanotubes, *Appl. Nanosci.*, 2020, **10**, 1115–1129.

64. A. Kumar, Adsorption of Methane on Activated Carbon by Volumetric Method, *Masters thesis, National Institute of Technology, Rourkela*, 2011.

65. International Organization for Standardization, ISO 697:1981, Surface Active Agents. Washing Powders, Determination of Apparent Density, Method by Measuring the Mass of a Given Volume, https://www.iso.org/standard/4897.html.

66. International Organization for Standardization, ISO 60:1977 Plastics. Determination of Apparent Density of Material that can be Poured from a Specified Funnel, https://www.iso.org/standard/3698.html.
67. K. Thu, Y. D. Kim, A. B. Ismil, B. B. Saha and K. C. Ng, Adsorption characteristics of methane on Maxsorb III by gravimetric method, *Appl. Therm. Eng.*, 2014, **72**, 200–205.
68. W. S. Loh, *et al.*, Improved isotherm data for adsorption of methane on activated carbons, *J. Chem. Eng. Data*, 2010, **55**, 2840–2847.
69. K. A. Rahman, *et al.*, Experimental adsorption isotherm of methane onto activated carbon at sub- and supercritical temperatures, *J. Chem. Eng. Data*, 2010, **55**, 4961–4967.
70. K. A. Rahman, *et al.*, Thermal enhancement of charge and discharge cycles for adsorbed natural gas storage, *Appl. Therm. Eng.*, 2011, **31**, 1630–1639.
71. J. C. Santos, F. Marcondes and J. M. Gurgel, Performance analysis of a new tank configuration applied to the natural gas storage systems by adsorption, *Appl. Therm. Eng.*, 2009, **29**, 2365–2372.
72. E. M. Strizhenov, *et al.*, Adsorption of methane on AU-5 microporous carbon adsorbent, *Prot. Met. Phys. Chem. Surf.*, 2013, **49**, 521–527.
73. M. S. Balathanigaimani, M. J. Lee, W. G. Shim, J. W. Lee and H. Moon, Charge and discharge of methane on phenol-based carbon monolith, *Adsorption*, 2008, **14**, 525–532.
74. N. Tzabar and H. J. M. ter Brake, Adsorption isotherms and Sips models of nitrogen, methane, ethane, and propane on commercial activated carbons and polyvinylidene chloride, *Adsorption*, 2016, **22**, 901–914.
75. K. Kiełbasa, J. Sreńscek-Nazzal and B. Michalkiewicz, Impact of tailored textural properties of activated carbons on methane storage, *Powder Technol.*, 2021, **394**, 336–352.
76. A. Ramesh, M. Jeyavelan, J. A. A. Rajju Balan, O. N. Srivastava and M. S. Leo Hudson, Supercapacitor and room temperature H, CO_2 and CH_4 gas storage characteristics of commercial nanoporous activated carbon, *J. Phys. Chem. Solids*, 2021, **152**, 109969.
77. L. Romagnuolo, E. Frosina, F. Fortunato, A. Andreozzi and A. Senatore, 1D model for *n*-butane adsorption and thermal variation for EVAP canister of gasoline-fueled vehicles: Validation with experimental results and DFSS optimization, *Appl. Therm. Eng.*, 2022, **209**, 118267.
78. UNECE. Global Technical Regulation No. 19 (EVAPorative emission test procedure for the Worldwide harmonized Light vehicle Test Procedure (WLTP EVAP)), 2004.
79. M. Zgrzebnicki, A. Kałamaga and R. Wrobel, Sorption and textural properties of activated carbon derived from charred beech wood, *Molecules*, 2021, **26**, 7604.
80. K. Sato, N. Kobayashi and M. Hasatani, Activated Carbon Performance Experiment, *SAE Tech. Pap. Ser.*, 2008, **1**, 409–416.
81. L. Romagnuolo, *et al.*, Experimental adsorption and desorption characterization of a gasoline-fueled vehicle carbon canister for European

application filled with *n*-butane and nitrogen mixtures, *E3S Web Conf.*, 2020, **197**, 06016.

82. H. M. Lee, *et al.*, Effects of Pore Structure on *n*-Butane Adsorption Characteristics of Polymer-Based Activated Carbon, *Ind. Eng. Chem. Res.*, 2019, **58**, 736–741.

83. S. B. Yahia and A. Ouederni, Hydrocarbons Gas Storage on Activated Carbons, *Int. J. Chem. Eng. Appl.*, 2012, 220–227.

84. X. Bai, *et al.* Adsorption/Desorption in a Carbon Canister, *SAE Technical Paper Series*, 2018.

85. L. Pingkuo and H. Xue, Comparative analysis on similarities and differences of hydrogen energy development in the World's top 4 largest economies: A novel framework, *Int. J. Hydrogen Energy*, 2022, **47**, 9485–9503.

86. M. Doğan, P. Sabaz, Z. Bicil, B. Koçer Kizilduman and Y. Turhan, Activated carbon synthesis from tangerine peel and its use in hydrogen storage, *J. Energy Inst.*, 2020, **93**, 2176–2185.

87. N. Kostoglou, *et al.*, Nanoporous polymer-derived activated carbon for hydrogen adsorption and electrochemical energy storage, *Chem. Eng. J.*, 2022, **427**, 131730.

88. N. Bader and A. Ouederni, Functionalized and metal-doped biomass-derived activated carbons for energy storage application, *J. Energy Storage*, 2017, **13**, 268–276.

89. M. Jordá-Beneyto, F. Suárez-García, D. Lozano-Castelló, D. Cazorla-Amorós and A. Linares-Solano, Hydrogen storage on chemically activated carbons and carbon nanomaterials at high pressures, *Carbon*, 2007, **45**, 293–303.

90. M. Mohan, V. K. Sharma, E. A. Kumar and V. Gayathri, Hydrogen storage in carbon materials, *Energy Storage*, 2019, 1–26.

91. H. T. Hwang and A. Varma, Hydrogen storage for fuel cell vehicles, *Curr. Opin. Chem. Eng.*, 2014, **5**, 42–48.

92. A. S. Oberoi, J. Andrews, A. L. Chaffee and L. Ciddor, Hydrogen storage capacity of selected activated carbon electrodes made from brown coal, *Int. J. Hydrogen Energy*, 2016, **41**, 23099–23108.

93. J. Andrews, R. Ojha, S. M. Rezaei Niya and S. Seibt, Electrochemical storage reactions of hydrogen in activated carbon from phenolic resin, *Catal. Today*, 2021, **397–399**, 155–164.

94. J. Andrews, S. M. Rezaei Niya and R. Ojha, Electrochemical hydrogen storage in porous carbons with acidic electrolytes: Uncovering the potential, *Curr. Opin. Electrochem.*, 2022, **31**, 100850.

95. S. Seifi and S. Masoum, Significantly enhanced electrochemical hydrogen storage performance of biomass nanocomposites from Pistacia Atlantica modified by CuO nanostructures with different morphologies, *Int. J. Hydrogen Energy*, 2021, **46**, 8078–8090.

96. S. H. Md Arshad, *et al.*, Preparation of activated carbon from empty fruit bunch for hydrogen storage, *J. Energy Storage*, 2016, **8**, 257–261.

97. Y. J. Heo and S. J. Park, Synthesis of activated carbon derived from rice husks for improving hydrogen storage capacity, *J. Ind. Eng. Chem.*, 2015, **31**, 330–334.

98. T. Ramesh, N. Rajalakshmi and K. S. Dhathathreyan, Activated carbons derived from tamarind seeds for hydrogen storage, *J. Energy Storage*, 2015, **4**, 89–95.

99. T. S. Blankenship, N. Balahmar and R. Mokaya, Oxygen-rich microporous carbons with exceptional hydrogen storage capacity, *Nat. Commun.*, 2017, **8**, 1545.

100. S. S. Samantaray, S. R. Mangisetti and S. Ramaprabhu, Investigation of room temperature hydrogen storage in biomass-derived activated carbon, *J. Alloys Compd.*, 2019, **789**, 800–804.

101. S. Hajialigol and S. Masoum, Promising electrochemical hydrogen storage properties of nano biomass derived from walnut shell, *Int. J. Hydrogen Energy*, 2019, **44**, 10713–10721.

102. G. Li, *et al.*, Preparation and characterization of the hydrogen storage activated carbon from coffee shell by microwave irradiation and KOH activation, *Int. Biodeterior. Biodegrad.*, 2016, **113**, 386–390.

103. S. Stelitano, *et al.*, Pinecone-derived activated carbons as an effective medium for hydrogen storage, *Energies*, 2020, **13**, 1–16.

104. H. Jin, Y. S. Lee and I. Hong, Hydrogen adsorption characteristics of activated carbon, *Catal. Today*, 2007, **120**, 399–406.

105. N. M. Musyoka and B. K. Mutuma, Onion-derived activated carbons with enhanced surface area for improved hydrogen storage and electrochemical energy application, *RSC Adv.*, 2020, **10**, 26928–26936.

106. X. Liu, C. Zhang, Z. Geng and M. Cai, High-pressure hydrogen storage and optimizing fabrication of corncob-derived activated carbon, *Microporous Mesoporous Mater.*, 2014, **194**, 60–65.

107. N. Bader and A. Ouederni, Optimization of biomass-based carbon materials for hydrogen storage, *J. Energy Storage*, 2016, **5**, 77–84.

108. G. Nazir, *et al.*, Heteroatoms-doped hierarchical porous carbons: Multifunctional materials for effective methylene blue removal and cryogenic hydrogen storage, *Colloids Surf., A*, 2021, **630**, 127554.

109. A. I. Sultana and M. T. Reza, Investigation of hydrothermal carbonization and chemical activation process conditions on hydrogen storage in loblolly pine-derived superactivated hydrochars, *Int. J. Hydrogen Energy*, 2022, **47**, 26422–26434.

110. Y. Xiao, *et al.*, Melaleuca bark based porous carbons for hydrogen storage, *Int. J. Hydrogen Energy*, 2014, **39**, 11661–11667.

111. M. Sevilla, A. B. Fuertes and R. Mokaya, Preparation and hydrogen storage capacity of highly porous activated carbon materials derived from polythiophene, *Int. J. Hydrogen Energy*, 2011, **36**, 15658–15663.

112. H. W. Kroto, J. R. Heath, S. C. O'Brien, R. F. Curl and R. E. Smalley, C_{60}: Buckminsterfullerene, *Nature*, 1985, **318**, 162–163.

113. P. Mauron, *et al.*, Reversible hydrogen absorption in sodium intercalated fullerenes, *Int. J. Hydrogen Energy*, 2012, **37**, 14307–14314.

114. L. S. Karpushenkava, G. J. Kabo and V. V. Diky, Thermodynamic properties and hydrogen accumulation ability of fullerene hydride $C_{60}H_{36}$, *Fullerenes, Nanotubes, Carbon Nanostruct.*, 2007, **15**, 227–247.

115. M. C. Tellez-Juárez, *et al.*, Hydrogen storage in activated carbons produced from coals of different ranks: Effect of oxygen content, *Int. J. Hydrogen Energy*, 2014, **39**, 4996–5002.

116. W. Zhao, *et al.*, Optimization of activated carbons for hydrogen storage, *Int. J. Hydrogen Energy*, 2011, **36**, 11746–11751.

117. W. Zhao, V. Fierro, N. Fernández-Huerta, M. T. Izquierdo and A. Celzard, Impact of synthesis conditions of KOH activated carbons on their hydrogen storage capacities, *Int. J. Hydrogen Energy*, 2012, **37**, 14278–14284.

118. M. Takht Ravanchi, T. Kaghazchi and A. Kargari, Application of membrane separation processes in petrochemical industry: a review, *Desalination*, 2009, **235**, 199–244.

119. N. Lamia, *et al.*, Adsorption of propane, propylene and isobutane on a metal-organic framework: Molecular simulation and experiment, *Chem. Eng. Sci.*, 2009, **64**, 3246–3259.

120. A. Ortiz, A. Ruiz, D. Gorri and I. Ortiz, Room temperature ionic liquid with silver salt as efficient reaction media for propylene/propane separation: Absorption equilibrium, *Sep. Purif. Technol.*, 2008, **63**, 311–318.

121. J. R. Alcántara-Avila, F. I. Gómez-Castro, J. G. Segovia-Hernández, K. I. Sotowa and T. Horikawa, Optimal design of cryogenic distillation columns with side heat pumps for the propylene/propane separation, *Chem. Eng. Process. Process. Intensif.*, 2014, **82**, 112–122.

122. S. Faramawy, T. Zaki and A. A. E. Sakr, Natural gas origin, composition, and processing: A review, *J. Nat. Gas Sci. Eng.*, 2016, **34**, 34–54.

123. B. N. Ho, *et al.*, Determination of methane, ethane and propane on activated carbons by experimental pressure swing adsorption method, *J. Nat. Gas Sci. Eng.*, 2021, **95**, 104124.

124. N. Tzabar, H. J. Holland, C. H. Vermeer and H. J. M. Ter Brake, Modeling the adsorption of mixed gases based on pure gas adsorption properties, *IOP Conf. Ser. Mater. Sci. Eng.*, 2015, **101**, 012169.

125. F. Birkmann, C. Pasel, M. Luckas and D. Bathen, Trace Adsorption of Ethane, Propane, and *n*-Butane on Microporous Activated Carbon and Zeolite 13X at Low Temperatures, *J. Chem. Eng. Data*, 2017, **62**, 1973–1982.

126. A. H. Johnstone, CRC Handbook of Chemistry and Physics, 69th Edition, Editor in Chief R. C. Weast, CRC Press Inc., Boca Raton, Florida, 1988, pp. 2400, price £57.50. ISBN 0-8493-0369-5, *J. Chem. Technol. Biotechnol.*, 2007, **50**, 294–295.

127. C. Mira-Hernández, J. A. Weibel, E. A. Groll and S. V. Garimella, Compressed-liquid energy storage with an adsorption-based vapor accumulator for solar-driven vapor compression systems in residential cooling, *Int. J. Refrig.*, 2016, **64**, 176–186.

CHAPTER 10

Trends and Perspectives Towards Activated Carbon and Activated Carbon-derived Materials in Environmental Catalysis Applications

YASMIN VIEIRA[a,b] AND GUILHERME LUIZ DOTTO*[a,b,c]

[a] Research Group on Adsorptive and Catalytic Process Engineering (ENGEPAC), Federal University of Santa Maria, Av. Roraima, 1000-7, 97105-900 Santa Maria, RS, Brazil; [b] Department of Chemistry, Federal University of Santa Maria, Av. Roraima, 1000-13, 97105-900 Santa Maria, RS, Brazil; [c] Department of Chemical Engineering, Federal University of Santa Maria, Av. Roraima, 1000-7, 97105-900 Santa Maria, RS, Brazil
*Email: guilherme_dotto@yahoo.com.br

10.1 Activated Carbons and Their Many Properties

Activated carbon (AC) is a versatile and environmentally friendly material with unique characteristics that can be tailored and designed for various specific purposes and applications.[1] Due to its intrinsic properties, such as large surface area, high porosity, electron conduction, and effective mass transport phenomenon, it has been widely explored in the areas of adsorption, electrochemistry, combustion enhancement, energy storage, and purification, as well as in many other chemical and engineering processes.[2,3] To date, many forms of AC have been developed, including powdered AC,

Activated Carbon: Progress and Applications
Edited by Chandrabhan Verma and Mumtaz A. Quraishi
© The Royal Society of Chemistry 2023
Published by the Royal Society of Chemistry, www.rsc.org

granular AC, graphite, carbon nanotubes (CNTs), and graphene (Figure 10.1). Furthermore, the environmental relevance of AC goes further than its application because the processes for its preparation can emerge from solving other environmental problems, such as residue recycling and reuse.[4]

The choices of precursors, the method of activation, and the control of processing conditions are the three main variables responsible for the characteristics of the final carbon product. Overall, AC and its forms can be prepared from organic materials rich in carbon, such as coal, lignite, wood, shells, peat, pitches, petrol residues, and biomass. Furthermore, the activation procedures fall within two major scopes: thermal and chemical activation. Finally, the control of processing is determined by the choice of carbonization method, which can include pyrolysis and hydrothermal carbonization, where the parameters temperature, residence time, gas atmosphere, pH, pressure, and heating rate can be modified to produce the desired form of AC.[5]

Even though carbonaceous materials have been employed in catalysis since the 1930s, their efficiency and critical role have not been studied in-depth or highlighted as it has been in recent years, resulting in the specific

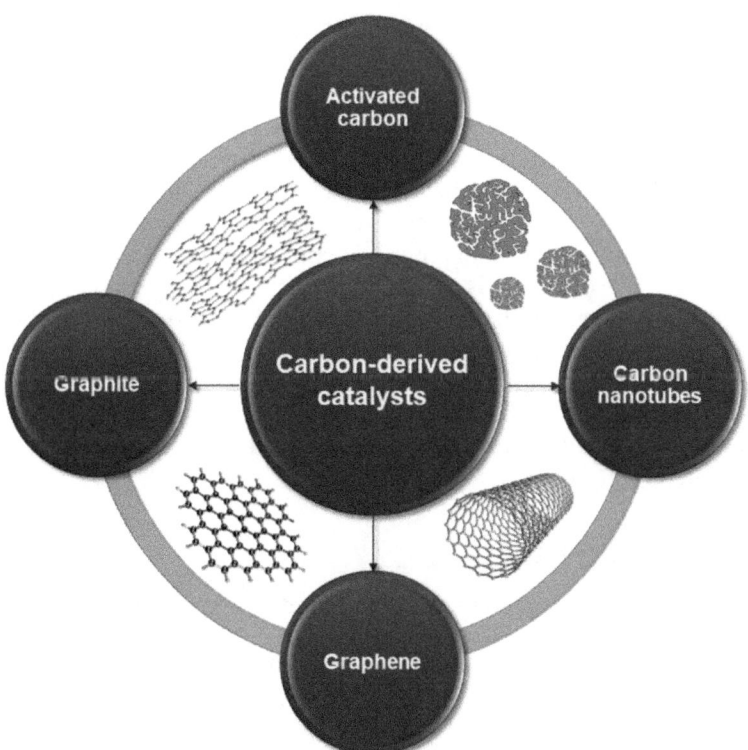

Figure 10.1 The main forms of carbonaceous materials in carbon-derived catalysts.

area nowadays known as carbocatalysis.[6] AC and AC-derived materials fall within the scope of heterogeneous catalysts, meaning that most of their catalytic efficiency arises from surface phenomena. Specifically, the catalytic process conducted using AC materials relies on the surface charges and functional groups that provide active sites on their surface, which are enhanced by the material's typically large surface area and porous structure.[2]

In synthetic procedures, AC has been applied successfully to dehydrogenation and alcohol dehydration.[7–9] However, the major applications of AC are focused on environmental chemistry, where its uses vary from adsorption to more specific catalytic reactions such as NO_x reduction,[10,11] SO_x oxidation,[12,13] ozonation,[14] and catalytic wet air oxidation.[15] Graphite has been used as a high yield catalyst in Friedel–Crafts acylation[16] and a substitute for strong Lewis acids in the alkylation of aromatic compounds and primary alcohols.[17] CNTs and modified CNTs have been employed for hydrogenation and dehydrogenation processes[18–21] and the decomposition and oxidation of NO, methane, and their derivatives.[22–24] Finally, graphene has been used to produce aldehydes and ketones from alcohols, alkenes, and alkynes, while graphene oxide (GO) has been found to promote four-electron oxygen transfer reactions.[25,26] In addition to the well-established applications described earlier, AC-derived catalysts are essential to the production of chlorination agents such as phosgene ($CoCl_2$) and thionyl chloride ($SOCl_2$) for the synthesis of glyphosate herbicide ($C_3H_8NO_5P$) and desulfurization of natural gas and universal oil, in addition to providing control of the reaction medium as well as being valuable synthesis building blocks.[27]

However, even though these compounds have been widely employed in the organic chemistry field, their mechanisms have not been as explored as it is nowadays, where novel applications have emerged. Recently, AC-derived catalysts have been employed for energy storage and conversion and in the degradation and remediation of emergent contaminants. They provided new angles on how these materials can be investigated. The next sections cover AC and AC-derived materials, including their different structures and forms, main characteristics, and preparation methods. Next, the most relevant and novel applications in the environmental catalysis field are summarized. Besides, a discussion is provided regarding future perspectives of its use by novel process intensification technologies, such as microwave (MW) and ultrasound (US) methods.

10.2 AC Catalysts

ACs are carbonaceous materials that have undergone some modification to reach the desired characteristics of high surface area and high porosity, consisting of at least 90% carbon.[1] The structural units of AC are composed of non-perfect carbon layer planes, which are cross-linked, resulting in an amorphous material with a highly accessible porous structure. The popular use and production of ACs date back many years, while the first known study reporting on the use of AC for adsorption purposes is from the 18th century.[28]

Even though initially ACs were obtained from coal and charcoal activation, the use of other starting materials rich in carbon, such as biomass residues, has become a common alternative.[5] The mechanisms behind AC catalysts have been correlated with several factors, ranging from surface area and functionality in heterogeneous catalysis to atmosphere composition in flow reactions.[29,30]

The production processes of AC are designed to produce granular, powdered, or shaped products, which can be divided into thermal activation and chemical activation. However, a carbonization methodology must be employed first to obtain the AC. The best methodology will depend on several characteristics of the starting material, such as its moisture, inorganic, and lignocellulosic content. Typically, pyrolysis and hydrothermal carbonization are the most used processes for carbonization.[31] Gasification can also yield AC, however, as a sub-product. In pyrolysis, as the name states, decomposing the organic constituents in an oxygen-free environment results in AC, where its specific properties can be controlled by altering the temperature, residence time, and heating ramp.[32] In hydrothermal carbonization, there is flexibility for using high moisture biomass. However, the morphology of the final product is not as controlled as it is in the product obtained by pyrolysis.[33]

After the carbonized material is obtained, its pore structure needs to be further enhanced by employing thermal or chemical activation. The activation is, in other words, a process to controllably destruct the carbonized structure by oxidation—successful carbonization results in a material with a high surface area and small pore volume. This process is similar to pyrolysis in thermal activation, but the atmosphere is oxygen-rich. Therefore, it is ideal for starting materials with low lignocellulosic content, such as natural coals. On the other hand, chemical activation involves use of strong dehydrating and oxidizing reagents, such as phosphoric acid and zinc chloride. Even though the temperature can vary, it is generally lower than those employed for the thermal process. Therefore, chemical activation is ideal for cellulose-rich materials, such as wood and sawdust, originating from a type of AC also known as biochar.[1,5] One of the advantages of chemical activation is that it allows using different oxidizing agents as specific morphology drivers to prepare selective catalysts. For example, it has been demonstrated that using HNO_3, H_2O_2, O_2, and N_2O acids to prepare ACs for dehydration and dehydrogenation reactions results in different ratios of obtained products for cyclohexanol conversion.[34]

Several studies report the successful conversion of different carbon sources to AC. Mesoporous and microporous ACs have been prepared by sol–gel and chemical activation methods employing olive stone biomass, and their performance was tested for the methoxylation of α-pinene. The prepared materials presented excellent conversion rates for the studied reaction, especially compared to a commercially available AC sample.[35] Residues of palm oil were used to produce sulfonated carbons, which acted as effective acid catalysts in the hydrolysis of cellobiose.[36] Acid-treated ACs have also been employed for other catalytic reactions, such as synthesizing *N*-alkylimidazoles and imidazolium ionic liquids.[37] Alkaline ACs have been used for radical species generation by ozone decomposition.[38] Despite being less common, the production of synthetic AC by polymerization of resorcinol and formaldehyde through carbonization is possible, and it is further subjected to thermal or chemical activation.[39] In the case of AC-produced materials, other pore-tailoring alternatives are possible, such as surfactants.[40] Amongst their possible uses, synthetic AC materials containing sulfonic acid groups have been reported as effective Friedländer catalysts, *i.e.*, for quinolone substitution reactions.[41]

10.2.1 AC-based Catalysts

To further enhance the catalytic properties of ACs, several research groups have explored their use as supports for carrying metallic particles of regular and nanometric sizes, resulting in AC-based catalysts.[42–45] The advantages of these combinations include an increase in the number of active sites, improvement in electron transfer phenomena, and smaller bandgap energies needed for their catalytic activation.[46–49] More specifically, in the case of AC-based composites, the high surface area of AC allows the composite to capture the pollutant to be degraded in areas nearer to the radical generation sites as a form of pre-concentration. There are many possible methods for loading metallic particles onto AC surfaces, such as impregnation, co-precipitation, the sol–gel method, the solvothermal method, polycondensation, hydrolysis, and methods assisted by MW and US technologies[50] (Figure 10.2). The choice of the method to be used will depend on the metallic particle chosen, as well as the stability and morphology of the AC since different methods and combinations of materials can result in novel composites with new unique properties.

In the impregnation method, the AC is put in contact with the desired metallic salt in a process reactor that can be dry, by mixing the powders, or in the aqueous phase. The obtained mixture is then calcined and dried, resulting in the final composite.[51] An acid-treated commercial AC impregnated with 5% (w/w) of iron was successfully employed for the production of phenol by benzene catalysis, resulting in a material with a surface area of 607.2 $m^2 g^{-1}$, capable of affording a product yield of 20%, which is about 10 times higher than the rates obtained by other catalysts such as Pd supported on SiO_2. The authors also investigated the effects of acid reflux and calcination temperature on iron incorporation and surface area. They observed that the acid reflux step was crucial to successfully incorporating the metallic particles into the AC. Besides, as the calcination temperature increased, the linking of iron to the AC possibly decreased, resulting in materials with larger surface areas and less catalytic activity.[52]

In the co-precipitation method, the metal salt and the AC are mixed in a solution that undergoes heating and then a strong base is added. During the process of lowering the pH, the electrostatic charges on the surface of the AC change, and the metal salts are precipitated from the solution, and attracted

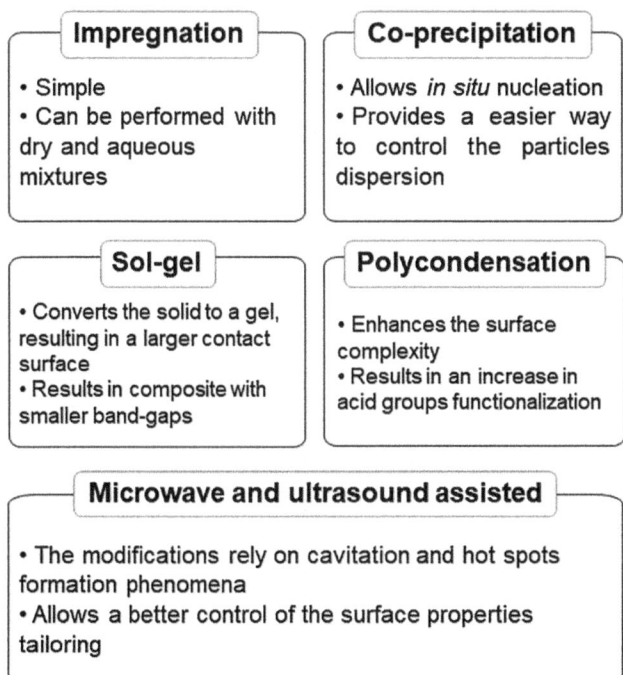

Impregnation

• Simple
• Can be performed with dry and aqueous mixtures

Co-precipitation

• Allows *in situ* nucleation
• Provides a easier way to control the particles dispersion

Sol-gel

• Converts the solid to a gel, resulting in a larger contact surface
• Results in composite with smaller band-gaps

Polycondensation

• Enhances the surface complexity
• Results in an increase in acid groups functionalization

Microwave and ultrasound assisted

• The modifications rely on cavitation and hot spots formation phenomena
• Allows a better control of the surface properties tailoring

Figure 10.2 Scheme representing the characteristics of the methods most commonly used in the production of carbon-based catalysts.

and linked to the AC surface by the difference in electrostatic potential.[51] Typically, this method produces magnetized AC, allowing magnetite to be synthesized directly onto the carbonaceous surface. In this process, the distribution of the metallic particles is more easily controlled than in other methods because the process is ruled by minor nucleation outbursts that kick-start a slow and controlled linking between carbon and metal. Even though this process seems simple, it is very efficient, successfully producing intricate supported AC materials, such as cross-linking magnetite to chitosan and biochar.[53]

The sol–gel method involves converting solids used to a gel during the synthetic process by employing an acid treatment to create an exfoliated surface capable of receiving the metallic particles, which are then calcined, returning to the powder state.[51,54] It is one of the most used methods for producing AC-based catalysts. It has been reported that this method optimized the bandgap of metallic catalysts already known for their low activation energy, such as copper ferrite. Furthermore, the combination of $CuFe_2O_4$ with AC produced with malt bagasse was able to decrease the electron/hole recombination, often observed for low bandgap catalysts, increasing the composite efficiency due to the synergic effects of charge dispersion and redistribution associated with a steric approximation of the molecules to be degraded to the radical species to be generated. In addition, the effects of the metallic catalyst to AC ratio were investigated. It was observed that the bandgap of the composite decreased as the AC content increased, which resulted in higher catalytic activity and efficiency.[49]

In the polycondensation process, AC is mixed with a polymeric precursor and metallic particles, which are then calcined, resulting in the desired composite.[51,54] This process is widely employed to produce metal-free AC materials, such as S-doped materials. It has been reported that different phosphorus oxyacids can act as polycondensation activators, influencing the addition content of acidic surface groups and the AC micropore volume. All the oxyacid activated materials presented high surface areas, up to 900 $m^2 g^{-1}$. The effects of orthophosphoric (H_3PO_4), pyrophosphoric ($H_4P_2O_7$), polyphosphoric ($H_6P_4O_{13}$), and phytic ($C_6H_{18}O_{24}P_6$) acids were investigated, and it was found that the acidic functionalities increased as the acid chain length increased.[55]

Finally, the methods based on US- and MW-assisted modifications rely on the effects of cavitation and hot spot formation, respectively, to modify the initial AC surface into receiving and linking itself with the metallic catalyst.[51,54] The use of US results in specific micro-jets and micro-liquid flows at high temperatures, enhancing the cavitation phenomenon resulting in larger surface areas and contributing to the formation of more stable suspensions of the metallic particles to be loaded onto the AC. By controlling time and frequency, it is also possible to control the regeneration capacity and the biological index of AC produced with biomass.[56] In addition to the impregnation of AC with metals, such as Mn,[57] US-assisted methods have been successfully employed for the S-doping of AC materials, resulting in composites with surface areas as high as 1500 $m^2 g^{-1}$.[58] On the other hand, the use of MWs provides specific heating spots and a more uniform heating rate, allowing better control for the tailoring of the surface properties.[59] Combinations of both methods are also possible, where US is used to clean surface impurities and enhance the dispersion and homogenization of the materials to be combined, and MWs are used for heating and surface linking.[60,61]

10.3 Graphite-based Catalysts

Graphite is the most stable form of naturally occurring carbon, consisting of stacked layers of single carbon atoms arranged in a two-dimensional lattice. Due to the weakness of the bonding between these layers, graphite is most commonly used as a precursor for graphene production by exfoliation. However, since natural graphite is found in large amounts worldwide, at very low costs, graphite can also provide a good hydrophobic surface which is very interesting for the catalysis of non-polar compounds. In addition to the possibility of its conversion to graphene, graphite can also be functionalized between layers and have its structure expanded to increase its porosity.[62] One downside of pristine graphite is the non-controlled and irregular surface morphology resulting from the weak interlayer bonding. However, it has been proposed that these structural defects can be corrected by Ar^+

sputtering followed by vacuum or oxygen annealing, resulting in highly oriented layered graphite.[63]

Pristine graphite presented excellent catalytic potential and selectivity in the Baeyer–Villiger oxidation. It was also superior to CNTs and AC in recyclability and reusability.[64] Graphite has also been successfully employed in other synthetic procedures, such as in Friedel–Crafts acylation of methoxybenzene, methylbenzene, and *o*-xylene, resulting in high yields.[65] Amongst the metal-supporting possibilities, Ni is the most explored metal for organic and organometallic reactions.[66] The methodology is simple, and the Ni nucleation onto the graphite surface can be achieved *in situ*.[67] The Ni–graphite composites have been used for the catalysis of several reactions, such as the reduction of aryl sulfonates, Suzuki coupling, as well as Negishi-coupling of vinylzirconocenes derived from terminal alkynes and aryl halides.[66]

Expanded graphite was prepared with flake graphite, concentrated sulfuric acid, and potassium permanganate, loaded with NiB. The catalyst was tested for the hydrogenation of sulfolene to sulfalone and compared to a standard Ni catalyst and NiB supported on MgO. The NiB supported on expanded graphite presented excellent catalytic conversion for the test reaction. Even though it contained over 6 times less Ni than the standard catalyst, it could reach the same results. Compared to the NiB supported on MgO, the NiB supported on expanded graphite showed superior conversion and selectivity rates, indicating that the expanded graphite plays a key role in the process.[62] Hydrogen generation on metal-supported graphite was also possible, demonstrating its high activity and durability.[68]

10.4 CNT-based Catalysts

CNTs consist of one or more layers of single carbon atoms arranged in a two-dimensional lattice rolled into a cylinder shape, with a typical diameter on the nanometric scale.[69] CNTs with a unique layer are named single-walled CNTs (SWCNTs).[70] When consisting of two or more layers, they are named multi-walled CNTs (MWCNTs).[71] The use of SWCNTs and MWCNTs due to their great electronic properties for electron transfer steps has been explored by employing these materials as metal-free catalysts. In addition, CNTs have shown even further interesting electronic effects differently from other carbon-based materials due to their tubular structure. Among these effects, it has been observed that the curvature of CNT walls results in a shift of π electron density flowing from the inner convex to its outer surface, culminating in a flowing shift of potential difference. Many studies have explored this specific property of CNTs, named confinement or tunnelling, by depositing metal particles inside and outside CNT walls.[72,73]

Moreover, CNTs also exhibit great chemical inertness and high oxidation stability. However, due to the high level of aggregation present in CNTs, some surface modifications are needed to reach their full catalytic performance. Even though in many cases the introduced surface groups seem to have become the active sites instead of the CNTs *per se*, the electronic mobility would not be as high as that if the CNT structure was not present.[74] CNTs as a support for other particles in catalysis were suggested shortly after their discovery, where Ru was supported onto nanotubes resulting in a catalyst used for cinnamaldehyde hydrogenation.[75] The doping of CNTs with less common metals, such as K, has been studied.[76,77]

Pt nanoparticles supported on SWCNTs successfully increased the oxidation performance of the electrocatalysis process for the oxidation of methanol and reduction of oxygen.[78] Similar results were also found for Pt supported onto MWCNTs.[79] Therefore, the combination of Pt and Ru has been explored to reduce Pt poisoning. In this case, the combined metals were supported onto CNTs, and their performance was investigated for electrocatalysis on a fuel cell. The synthetic procedure consisted of impregnation followed by thermal decomposition. The combination of materials resulted in composites with high surface area and increased power densities compared to the pristine materials.[80] CNTs have been successfully employed in organic chemistry for oxidative dehydrogenation reactions, such as conversion of ethylbenzene to styrene[81] and ethylbenzene to aceto-phenone,[82] hydrogenation of cyclohexene,[83] cyanosilylation of aldehydes,[84] and enantioselective hydrogenation of methyl 2-acetamidoacrylate and α-acetamidocinnamic acid, only to name a few.[85]

10.5   Graphene-based Catalysts

Undoubtedly, graphene is the most explored form of carbon-derived material in catalysis. Its use has emerged and been sustained in several research fields, being applied from energy conversion and storage to even optimization of resistance and weight of many materials, especially the ones used by armoured forces.[86–89] Graphene, which is the smallest structural unit from which all carbonaceous materials are made, consists of a two-dimensional sheet with a thickness of only one atom, where sp^2 bonded carbon atoms are packed and oriented in a honeycomb lattice.[90] Although thin, the graphene structure combines excellent mechanical strength and elasticity and exceptionally high electronic and thermal conductivities. It can be obtained by mechanical exfoliation, liquid-phase exfoliation, molecular self-assembly, and chemical vapor deposition. The choice will rely on the type of application and the limits of defects in the material for that purpose.[91]

Additionally, graphene can be converted to GO or reduced GO (rGO) to modify its electrical conductivity, hydrophilic behavior, mechanical strength, and dispersibility as desired and needed. The main difference between GO and rGO is the number of surface groups rich in oxygen, which is much higher for GO than for rGO, resulting in the former being more hydrophilic than the latter. As a result, the surface area of GO tends to be much smaller than rGO's, which is almost as high as that of pristine graphene. In addition, it has been observed that by converting GO to rGO, significant increases occur in electrical conductivity. However, even though the reduction process can significantly decrease the amount of oxygen-containing surface groups, it can never return to pristine graphene properties due to the inevitable presence of reduction residues.[92,93]

This diversity of characteristics caused by the different surface functionalities on so-called defects allows graphene to catalyze reactions as efficiently as reported. In summary, in addition to supporting oxygenated functional groups, carbon vacancies and holes, edge effects, and the presence of dopant elements, the catalytic activity of graphene, GO, and rGO has been attributed to their ability to adsorb molecular compounds and activate them by charge transfer. Compared to the previously cited carbon-derived supports for metallic particles, it is needless to say that the performance of graphene and its derivatives is far superior.[94]

However, in comparison to GO and rGO, pure and pristine graphene is somewhat limited and restricted for several catalytic applications.[95] From an electronic perspective, graphene is a zero bandgap catalyst with conduction and valence bands in the Brillouin zone, *i.e.*, the highest symmetry possible. Therefore, compared to GO and rGO, graphene is highly susceptible to losses due to electron/hole recombination.[96] Still, some processes such as heteroatom doping have been proven to be successful modifications for graphene. The doping of graphene with N atoms, for example, has also been proven to improve significantly common problems when working with Pt catalysts, such as catalyst poisoning by CO.[97] Graphene doped with B and N has been prepared by thermal annealing with boric acid and ammonia, resulting in a catalyst with superior activity compared to its commercial competitor composed of Pt and C, even in an alkaline environment.[98] In another study, the mechanisms behind N-doped graphene efficiency in alkaline media have been attributed to four- and two-electron reduction pathways.[25] Doping with S atoms has also been reported, demonstrating good results for hydrolysis, dehydration, and esterification of several organic compounds.[99–101]

It has been demonstrated that GO is an excellent catalyst for the oxidation of several alcohols and (Z)-stilbene and the hydration of various alkynes under mild conditions, reaching high yields of the desired products.[102] Furthermore, the impregnation of GO with Pd nanoparticles by simultaneous co-reduction resulted in a hybrid composite promoting Suzuki–Miyaura coupling reactions with less electrophilicity for the aryl bromide at lower Pd concentrations than conventionally used.[103] Monodispersion and immobilization of Pd nanoparticles in GO was also evaluated as an effective supporting methodology, resulting in size-dependent composites with high catalytic activity for liquid-phase hydrogenations of 3-hexyne and 4-octyne.[104] Similar efficiency results for the same catalytic reaction were obtained by employing the precursor $Pd(NH_3)_4(NO_3)_2$ as a substitute of Pd nanoparticles for GO impregnation through ion exchange, followed by a reduction in H_2 flow.[105] Moreover, due to its acidic groups, GO has been successfully employed in the Friedel–Crafts alkylation of indoles with Michael-acceptors,[106] aza-Michael addition of amines,[107] polymerization of cyclic-lactones,[108] oxidation of alcohols,[102] and photo-oxidation of tertiary amines.[109]

Contrarily to GO, the efficiency of rGO is attributed to diverse effects depending on the reaction. The reduction of nitrobenzenes, for example, has been attributed to the edge defects observed in the prepared rGO.[110] In contrast, the hydrolysis of the ester by sulfonated rGO was attributed to the presence of acidic groups.[101] Nitrobenzene reduction has also been reported with hydrazine hydrate catalyzed by rGO.[111] The hydrogenation of acetylene and alkenes in the petrochemical industry has been demonstrated even in ethylene-rich environments.[112] The metal oxides supported on rGO have further been investigated for substitution reactions of indole derivatives, such as 3-substituted indoles, with excellent selectivity for aryl iodide to ensure the formation of a single indole regioisomer.[113–115] A ZnO/rGO hybrid composite was prepared by self-assembly and *in situ* photoreduction to synthesize biologically active 3-substituted indoles, and it was found to be a highly recyclable heterogeneous catalyst with excellent yields in water.[116]

10.6   Environmental Catalysis Applications

Even though AC-derived catalysts have been widely employed since 1990 for organic catalysis, as demonstrated earlier, their use for environmental purposes started only around 20 years later, and the increase in the number of studies over the years is shown in Figure 10.3. In the environmental catalysis field, metal-based catalysis was very common due to the popularity of common reactions for pollutant degradation, such as Fenton-type reactions, which are very well established and effective. Since these metal-catalyzed reactions attended very well to the demands, there was no significant need for innovation in this field. However, scientific studies have demonstrated

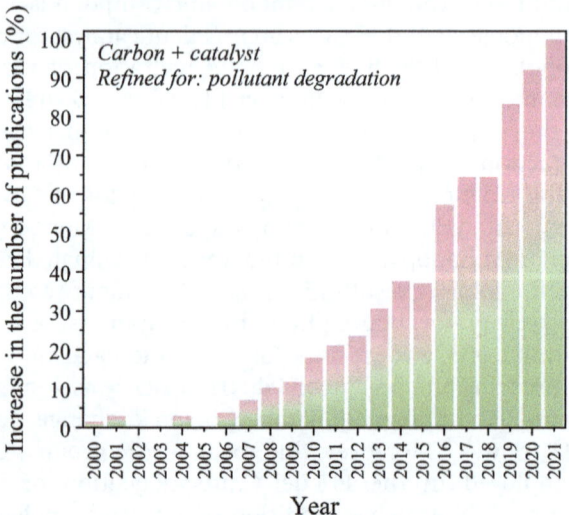

Figure 10.3   Interest increased from the year 2000 to 2021 in the field of carbocatalysis for environmental remediation applications.

the disadvantages of using pure metallic catalysts in the past years, such as the loss of efficiency due to the recombination of electron/hole pairs.

In addition to the scientific aspects that have boosted the development of novel catalytic processes for pollutant remediation, the current legislations in different parts of the world for the release of substances with harmful potential to the environment are becoming more rigorous, allowing each time fewer amounts of metallic ions in the final effluent. In this context, the development of catalysts with low toxicity and low metal leachate potential has emerged as an innovation trend, which has resulted in the development of metal-free carbon catalysts.

The following sections present the applications of such novel catalysts in environmental catalysis, emphasizing emerging contaminants and recalcitrant pollutant molecules, such as PPCPs, phenolic compounds, pesticides, and even microplastics. Although, over the years, several studies have been devoted to the degradation of common organic molecules, such as dyes, which is a theme already widely known and discussed, these pollutants will not be discussed here.

10.6.1 PPCPs

PPCPs include all compounds specifically designed to control any biological aspects for pharmaceuticals or in daily use products, such as fragrances and sunscreens. Several studies have shown how these substances have become distributed in various environmental compartments worldwide, including surface water, sea water, groundwater, soil, and biosolids. In addition to the risks due to its toxicity even at low concentrations, PPCPs can also bioaccumulate and, therefore, persist in the environment, modifying the natural regulation mechanisms.[117] Over the years, several studies have used carbon-derived materials as adsorbents to remove PPCPs. The use of metallic particles supported on different carbons has also been explored. Yet, the use of intrinsic metal-free carbon catalysts is a field in expansion.

MWCNTs modified with different levels of acidity/basicity were prepared and tested in the ozonation of sulfamethoxazole, an antibiotic widely used against both Gram-negative and -positive bacteria, which has been widely detected in various samples of water resources, posing risks to the environment equilibria. The catalytic performance of the prepared materials was compared to commercially available MWCNTs. The highest mineralization rate for sulfamethoxazole was achieved by the sample modified with nitric acid and carbonized at 900 °C. In comparison to the unmodified MWCNTs, the specific surface area of this sample increased from 331 to 529 $m^2 g^{-1}$. The authors linked the efficiency to the surface chemistry of the samples, and in the absence of oxygen-containing groups, an increase in degradation was observed.[118]

The catalytic ozonation using rGO with low structural defects was tested to degrade *p*-hydroxybenzoic acid, a paraben used in several personal care formulations such as sunscreens, repellents, and toothpaste. The influence of temperature used in the carbonization of the catalysts was also investigated, which showed that materials with surface areas of 305.4 and 265.4 $m^2 g^{-1}$ were obtained for 300 and 700 °C, respectively. However, the adsorption for both materials was negligible. The mechanistic studies performed by the authors revealed that the rGO catalysts converted the ozone molecules mainly to $^\bullet O_2^-$ and $^1O_2$. After approximately 15 min, both catalysts promoted the complete degradation of a total concentration of 20 $mg\,L^{-1}$ of *p*-hydroxybenzoic acid under low-dose ozone conditions, which was about 4 times faster compared to high-dose ozone alone.[119]

Diverse nanoscale carbon-derived catalysts with different dimensions, carbon types, and functional groups were tested for the oxidation of sulfachloropyridazine, an antimicrobial agent used to treat urinary tract infections. The carbon catalysts used were MWCNTs, annealed diamond nanopowders, GO, and rGO. The degradation of 20 $mg\,L^{-1}$ of sulfachloropyridazine was readily reached after approximately 20 min of reaction employing rGO as a catalyst. In addition, through radical quenching and theoretical modeling, the authors demonstrated that the pathways for pollutant degradation induced by carbocatalysis include much more non-radical-dependent oxidation routes, resulting in a globally more efficient process than most metallic catalysts.[120]

10.6.2   Phenol and Phenolic Compounds

Phenol is an important industrial commodity used as a precursor in preparing several other substances and products, such as bisphenol-A, aspirin, and herbicides. Even though it is necessary, the use of phenol in industry produces a large quantity of phenol-containing wastes, which can cause severe damage to the environment, especially aquatic ecosystems.[121] The toxicological reports for phenol show that it is not safe for animals or human beings, and poisoning can occur through skin absorption or ingestion. Despite the exposure route, the contact can culminate in detrimental health effects.[122,123] Thus, since 1977, phenol has been considered as a toxic substance and a potential chemical hazard.[121] However, in the last few years, the harmful effects of its accumulation have become more explicit, resulting in the need for its effective degradation to non-harmful compounds.[124]

The pristine carbon-derived catalysts SWCNTs, MWCNTs, rGO, mesoporous AC, fullerenes, nanodiamond powders, and graphitic carbon nitride were investigated to degrade phenol under persulfate-activated catalysis. The surface areas and oxygen contents of the used materials were 385.2 $m^2 g^{-1}$ and 1.6%, 253.6 $m^2 g^{-1}$ and 2.1%, 462.2 $m^2 g^{-1}$ and 6.0%, 1071.9 $m^2 g^{-1}$ and 8.8%, 4.8 $m^2 g^{-1}$ and 9.1%, 315.5 $m^2 g^{-1}$ and 4.1%, and 11.7 $m^2 g^{-1}$ and 1.1%, respectively. The adsorption contribution to the efficiency of the process increased as the dimensions of the structure of the carbons used increased, *i.e.*, from 0D to 3D. A higher oxidative potential was obtained for the reaction catalyzed with rGO, followed by the mesoporous-AC, SWCNTs, and MWCNTs. Lower efficiencies were obtained for the nanodiamond powders, followed by graphitic carbon nitride and fullerenes.[125]

Phenol degradation is also often commonly employed as a probe reaction to verify catalysis efficiency and mechanisms, not only as a target pollutant molecule to be degraded. According to theoretical calculations, the specific catalytic sites found on graphene favor the generation of persulfate radicals over other reactive species by activating the O–O bond in peroxymonosulfate. The efficiency related to this process is intrinsically linked to the absence of O-containing surface groups attached to graphene, GO, or rGO. It has been demonstrated that the removal of O% can be achieved by a simple thermal treatment, where temperatures between 700 and 900 °C seem to attain higher efficiencies. Higher defect degrees also play an important role in persulfate activation. This property can also be increased by using higher carbonization temperatures, which will cause the structure of GO and rGO to collapse, resulting in a smaller surface area. Therefore, this process needs optimization.[126]

It has been suggested that simple modifications such as N-doping can significantly improve the catalytic performance of carbon-derived catalysts. The addition of a quaternary N atom can dramatically increase the electron transfer capacity of graphene, for example, in addition to lowering the energy needed for surface interactions to occur.[127] In addition to N-doping, other elements have been explored, such as B and P; however, no positive effects were observed toward catalytic activation for phenol degradation. It has been reported that doping with B or P resulted in materials with a catalytic removal capacity similar to the adsorption efficiency of N-doped graphene, suggesting that doping with B and P cannot improve the electron transfer capacity of the structural carbon net.[128] Contrarily to the capacity of the conventional structure for enhancing persulfate radical generation, the delocalization of electrons found in graphene and its derivative structures when N-doped has been reported to enhance conventional oxidative processes for the generation of the reactive species $^\bullet OH$ and $O_2^{\bullet -}$ toward phenol degradation. Compared with GO and rGO, graphitic carbon nitride showed superior ozone decomposition capacity, resulting in a higher capacity for generating reactive oxygen species.[129]

Annealed GO doped with N and subsequently reduced, resulting in N-doped rGO, was investigated for phenol degradation in an aqueous system. The catalyst was prepared by a modified Hummers' method with melamine as the N precursor. After carbonization at 700 °C, the material reached a surface area of 227.5 $m^2 g^{-1}$, with micro- and mesoporous structures. The catalytic activation of the material was done using peroxymonosulfate (2 $g L^{-1}$), for an initial phenol concentration of 20 $mg L^{-1}$ and a catalyst dosage of 0.1 $g L^{-1}$. If compared to other carbon catalysts (including graphene nanosheets, SWCNTs, pristine rGO, as well as rGO and MWCNTs doped with N from ammonium nitrate precursors), the N-doped rGO prepared with melamine as N source presented much higher catalytic potential for phenol removal, reaching total degradation after approximately 20 min. The other materials reached a maximum of 60% of removal after 180 min. The N-doped rGO was even compared to a typical metal catalyst

(Co_3O_4), which took about 7.5 times longer to reach the same degradation level. Furthermore, the N-doped rGO exhibited a significant efficiency increase of approximately 80 times compared to pristine rGO, with excellent reusability and stability.[127]

The use of different N precursors in doping graphene-derived materials was investigated in catalytic wet air oxidation and ozonation employing phenol degradation. The authors used melamine and urea as the N sources, which resulted in catalysts with surface areas of 102 and 47 mg g^{-1}, respectively. The sample prepared with melamine reached an N content of 9.3%, while the sample prepared with urea reached 7.5%. The samples did not show significant differences in O% content. In both catalytic procedures employed, the sample prepared with melamine reached higher efficiencies than the one prepared with urea. Therefore, the results conclude that the N precursor can influence the reaction rate by modifying the pyrrolic and pyridinic content in the doping procedure. In contrast, the higher the pyridine/pyrrole ratio, the higher the catalytic activity.[130]

Even though graphene and its derivatives have several clear advantages over metallic catalysts, their major drawbacks are their high costs and the time needed for their carbonization. Typically, a standard carbonization procedure takes several hours due to the several stages of heating and cooling and the residence time. However, alternative methodologies such as MW-assisted carbonization can reduce the total time taken from 3 h to 1 min, resulting in higher production capacity. Therefore, two samples were prepared by employing the conventional and alternative methods, and were tested for the catalytic ozonation of a phenolic compound to verify the influence of the carbonization procedure on the morphology of N-doped rGO. The catalyst prepared by MW heating demonstrated a higher degree of graphitic character, a higher N-doping level, and a higher surface area than the sample prepared through conventional heating. These effects were attributed to the specific heating effects produced by MWs, which will be discussed in detail in the next section. Consequently, the pyridinic/pyrrolic N to graphitic N ratio tends to increase, resulting in a higher catalytic performance toward converting ozone to reactive oxygen species. After 15 min, complete degradation of the studied phenolic compound was achieved by the MW-carbonized N-doped rGO catalyst, with a total organic carbon conversion of 95%.[131]

In addition to the possibility of doping, co-doping (*i.e.*, doping with more than one heteroatom) strategies have also been explored for phenol degradation. Even though the co-doping with combinations of N and B or P or B and P was ineffective, the combination of N and S atoms onto the rGO lattice reduced the carbon inertness, improving its catalytic activity. After 20 min, total degradation considering an initial phenol concentration of 20 mg L^{-1}, was achieved by adding 0.4 g L^{-1} of the N and S co-doped catalyst. Still, a critical parameter observed for the material's performance was the optimal dosage of S loading, with improvement in the range of 0.05 to 0.3 g g^{-1} of the S precursor to GO. For amounts higher than 0.5 g g^{-1}, inverse effects were observed, supported by experimental and theoretical calculations.[132] Thus, when doping with non-metallic heteroatoms, the key to obtaining highly effective materials lies in the precursor source and the optimal dosage used in the synthetic procedure.

10.6.3 Pesticides

Pesticides are substances designed to control several pests, ranging from weeds to fungi or insects. The use of pesticides is necessary to attend to demands for food production, although it is widely known that these substances are harmful to all forms of life by triggering various deleterious mechanisms that include oxidative stress, growth alterations, metabolic inhibition, and mutagenic and genotoxic effects.[133,134] It is estimated that, without the use of pesticides, the production would decrease by 32%, 54%, and 78% for fruits, vegetables, and cereals, respectively.[135] Therefore, advanced and effective remediation technologies need to be implemented in addition to more controlled use. Furthermore, most pesticide molecules behave either as being devoid of charges or as zwitterions, *i.e.*, containing both positive and negative charges over a wide pH range. Therefore, it is very hard to control the interaction mechanisms. In this context, carbon-derived catalysts are promising alternatives due to their easily tailored hydrophobic or hydrophilic surface that can enhance their performance toward these types of molecules.

The combination of ozone and GO was investigated for *N,N*-diethyl-*m*-toluamide degradation, a process that typically presents some limitations, because the most effective degradation of this compound occurs by $^{\bullet}$OH attacks. GO dosage, pH variation, and common anions and matrices in reclaimed water were also investigated. The adsorption effects were found to be limited. After only 10 min, the addition of GO to the system was able to improve the ozonation efficiency from 40% to 95%, increasing the kinetic rate constant from 0.059 to 0.399 min^{-1}. Through radical quenching experiments, it was demonstrated that the yields of $^{\bullet}$OH generation dramatically increased with the addition of limited amounts of GO to the system, which did not improve as the catalyst dosage increased over 20 mg L^{-1}. The pH was also critical to the reaction, which was inhibited under acidic conditions. The presence of humic acid and bicarbonate in moderate amounts did not significantly alter the efficiency for *N,N*-diethyl-*m*-toluamide degradation, which was proven also considering real reclaimed water samples.[136]

Even though carbon-derived materials are usually presented as supports in composites, their highly ordered structure allows some of their shapes to be formed *in situ*, like carbon nanofibers. The combination of a ceramic monolith with nano-carbon structures was investigated to degrade atrazine and metolachlor herbicides in a batch system containing pharmaceuticals and phenolic compounds. Even though it was found that this combination of materials decreased the overall removal efficiency of the systems by decreasing the reaction rate, the mineralization of the pollutants increased in the global process.[137] The authors also investigated the efficiency of the carbon nanofibers supported on a cordierite monolith in continuous ozonation modes, showing that in a system containing metolachlor alone, the total organic carbon removal increased by 30%. At the same time, the toxicity of the final solution decreased by 15%.[138] Finally, n-doping was investigated as a final solution for the problems presented by carbon nanofibers grown on monoliths. This modification increased the process efficiency altogether, enhancing the carbon nanofiber activation, total organic carbon removal, and final toxicity decrease.[139]

The simultaneous degradation of the pesticides acetochlor, pretilachlor, butachlor, florasulam, clodinafop-propargyl, diuron, nicosulfuron, pyrazosulfuron-ethyl, halosulfuron-methyl, and rimsulfuron was investigated in a photocatalytic system with graphitic carbon nitride. The pesticide concentration in the system was 2.0 mg L^{-1}, while the catalyst dosage was 1 g L^{-1}, and the pH was set as 7. Even though the bandgaps obtained for all the catalyst samples were relatively high, ranging from 2.75 to 2.64 eV, their activation was achieved under visible light irradiation, resulting in significant degradation of the pesticides after 180 min. Furthermore, experiments including the presence of soil and microplastics were carried out to simulate the conditions under real environmental conditions. In an environment containing soil considering a light and dark cycle of 12 h, the degradation of the pesticides was reached after 5 days, indicating that the designed catalyst is effective. However, microplastics in the simulated system hindered the catalytic activity.[140]

10.6.4 Microplastics

Microplastics are plastic particles smaller than 5 μm, resulting from their specifical production for use in PPCPs or from the natural fragmentation that polymers undergo in the environment. However, the large amounts of plastics produced over the years associated with non-existing recycling policies and incorrect disposal resulted in the pollution that includes several complex interactions, such as interference in light penetration and adsorption of other contaminants present in the same medium.[141] Despite the mechanisms behind microplastic degradation and environmental interactions involved being complex and not yet fully disclosed, it is now better understood how extensive the contamination with these particles is. Furthermore, these particles have already found their way into the food chain. Therefore, the ingestion of microplastics is already a reality.[142]

In addition to several studies showing that the mobility of microplastics as vectors for carrying several pollutants through the environment is very high, its mobility inside several organisms has been investigated and proven to be true. Even though microplastics are solid particles, strikingly, studies show that plastic particles have the potential to permeate several lipidic membranes of cells, entering organisms and causing oxidative or inflammatory stresses at levels still unknown.[143] The translocation of plastic particles to and their deposition in fetal tissue previously demonstrated in rats was later observed in humans, where it was found in all related placental samples, including maternal, fetal, and amniochorial membranes. More recently, plastic particles smaller than 700 nm have been found in human blood samples, with a mean concentration of 1.6 $\mu g\,mL^{-1}$ of blood.[144]

It is estimated that, by 2050, there will be more plastic than fish in the ocean; therefore, its recovery and efficient methods for controlling this type of pollution need to emerge and be integrated as part of a long-term sustainable economy.[145,146] Due to the complexity of the polymeric chains of plastics, the use of carbon catalysts for their degradation has been somewhat limited. Most studies focus on using carbonaceous bases to improve the catalytic properties of metallic particles, such as MWCNTs impregnated with TiO_2 for polyethylene degradation. The combination improved the optical bandgap of pristine TiO_2, as expected, and improved its stability and dispersion in solution.[147] The combination of an N-enriched carbon source with metallic particles was also investigated through the developed nitride/nickel phosphide composite, which was employed for polyethylene terephthalate and polylactic acid photoreforming under alkaline conditions.[148]

Other mechanisms were also briefly explored, such as the potential application of peroxymonosulfate activation with CNTs with high-level nitrogen dopants and encapsulated metal nanoparticles. For polyethylene degradation, N-doped CNTs with Mn particles proved to exhibit superior catalytic performance to GO, pristine N-doped CNTs, and pristine MnO_2 particles. In addition, the composite presented low toxicity, and the fragments generated over the reaction course could be used as a carbon source for waterborne algae.[149] Moreover, the effects that the presence of carbon-derived materials have on plastic photodegradation are still under investigation and need to be explored before it is possible to say that carbon catalysts are a viable option in this case further.

10.7 Trending Process Intensification Technologies: The Use of MWs and US as Enhancing and Synergic Energy Sources for Carbocatalysis

In the "as fast as possible" era, where we are experiencing the greatest rates of scientific production and acquisition of knowledge and information, it can be stated that time is the key variable when choosing the best intensification energy source for any chemical process. In this context, MW and US energies have much to offer. Both technologies are already widely used for organic and material synthesis and have been explored for advanced oxidation processes. In the carbocatalysis scenario, each energy source can provide specific effects that can activate carbon-derived materials in different forms, resulting in the enhancement of reaction rates as well as critical reduction of the total time needed for the process to occur even if assisted with other energy sources, such as light.

MWs are found between infrared and radio waves in the electromagnetic spectrum. MW technology is used daily, such as in heating food in a domestic MW oven and telecommunication services. The transfer of energy provided by MWs occurs through the dielectric interaction of the waves with the polar moments of molecules, resulting in oscillations and electrical resistance that dissipate in the form of heat. As a result, it is possible to observe a fast and uniform overall heating rate.[150] The catalyst used needs to have, at the same time, a moderate dielectric constant (ε') and a high dielectric loss factor (ε''), where the former component represents how much energy is absorbed or reflected and the latter is how much the electrical energy is converted to heat. The ratio between $\varepsilon''/\varepsilon'$ is named the dielectric loss tangent ($\tan \delta$), which needs to be high to reflect the MW absorbing properties of the material.[151]

Even though it may seem that carbon-derived materials would not be good MW absorbers, their structural organization rich in delocalized π-electrons and the associated edge defects allow the unique phenomenon to occur. It has been observed that the $\tan \delta$ value increases as the stiffness of the sample decreases. For example, for coal, the $\tan \delta$ value ranges from 0.02 to 0.08, while for ACs, it varies from 0.57 to 0.8, being even higher as the temperature increases. Due to the high electronic mobility, the interaction results in a high increase in the kinetic energy of the electrons, resulting in localized ionization sites known as hot spots. Hot spots can sometimes be perceived visually as sparks. However, this phenomenon consists of the formation of microplasmas, whose temperature is incredibly higher than that in the remaining solution. Therefore, because of these specific interactions, carbon-derived catalysts have been employed for soil remediation, biomass pyrolysis, organic waste conversions, and environmental catalysis, where the most applications can be found.[152]

Several studies show that combining MWs and pristine carbon-derived materials can improve the efficiency of removing and degrading several pollutants, such as phenolic compounds, dyes, and pesticides. For example, the degradation of *p*-nitrophenol was demonstrated in a fixed bed reactor irradiated with MWs where AC was used as a catalyst. Considering an initial pollutant concentration of 1330 $mg\,L^{-1}$ and a MW power of 500 W, the system obtained a total removal of the starting compound of 90% and a total organic carbon removal of 80%.[153] In addition, the combination of AC and H_2O_2 was investigated for nitrobenzene degradation. A severe increase in reactive oxygen species production was reported when the reaction was assisted by MW irradiation.[154] The degradation of congo red[155] and acid orange[156] was also successfully achieved by employing the combination of AC powders and MW irradiation. According to these studies, the critical factors for dye degradation were catalyst dosage and irradiation time.[155,156]

The different structural factors regarding CNTs were investigated by studying structures with different diameters and evaluating their efficiency toward the degradation of the dyes methyl orange and methylene blue, the endocrine disruptors sodium dodecylbenzene sulfonate and bisphenol-A, and the pesticide methyl parathion. It was reported that as the diameter of the CNTs increased, the catalytic efficiency decreased. In a system considering $0.48 \, g \, L^{-1}$ of CNTs with diameters of 10–20 nm and a MW power of 500 W, the kinetic rate constants for the studied pollutants were calculated to be 0.334, 0.679, 0.726, 0.168, and 0.463 min^{-1} for 30.44 $\mu mol \, L^{-1}$ of methyl orange, methylene blue, sodium dodecyl benzene sulfonate, bisphenol-A, and methyl parathion, which are approximately 3 and 1.5 times superior to the kinetic rates calculated for the CNTs with diameters of 20–40 and 40–60 nm.[157]

Modifications of carbon-derived materials with metallic particles have been explored with MW irradiation. Carbon has been mostly employed as support, resulting in the need for fewer amounts of metal and, therefore, better cost-effectiveness. From the perspective of the intrinsic metal catalyst, the supporting of carbon-derived materials results in a composite with a lower bandgap, which is more suitable for MW applications since MWs are waves of low electromagnetic energy. On the other hand, if we consider the improvement of the carbonaceous structure as the standpoint, its enrichment with metallic particles can create interesting electron mobility sites. Then, the electrons from carbon can participate in the oxidation and reduction of the metallic species.

The combinations of various metallic oxides and sulfate forms of titanium, copper, cobalt, iron, and cerium with AC, MWCNTs, and rGO have been prepared and tested under MW assistance for the degradation of many pollutants.[158–161] For example, the combination of TiO_2 with AC and MWCNTs was successfully employed for methyl orange degradation,[160] while crystal violet and humic acid were degraded using copper, cerium, cobalt, and AC composites.[158,159] According to quenching effects, even though it would be expected that several reactive oxygen species would be generated, the predominance of $^{\bullet}OH$ species occurred in all systems, which is attributed to the intrinsic interactions of MW irradiation with H_2O_2 or the catalyst itself. It has also been shown that improvements obtained in catalytic activity under MW irradiation can even allow composites to reach significant pollutant degradation rates for real wastewater samples, as demonstrated by rGO and $CuFeS_2$. For example, due to synergistic effects, the total elapsed time for the degradation of direct black 22 dye and the removal of total organic carbon from a textile wastewater sample is approximately 7 min under MW irradiation. In addition to the ultra-fast treatment of a real wastewater sample, it was also reported that the phytotoxicity of the samples irradiated with MWs was much lower than for the samples where the catalytic activation occurred through heat or light.[161]

Contrarily to MWs, which are electromagnetic waves, US is, as the name states, composed of sound waves with frequencies over 20 kHz, which are capable of promoting specific interaction effects that are very interesting for the enhancement of catalytic processes. For example, the use of US in advanced oxidation processes demonstrates that US is a powerful tool for improving the partial oxidation of certain compounds, generating reactive oxygen species, and promoting cavitation effects that can be extended to the catalyst surface well.[162] On the other hand, the use of US is somewhat limited when it comes to carbocatalysis. Even though no specific effects are related to carbon-derived materials and US as in MW applications, US can promote the carbon surface's regeneration and help reactivate the so-called poisoned catalytic sites, which is very interesting for the recycling and reuse of catalysts.[163,164]

References

1. F. Çeçen, *Kirk-Othmer Encycl. Chem. Technol.*, 2014, 1–34.
2. I. Matos, M. Bernardo and I. Fonseca, *Catal. Today*, 2017, **285**, 194–203.
3. W. Xin and Y. Song, *RSC Adv.*, 2015, **5**, 83239–83285.
4. M. M. Titirici, R. J. White, N. Brun, V. L. Budarin, D. S. Su, F. Del Monte, J. H. Clark and M. J. MacLachlan, *Chem. Soc. Rev.*, 2014, **44**, 250–290.
5. Y. Vieira, J. M. N. dos Santos, J. Georgin, M. L. S. Oliveira, D. Pinto and G. L. Dotto, *Gondwana Res.*, 2022, **110**, 393–420.
6. D. R. Dreyer and C. W. Bielawski, *Chem. Sci.*, 2011, **2**, 1233–1240.
7. C. De, A. Holmen, Z. Sui and X. Zhou, *Chin. J. Catal.*, 2014, **35**, 824–841.
8. J. Bedia, R. Barrionuevo, J. Rodríguez-Mirasol and T. Cordero, *Appl. Catal., B*, 2011, **103**, 302–310.
9. C. Moreno-Castilla, F. Carrasco-Marín, C. Parejo-Pérez and M. V. López Ramón, *Carbon*, 2001, **39**, 869–875.
10. S. Raja, M. S. Alphin and L. Sivachandiran, *Catal. Sci. Technol.*, 2020, **10**, 7795–7813.
11. J. Li, D. Yin, D. Long, J. Wang, L. Ling and W. Qiao, *RSC Adv.*, 2016, **6**, 27272–27281.
12. Z. C. Kampouraki, D. A. Giannakoudakis, K. S. Triantafyllidis and E. A. Deliyanni, *Green Chem.*, 2019, **21**, 6685–6698.
13. J. S. Valente and R. Quintana-Solorzano, *Energy Environ. Sci.*, 2011, **4**, 4096–4107.
14. P. C. C. Faria, J. J. M. Órfão and M. F. R. Pereira, *Ind. Eng. Chem. Res.*, 2006, **45**, 2715–2721.
15. A. Rey, M. Faraldos, A. Bahamonde, J. A. Casas, J. A. Zazo and J. J. Rodríguez, *Ind. Eng. Chem. Res.*, 2008, **47**, 8166–8174.
16. M. Kodomari, Y. Suzuki and K. Yoshida, *Chem. Commun.*, 1997, 1567–1568.
17. G. A. Sereda, V. B. Rajpara and R. L. Slaba, *Tetrahedron*, 2007, **63**, 8351–8357.
18. Z. J. Liu, Z. Xu, Z. Y. Yuan, D. Lu, W. Chen and W. Zhou, *Catal. Lett.*, 2001, **72**, 203–206.
19. R. Giordano, P. Serp, P. Kalck, Y. Kihn, J. Schreiber, C. Marhic and J. L. Duvail, *Eur. J. Inorg. Chem.*, 2003, **2003**, 610–617.
20. A. Chambers, T. Nemes, N. M. Rodriguez and R. T. K. Baker, *J. Phys. Chem. B*, 1998, **102**, 2251–2258.
21. N. S. Fígoll, M. R. Sad, J. N. Beltramini, E. L. Jablonskl and J. M. Parera, *Ind. Eng. Chem. Prod. Res. Dev.*, 2002, **19**, 545–551.

22. J. Z. Luo, L. Z. Gao, Y. L. Leung and C. T. Au, *Catal. Lett.*, 2000, **66**(1), 91–97.
23. S. K. Shaikhutdinov, L. B. Avdeeva, B. N. Novgorodov, V. I. Zaikovskii and D. I. Kochubey, *Catal. Lett.*, 1997, **47**(1), 35–42.
24. C. A. Bessel, K. Laubernds, N. M. Rodriguez and R. T. K. Baker, *J. Phys. Chem. B*, 2001, **105**, 1121–1122.
25. S. Yasuda, L. Yu, J. Kim and K. Murakoshi, *Chem. Commun.*, 2013, **49**, 9627–9629.
26. C. Su and K. P. Loh, *Acc. Chem. Res.*, 2013, **46**, 2275–2285.
27. Y. Zhai, Z. Zhu and S. Dong, *ChemCatChem*, 2015, 7, 2806–2815.
28. P. Walden, *J. Chem. Educ.*, 1951, 304–308.
29. G. C. Grunewald and R. S. Drago, *J. Am. Chem. Soc.*, 1991, **113**, 1636–1639.
30. F. Rodríguez-Reinoso, *Carbon*, 1998, **36**, 159–175.
31. J. S. Cha, S. H. Park, S. C. Jung, C. Ryu, J. K. Jeon, M. C. Shin and Y. K. Park, *J. Ind. Eng. Chem.*, 2016, **40**, 1–15.
32. J. Zhang, J. Liu and R. Liu, *Bioresour. Technol.*, 2015, **176**, 288–291.
33. B. Hu, S. H. Yu, K. Wang, L. Liu and X. W. Xu, *Dalton Trans.*, 2008, 5414–5423.
34. I. F. Silva, J. Vital, A. M. Ramos, H. Valente, A. M. Botelho Do Rego and M. J. Reis, *Carbon*, 1998, **36**, 1159–1165.
35. I. Matos, M. F. Silva, R. Ruiz-Rosas, J. Vital, J. Rodríguez-Mirasol, T. Cordero, J. E. Castanheiro and I. M. Fonseca, *Microporous Mesoporous Mater.*, 2014, **199**, 66–73.
36. A. do, C. Fraga, C. P. B. Quitete, V. L. Ximenes, E. F. Sousa-Aguiar, I. M. Fonseca and A. M. B. Rego, *J. Mol. Catal. A: Chem.*, 2016, **422**, 248–257.
37. C. J. Durán-Valle, M. Madrigal-Martínez, M. Martínez-Gallego, I. M. Fonseca, I. Matos and A. M. Botelho Do Rego, *Catal. Today*, 2012, **187**, 108–114.
38. P. C. C. Faria, J. J. M. Órfão and M. F. R. Pereira, *Ind. Eng. Chem. Res.*, 2006, **45**, 2715–2721.
39. C. Lin and J. A. Ritter, *Carbon*, 1997, **35**, 1271–1278.
40. I. Matos, S. Fernandes, L. Guerreiro, S. Barata, A. M. Ramos, J. Vital and I. M. Fonseca, *Microporous Mesoporous Mater.*, 2006, **92**, 38–46.
41. J. Lõpez-Sanz, E. Pérez-Mayoral, E. Soriano, D. Omenat-Morán, C. J. Durán, R. M. Martín-Aranda, I. Matos and I. Fonseca, *ChemCatChem*, 2013, **5**, 3736–3742.
42. P. A. Ajibade and E. C. Nnadozie, *ACS Omega*, 2020, **5**, 32386–32394.
43. Y. Li, J. Chen, J. Liu, M. Ma, W. Chen and L. Li, *J. Environ. Sci.*, 2010, **22**, 1290–1296.
44. N. Welter, J. Leichtweis, S. Silvestri, P. I. Z. Sánchez, A. C. C. Mejía and E. Carissimi, *J. Alloys Compd.*, 2022, **901**, 163758.
45. J. Leichtweis, S. Silvestri, N. Stefanello and E. Carissimi, *Chemosphere*, 2021, **281**, 130987.
46. Z. He, S. Yang, Y. Ju and C. Sun, *J. Environ. Sci.*, 2009, **21**, 268–272.

47. P. Raizada, P. Singh, A. Kumar, G. Sharma, B. Pare, S. B. Jonnalagadda and P. Thakur, *Appl. Catal., A*, 2014, **486**, 159–169.
48. J. Leichtweis, S. Silvestri and E. Carissimi, *Biomass Bioenergy*, 2020, **140**, 105648.
49. J. Leichtweis, S. Silvestri, N. Welter, Y. Vieira, P. I. Zaragoza-Sánchez, A. C. Chávez-Mejía and E. Carissimi, *Process Saf. Environ. Prot.*, 2021, **150**, 497–509.
50. M. M. Mian and G. Liu, *RSC Adv.*, 2018, **8**, 14237–14248.
51. F. Pinna, *Catal. Today*, 1998, **41**, 129–137.
52. J. S. Choi, T. H. Kim, K. Y. Choo, J. S. Sung, M. B. Saidutta, S. O. Ryu, S. D. Song, B. Ramachandra and Y. W. Rhee, *Appl. Catal., A*, 2005, **290**, 1–8.
53. Y. Vieira, E. C. Lima and G. L. Dotto, *Nano-Biosorbents Decontam. Water, Air, Soil Pollut.*, 2022, 551–568.
54. M. E. Khan, *Nanoscale Adv.*, 2021, **3**, 1887–1900.
55. C. Cheng, J. Zhang, Y. Mu, J. Gao, Y. Feng, H. Liu, Z. Guo and C. Zhang, *J. Anal. Appl. Pyrolysis*, 2014, **108**, 41–46.
56. Y. Fu, X. Ding, J. Zhao and Z. Zheng, *Ultrason. Sonochem.*, 2020, **69**, 104921.
57. Y. F. Qu, J. X. Guo, Y. H. Chu, M. C. Sun and H. Q. Yin, *Appl. Surf. Sci.*, 2013, **282**, 425–431.
58. Z. B. Zhang, X. Y. Liu, D. W. Li, T. T. Gao, Y. Q. Lei, B. G. Wu, J. W. Zhao, Y. K. Wang and L. Wei, *New Carbon Mater.*, 2018, **33**, 409–416.
59. W. Ao, J. Fu, X. Mao, Q. Kang, C. Ran, Y. Liu, H. Zhang, Z. Gao, J. Li, G. Liu and J. Dai, *Renewable Sustainable Energy Rev.*, 2018, **92**, 958–979.
60. S. Cheng, L. Zhang, H. Xia, J. Peng, J. Shu and C. Li, *RSC Adv.*, 2016, **6**, 78936–78946.
61. X. Jiang, H. Xia, L. Zhang, J. Peng, S. Cheng, J. Shu, C. Li and Q. Zhang, *Powder Technol.*, 2018, **338**, 857–868.
62. W. Li, C. Han, W. Liu, M. Zhang and K. Tao, *Catal. Today*, 2007, **125**, 278–281.
63. A. V. Nartova, A. V. Bukhtiyarov, R. I. Kvon, E. M. Makarov, I. P. Prosvirin and V. I. Bukhtiyarov, *Surf. Sci.*, 2018, **677**, 90–92.
64. Y. F. Li, M. Q. Guo, S. F. Yin, L. Chen, Y. B. Zhou, R. H. Qiu and C. T. Au, *Carbon*, 2013, **55**, 269–275.
65. M. Kodomari, Y. Suzuki and K. Yoshida, *Chem. Commun.*, 1997, 1567–1568.
66. T. A. Butler, E. C. Swift and B. H. Lipshutz, *Org. Biomol. Chem.*, 2007, **6**, 19–25.
67. B. H. Lipshutz, B. A. Frieman, T. Butler and V. Kogan, *Angew. Chem., Int. Ed.*, 2006, **45**, 800–803.
68. Y. Liang, H. Bin Dai, L. P. Ma, P. Wang and H. M. Cheng, *Int. J. Hydrogen Energy*, 2010, **35**, 3023–3028.
69. W. R. Davis, R. J. Slawson and G. R. Rigby, *Nature*, 1953, **171**(4356), 756.
70. S. Iijima and T. Ichihashi, *Nature*, 1993, **363**(6430), 603–605.
71. S. Iijima, *Nature*, 1991, **354**(6348), 56–58.

72. J. Xiao, X. Pan, S. Guo, P. Ren and X. Bao, *J. Am. Chem. Soc.*, 2015, **137**, 477–482.
73. X. Pan and X. Bao, *Acc. Chem. Res.*, 2011, **44**, 553–562.
74. M. Melchionna, S. Marchesan, M. Prato and P. Fornasiero, *Catal. Sci. Technol.*, 2015, **5**, 3859–3875.
75. N. Coustel, B. Coq, V. Brotons, P. S. Kumbhar, R. Dutartre, P. Geneste, J. M. Planeix, P. Bernier and P. M. Ajayan, *J. Am. Chem. Soc.*, 1994, **116**, 7935–7936.
76. M. Bockrath, J. Hone, A. Zettl, P. L. McEuen, A. G. Rinzler and R. E. Smalley, *Phys. Rev. B*, 2000, **61**, R10606.
77. J. Kong, C. Zhou, E. Yenilmez and H. Dai, *Appl. Phys. Lett.*, 2000, **77**, 3977.
78. G. Girishkumar, K. Vinodgopal and P. V. Kamat, *J. Phys. Chem. B*, 2004, **108**, 19960–19966.
79. M. Karimi, M. Mirzaei and M. H. Ahmadieh, *Iran. J. Allergy, Asthma Immunol.*, 2006, **5**, 63–67.
80. M. Carmo, V. A. Paganin, J. M. Rosolen and E. R. Gonzalez, *J. Power Sources*, 2005, **142**, 169–176.
81. G. Mestl, N. I. Maksimova, N. Keller, V. V Roddatis and R. Schlögl, *Angew. Chem., Int. Ed.*, 2001, **40**, 2066–2068.
82. J. Luo, F. Peng, H. Yu, H. Wang and W. Zheng, *ChemCatChem*, 2013, **5**, 1578–1586.
83. S. Banerjee and S. S. Wong, *J. Am. Chem. Soc.*, 2002, **124**, 8940–8948.
84. C. Baleizão, B. Gigante, H. Garcia and A. Corma, *J. Catal.*, 2004, **221**, 77–84.
85. C. C. Gheorghiu, B. F. Machado, C. Salinas-Martínez De Lecea, M. Gouygou, M. C. Román-Martínez and P. Serp, *Dalton Trans.*, 2014, **43**, 7455–7463.
86. X. Li and L. Zhi, *Chem. Soc. Rev.*, 2018, **47**, 3189–3216.
87. M. Pumera, *Energy Environ. Sci.*, 2011, **4**, 668–674.
88. T. Nieberle, S. R. Kumar, A. Patnaik and C. Goswami, *Lect. Notes Mech. Eng.*, 2021, 239–248.
89. J. Naveen, M. Jawaid, K. L. Goh, D. M. Reddy, C. Muthukumar, T. M. Loganathan and K. N. G. L. Reshwanth, *Nanomaterials*, 2021, **11**, 1239.
90. A. K. Geim and K. S. Novoselov, *Nat. Mater.*, 2007, **6**(3), 183–191.
91. S. Park and R. S. Ruoff, *Nat. Nanotechnol.*, 2009, **4**(4), 217–224.
92. R. K. Singh, R. Kumar and D. P. Singh, *RSC Adv.*, 2016, **6**, 64993–65011.
93. O. C. Compton and S. T. Nguyen, *Small*, 2010, **6**, 711–723.
94. S. Navalon, A. Dhakshinamoorthy, M. Alvaro, M. Antonietti and H. García, *Chem. Soc. Rev.*, 2017, **46**, 4501–4529.
95. C. Biswas and Y. H. Lee, *Adv. Funct. Mater.*, 2011, **21**, 3806–3826.
96. R. Dalven, *Introduction to Applied Solid State Physics*, Springer, US, 1990.
97. X. K. Kong, C. Le Chen and Q. W. Chen, *Chem. Soc. Rev.*, 2014, **43**, 2841–2857.
98. S. Wang, L. Zhang, Z. Xia, A. Roy, D. W. Chang, J. B. Baek and L. Dai, *Angew. Chem., Int. Ed.*, 2012, **51**, 4209–4212.

99. E. Lam, J. H. Chong, E. Majid, Y. Liu, S. Hrapovic, A. C. W. Leung and J. H. T. Luong, *Carbon*, 2012, **50**, 1033–1043.

100. F. Liu, J. Sun, L. Zhu, X. Meng, C. Qi and F. S. Xiao, *J. Mater. Chem.*, 2012, **22**, 5495–5502.

101. J. Ji, G. Zhang, H. Chen, S. Wang, G. Zhang, F. Zhang and X. Fan, *Chem. Sci.*, 2011, **2**, 484–487.

102. D. R. Dreyer, H.-P. Jia and C. W. Bielawski, *Angew. Chem.*, 2010, **122**, 6965–6968.

103. G. M. Scheuermann, L. Rumi, P. Steurer, W. Bannwarth and R. Mülhaupt, *J. Am. Chem. Soc.*, 2009, **131**, 8262–8270.

104. Á. Mastalir, Z. Király, M. Benkő and I. Dékány, *Catal. Lett.*, 2008, **124**, 34–38.

105. Á. Mastalir, Z. Király, Á. Patzkó, I. Dékány and P. L'Argentiere, *Carbon*, 2008, **46**, 1631–1637.

106. A. Vijay Kumar and K. Rama Rao, *Tetrahedron Lett.*, 2011, **52**, 5188–5191.

107. S. Verma, H. P. Mungse, N. Kumar, S. Choudhary, S. L. Jain, B. Sain and O. P. Khatri, *Chem. Commun.*, 2011, **47**, 12673–12675.

108. D. R. Dreyer, K. A. Jarvis, P. J. Ferreira and C. W. Bielawski, *Polym. Chem.*, 2012, **3**, 757–766.

109. Y. Pan, S. Wang, C. W. Kee, E. Dubuisson, Y. Yang, K. P. Loh and C. H. Tan, *Green Chem.*, 2011, **13**, 3341–3344.

110. Y. Gao, D. Ma, C. Wang, J. Guan and X. Bao, *Chem. Commun.*, 2011, **47**, 2432–2434.

111. S. Wu, G. Wen, X. Liu, B. Zhong and D. S. Su, *ChemCatChem*, 2014, **6**, 1558–1561.

112. A. Primo, F. Neatu, M. Florea, V. Parvulescu and H. Garcia, *Nat. Commun.*, 2014, **5**(1), 1–9.

113. M. C. Dobish and J. N. Johnston, *Org. Lett.*, 2010, **12**, 5744–5747.

114. T. Jensen, H. Pedersen, B. Bang-Andersen, R. Madsen and M. Jørgensen, *Angew. Chem., Int. Ed.*, 2008, **47**, 888–890.

115. T. P. Singh and O. M. Singh, *Mini-Rev. Med. Chem.*, 2018, **18**, 9–25.

116. U. C. Rajesh, J. Wang, S. Prescott, T. Tsuzuki and D. S. Rawat, *ACS Sustainable Chem. Eng.*, 2015, **3**, 9–18.

117. A. S. Adeleye, J. Xue, Y. Zhao, A. A. Taylor, J. E. Zenobio, Y. Sun, Z. Han, O. A. Salawu and Y. Zhu, *J. Hazard. Mater.*, 2022, **424**, 127284.

118. A. G. Gonçalves, J. J. M. Órfão and M. F. R. Pereira, *Catal. Commun.*, 2013, **35**, 82–87.

119. Y. Wang, Y. Xie, H. Sun, J. Xiao, H. Cao and S. Wang, *ACS Appl. Mater. Interfaces*, 2016, **8**, 9710–9720.

120. X. Duan, Z. Ao, L. Zhou, H. Sun, G. Wang and S. Wang, *Appl. Catal., B*, 2016, **188**, 98–105.

121. H. Babich and D. L. Davis, *Regul. Toxicol. Pharmacol.*, 1981, **1**, 90–109.

122. R. M. Bruce, J. Santodonato and M. W. Neal, *Toxicol. Ind. Health*, 1987, **3**, 535–568.

123. M. Honda and N. Suzuki, *Int. J. Environ. Res. Public Heal.*, 2020, **17**, 1363.
124. S. Mohammadi, A. Kargari, H. Sanaeepur, K. Abbassian, A. Najafi and E. Mofarrah, *New pub Balaban*, 2015, **53**, 2215–2234.
125. X. Duan, H. Sun, J. Kang, Y. Wang, S. Indrawirawan and S. Wang, *ACS Catal.*, 2015, **5**, 4629–4636.
126. X. Duan, H. Sun, Z. Ao, L. Zhou, G. Wang and S. Wang, *Carbon*, 2016, **107**, 371–378.
127. X. Duan, Z. Ao, H. Sun, S. Indrawirawan, Y. Wang, J. Kang, F. Liang, Z. H. Zhu and S. Wang, *ACS Appl. Mater. Interfaces*, 2015, **7**, 4169–4178.
128. X. Duan, S. Indrawirawan, H. Sun and S. Wang, *Catal. Today*, 2015, **249**, 184–191.
129. Z. Song, Y. Zhang, C. Liu, B. Xu, F. Qi, D. Yuan and S. Pu, *Chem. Eng. J.*, 2019, **357**, 655–666.
130. R. P. Rocha, A. G. Gonçalves, L. M. Pastrana-Martínez, B. C. Bordoni, O. S. G. P. Soares, J. J. M. Órfão, J. L. Faria, J. L. Figueiredo, A. M. T. Silva and M. F. R. Pereira, *Catal. Today*, 2015, **249**, 192–198.
131. Y. Wang, H. Cao, C. Chen, Y. Xie, H. Sun, X. Duan and S. Wang, *Chem. Eng. J.*, 2019, **355**, 118–129.
132. X. Duan, K. O'Donnell, H. Sun, Y. Wang and S. Wang, *Small*, 2015, **11**, 3036–3044.
133. S. Shukla, R. C. Jhamtani, M. S. Dahiya and R. Agarwal, *Toxicol. Rep.*, 2017, **4**, 240–244.
134. Ö. Demirci, K. Güven, D. Asma, S. Öğüt and P. Uğurlu, *Ecotoxicol. Environ. Saf.*, 2018, **147**, 749–758.
135. M. Tudi, H. D. Ruan, L. Wang, J. Lyu, R. Sadler, D. Connell, C. Chu and D. T. Phung, *Int. J. Environ. Res. Public Heal.*, 2021, **18**, 1112.
136. J. N. Liu, Z. Chen, Q. Y. Wu, A. Li, H. Y. Hu and C. Yang, *Sci. Rep.*, 2016, **6**(1), 1–9.
137. J. Restivo, J. J. M. Órfão, M. F. R. Pereira, E. Garcia-Bordejé, P. Roche, D. Bourdin, B. Houssais, M. Coste and S. Derrouiche, *Chem. Eng. J.*, 2013, **230**, 115–123.
138. J. Restivo, J. J. M. Órfão, S. Armenise, E. Garcia-Bordejé and M. F. R. Pereira, *J. Hazard. Mater.*, 2012, **239–240**, 249–256.
139. J. Restivo, E. Garcia-Bordejé, J. J. M. Órfão and M. F. R. Pereira, *Chem. Eng. J.*, 2016, **293**, 102–111.
140. X. Liu, C. Li, Y. Zhang, J. Yu, M. Yuan and Y. Ma, *Appl. Catal. B Environ.*, 2017, **219**, 194–199.
141. Y. Vieira, E. C. Lima, E. L. Foletto and G. L. Dotto, *Sci. Total Environ.*, 2020, 141981.
142. Q. F. Khan, S. Anum, F. Sharif, M. Farhan, H. A. Sakandar, R. Rasheed and L. Shahzad, in *Emerging Contaminants and Associated Treatment Technologies*, ed. M. Z. Hashmi, Springer, Cham, 2022, pp. 67–76.
143. J.-B. Fleury and V. A. Baulin, *Proc. Natl. Acad. Sci. U. S. A.*, 2021, **118**, 1–8.
144. H. A. Leslie, M. J. M. van Velzen, S. H. Brandsma, A. D. Vethaak, J. J. Garcia-Vallejo and M. H. Lamoree, *Environ. Int.*, 2022, **163**, 107199.

145. A. A. Adelodun, *Front. Environ. Sci.*, 2021, **9**, 264.
146. World Economic Forum, *The New Plastics Economy Rethinking the Future of Plastics*, Switzerland, 2016.
147. Y. An, J. Hou, Z. Liu and B. Peng, *Mater. Chem. Phys.*, 2014, **148**, 387–394.
148. T. Uekert, H. Kasap and E. Reisner, *J. Am. Chem. Soc.*, 2019, **141**, 15201–15210.
149. J. Kang, L. Zhou, X. Duan, H. Sun, Z. Ao and S. Wang, *Matter*, 2019, **1**, 745–758.
150. R. R. Mishra and A. K. Sharma, *Composites, Part A*, 2016, **81**, 78–97.
151. R. Wei, P. Wang, G. Zhang, N. Wang and T. Zheng, *Chem. Eng. J.*, 2020, **382**, 122781.
152. J. A. Menéndez, A. Arenillas, B. Fidalgo, Y. Fernández, L. Zubizarreta, E. G. Calvo and J. M. Bermúdez, *Fuel Process. Technol.*, 2010, **91**, 1–8.
153. L. Bo, X. Quan, S. Chen, H. Zhao and Y. Zhao, *Water Res.*, 2006, **40**, 3061–3068.
154. D. Tan, H. Zeng, J. Liu, X. Yu, Y. Liang and L. Lu, *J. Environ. Sci.*, 2013, **25**, 1492–1499.
155. Z. Zhang, Y. Shan, J. Wang, H. Ling, S. Zang, W. Gao, Z. Zhao and H. Zhang, *J. Hazard. Mater.*, 2007, **147**, 325–333.
156. S. Yang, P. Wang, X. Yang, G. Wei, W. Zhang and L. Shan, *J. Environ. Sci.*, 2009, **21**, 1175–1180.
157. J. Chen, S. Xue, Y. Song, M. Shen, Z. Zhang, T. Yuan, F. Tian and D. D. Dionysiou, *J. Hazard. Mater.*, 2016, **310**, 226–234.
158. X. Yao, Q. Lin, L. Zeng, J. Xiang, G. Yin and Q. Liu, *Chem. Eng. J.*, 2017, **330**, 783–791.
159. J. Yin, J. Cai, C. Yin, L. Gao and J. Zhou, *J. Environ. Chem. Eng.*, 2016, **4**, 958–964.
160. Z. Zhang, Y. Xu, X. Ma, F. Li, D. Liu, Z. Chen, F. Zhang and D. D. Dionysiou, *J. Hazard. Mater.*, 2012, **209–210**, 271–277.
161. Y. Vieira, M. B. Ceretta, E. L. Foletto, E. A. Wolski and S. Silvestri, *J. Water Process Eng.*, 2020, **36**, 101397.
162. N. H. Ince, *Ultrason. Sonochem.*, 2018, **40**, 97–103.
163. J. L. Lim and M. Okada, *Ultrason. Sonochem.*, 2005, **12**, 277–282.
164. C. Liu, Y. Sun, D. Wang, Z. Sun, M. Chen, Z. Zhou and W. Chen, *Ultrason. Sonochem.*, 2017, **34**, 142–153.

CHAPTER 11

Advanced Applications of Activated Carbon: Catalysis and Engineering

M. S. F. FAZLI-KU AND C. T. TYE*

School of Chemical Engineering, Engineering Campus, Universiti Sains Malaysia, 14300 Nibong Tebal, Pulau Pinang, Malaysia
*Email: chcttye@usm.my

11.1   Introduction

Activated carbon (AC) is an amorphous carbon substance with a large specific surface area (typically greater than 500 $m^2 g^{-1}$) that is derived from various carbon sources and has many applications, particularly in absorption.[1] All natural organics carbon-rich materials and fossil-derived carbonaceous materials can be converted into AC. AC can be made from wood, nut shells, coconut shells, peat, lignite, coal (anthracite or brown coal), petroleum coke, and other natural and synthetic high polymer raw materials.[2]

In the production of AC, carbonaceous waste is widely used as the raw material (precursor). It includes olive stones, biomass leftovers, rice husks, corn cobs, bagasse, hard apricot stone shells, and almond, walnut, and hazelnut shells.[3] Waste tyres, pulp-mill residues, phenol formaldehyde resins, bones, and coffee beans are also used.[4–8] On the other hand, large volumes of sludge produced in municipal and industrial wastewater treatment plants also have the potential for use in AC production because they are mainly composed of organic materials.[9]

Activated Carbon: Progress and Applications
Edited by Chandrabhan Verma and Mumtaz A. Quraishi
© The Royal Society of Chemistry 2023
Published by the Royal Society of Chemistry, www.rsc.org

The raw carbonaceous materials used in the production of AC must fulfil a few characteristics, including being low in inorganic components, widely available, and inexpensive.[10] Furthermore, the raw material should be easily activatable. There are various instances of AC production from a variety of natural or waste materials in the literature. Carbonization and activation are the two main steps in the production of AC.[11] The properties of the final product vary based on the source of material used, as well as the carbonization and activation conditions employed.

It is important to select a cost-effective material for AC preparation due to the high costs involved with AC production and regeneration processes. Choosing a raw material with low ash, high carbon content, high volatile content, low inorganic content, and low cost is essential for producing AC with superior properties.[10] The most prevalent waste sources for AC synthesis are coconut shells, fruit stones, and palm oil shells.

AC has wide applications, particularly in wastewater treatment, the food industry, and the medical field. In recent years, many different forms of carbon, such as ACs and carbon blacks, have been used as support materials for the development of heterogeneous catalysts.[12] The structure, morphology, and characteristics of AC appear to favour the catalyst performance as the support material. Rather than the conventional application of AC as an adsorbent, this chapter provides an overview of the advanced application of AC as a catalyst or catalyst support in several major reactions and their recent developments, such as catalytic CH_4 reforming with CO_2, transesterification for production of biodiesel, catalytic cracking of vegetable oil, and isomerization of glucose. In addition, the modification of AC for improved catalytic activity is discussed.

11.2   Characteristics of AC as a Catalyst

AC comes in a variety of forms and sizes. The physicochemical properties of AC have a direct impact on its performance as a catalyst, and these are derived from its precursor and production conditions. As noted, AC may be produced from any raw material with high carbon content, including inexpensive waste items. These carbon-based materials are converted to AC through a series of steps which include washing to remove impurities, drying to remove free moisture which may influence the subsequent steps, milling and sieving the carbon precursors, and demineralization/deashing using acidic or basic solutions. The materials would then be carbonized to remove non-carbon (volatile) components. The organic material is then converted to elemental carbon. For AC to serve as a catalyst, it should contain only the least amounts of ash content and minerals.[13]

The porosity of AC typically varies depending on the type of raw material used and the activation process. The internal structures and adsorption capabilities of AC have been described by the pore size distribution. Due to different preparation procedures, the pore diameters of AC can be categorized into micropores (dia. < 2 nm), mesopores (dia. between 2 and 50 nm), or macropores (dia. > 50 nm).[14]

The porous structure of AC resulting from the carbonization step can be developed further *via* the activation step. The activation step increases its pore volume, pore diameter, and surface area. The activation step can be performed by two techniques: physical and chemical. During the activation step, removing the disordered carbon and decomposition of the lignin in the presence of activating agent will enhance the microporous structure. When the walls between the pores are burned, the pores generally expand.[15] Therefore, the provisional pores and microporosity increase following the step. As a result, the activation step is regarded as the crucial one among the preparation steps of a porous carbon catalyst.

There are three types of oxygen-containing surface groups on AC: acidic, basic, and neutral. Most oxygen complexes are found as acidic surface oxides such as carboxyls, lactones, and phenols. The carbon surface becomes hydrophilic and polar due to the presence of acidic surface oxides, and thus enhances the adsorption of polar and ionic species, like heavy metals. The basic surface oxygen groups are relatively much less understood. In general, the nature and concentration of surface functional groups can be altered by thermal or chemical treatment with agents such as carbon dioxide (CO_2), ammonia, ozone, nitric acid, and hydrogen peroxide.[16]

AC has a large surface area with various functional groups that function as active sites. Due to this surface property and relative hardness, AC can often serve as a catalyst.[17] The surface chemistry of AC is important in catalysis. If the original AC's catalytic activity is insufficient for specific applications in industrial processes, the surface properties are often modified by impregnation with metals and their oxides. A recent study reported that minimizing catalyst size and providing mesoporous and microporous systems are viable strategies for overcoming microporosity limitations while retaining structure and acid properties.[18] The AC catalyst activity and product selectivity would improve as a result. In general, improvements and modifications to AC directly affect its catalytic performance, which may be due to differences in the degree of metal dispersion, stability of metal crystallites, and/or mechanical properties.

In comparison to alternative support materials, using AC as a metal catalyst support offers several advantages. The pore size distribution and surface chemical characteristics (acidity, polarity and hydrophobicity) may be modified based on the intended application.[19] In addition, carbon support can be burned to extract metal particles. With its amorphous phase which can be modified to form a highly porous structure, with changeable surface chemistry properties leading to high reactivity, as well as its large specific surface area (up to 3000 $m^2 g^{-1}$), AC is gaining attention as a catalyst or catalyst support.[20]

11.3 Application of AC as a Catalyst

Catalysts mainly based on AC have been developed for several processes.

11.3.1 Catalytic CH_4 Reforming with CO_2

In the next fifteen years, natural gas is expected to be the fastest-growing fossil fuel resource.[21] The most abundant component of natural gas is methane (CH_4). Apart from being burned directly as fuel, CH_4 is also widely used as the feedstock for reforming. Methane reforming with CO_2 is also known as dry reforming. This reaction has the advantage of producing a suitable H_2/CO ratio for the synthesis of liquid hydrocarbons with higher market value and a high efficiency of chemical energy transfer.[22] It is endothermic and is conducted at a high temperature (>600 °C). Carbon normally forms on catalysts and deactivates them in most situations, posing a severe impediment to commercial application. To overcome this problem, various compounds have been developed to serve as catalysts for this re-action, including AC.[23] Nevertheless, the literature on using carbon catalysts for dry reforming is rather limited relative to metal catalysts.[24]

For thermal dry reforming, both CH_4 and CO_2 conversions are relatively low in the temperature range from 700 °C to 1000 °C.[25] AC catalysts exhibit strong initial catalytic activity, according to the findings of many research groups; however, they are rapidly deactivated due to carbonaceous deposits.[26] According to Moliner *et al.*,[27] the concentration of surface oxy-genated groups on the AC surface is related more to the initial conversion, and mesoporous carbons show more consistent activity than microporous carbons.

A higher temperature is beneficial for the endothermic dry-reforming re-action. Thermodynamically, at higher temperatures, this promotes both the reverse Boudouard reaction ($CO_2 + C \rightarrow 2CO$) and the reverse water–gas shift reaction ($CO_2 + H_2 \rightarrow CO + H_2O$), which consume CO_2, resulting in a lower conversion (<80%) of CH_4 than CO_2.[22]

Relatively high conversions of CH_4 and CO_2 were observed for dry re-forming using AC as a catalyst. The conversions of CH_4 and CO_2 using AC catalyst were reported to be as high as 92% and 98%, respectively, at 1000 °C;[25] the conversions of CH_4 and CO_2 were found to be 92% and 95%, respectively, at 950 °C;[28] and the conversions of CH_4 and CO_2 were 80.3% and 97.5%, respectively, at 900 °C.[29] Based on these studies, a high tem-perature is required for the catalytic effect during dry reforming by using AC. The use of an AC catalyst improved the conversions of both gases con-siderably, suggesting that AC is a potential catalyst for dry reforming.

On the other hand, to enhance the reaction and increase the conversion rate, noble metals have been introduced. Noble metals are loaded onto AC to form AC-supported metal catalysts for dry reforming.[30] Platinum, tungsten carbide (WC), cobalt, and cerium have all been tested as metal catalysts in dry reforming. The metal catalysts used in catalytic CH_4 reforming with CO_2 are listed in Table 11.1.

The conversion using metal catalysts is less promising when compared to AC-based catalysts. AC provides various benefits over metal catalysts in-cluding low cost,[31] large specific surface area, high sulfur tolerance,[32] and

Table 11.1 Common metal catalysts used in catalytic CH_4 reforming with CO_2.

Catalyst	Reaction	Outcome	Ref.
PtAl	Tubular quartz reactor, 500–900 °C, 20 mg catalyst	22.0% CH_4 conversion 28.3% CO_2 conversion	33
PtZr		13.9% CH_4 conversion 21.5% CO_2 conversion	
Pt/Al_2O_3	800 °C	65.0% CH_4 conversion 79.0% CO_2 conversion	34
Co–Ce/WC–AC	55 hours reaction time, tubular quartz reactor, 2.5 g catalyst, 800 °C	91.0% CH_4 conversion 93.0% CO_2 conversion	35
Co/AC	Tubular quartz reactor, 10 g catalyst, 900 °C	92.0% CH_4 conversion 98.0% CO_2 conversion	36

the flexibility to modify pore size distribution and surface chemistry. AC is a promising least expensive and environmentally friendly dry reforming catalyst.

11.3.2 Transesterification for Production of Biodiesel

Fatty acid methyl esters (FAME), also known as biodiesel, are gaining prominence as a sustainable alternative to diesel derived from petroleum. They are generated from renewable feedstocks such as plant/vegetable oil or animal fats and are blended in the fuel used in compression ignition engines. Though the most preferable technology for producing biodiesel is homogeneous catalytic transesterification in a stirred vessel, the technology involves high purification costs and the generation of effluent, as well as a series of technical issues such as mass transfer limitations due to the immiscibility of oil in methanol,[37] non-uniform product specifications, higher energy consumption, high alcohol usage, and soap production during the process.[38]

In a heterogeneous catalytic system, the catalyst could be easily separated from the produced biodiesel for subsequent use. To make up for the lower transesterification rate in a heterogeneous system relative to a homogeneous system due to the mass transfer limitations in three phases, catalysts that can provide more specific surface area and pores for active species to anchor and react with large triglyceride molecules are required.[39] AC is a highly effective catalyst support in the transesterification process when compared to other existing supports. AC has high physical and thermal stability, as well as a large specific surface area and an inert carbon structure.[40] It functions effectively in both liquid and vapour phase reactions as a catalyst support.[41]

AC is an excellent catalyst support for transesterification reactions due to its properties. Owing to the large specific surface area of AC, the reactants can spread widely and effectively. Transesterification is unaffected by the chemical structure of AC. Furthermore, the AC's low ash content can boost the transesterification reaction. At high temperatures and pressures, the surface properties of AC remain stable. Table 11.2 shows some ACs used as catalysts in the production of biodiesel *via* transesterification.

Table 11.2   ACs used as catalysts in the transesterification of biodiesel.

AC material	Feed	Reaction conditions	Outcome	Ref.
Coconut shells	Chicken fat and skin	Batch; 90 °C; 1 hour reaction time; 3 wt% catalyst	93% conversion	42
Charcoal	Waste cooking oil	Batch; 100 °C; 2 hours reaction	94.95% conversion	43
	Jatropha oil	time; 1 wt% catalyst	93.30% conversion	
Beech tree	Waste corn oil	Batch; 62.5 °C; 1 hour reaction time; 0.5 wt% catalyst	92 wt% liquid yield	44
Palm shells	Palm oil	Packed bed membrane; 70 °C; 157.04 g catalyst	94% conversion	40
Palm shells	Waste cooking oil	Fixed bed reactor; 60 °C; 8 hours reaction time	94% conversion	45

Other catalysts, in addition to ACs, are widely used as catalysts for transesterification. These include sodium oxide and potassium oxide. Using calcium oxide as a catalyst, Kouzu *et al.*[46] reported that the yield of biodiesel was greater than 99% after 2 hours of reaction time even though a portion of the catalyst was converted to calcium soap by interacting with free fatty acids in the waste cooking oil at the early stage of the transesterification. Rashid *et al.*[47] investigated the transesterification of rapeseed oil with KOH as an alkaline catalyst. The biodiesel yield was 95–96% with a reaction temperature of 65 °C and 2 hours reaction time.[47] However, the flash and combustion points are high when compared to the common biodiesel produced.

In general, various AC, feed, and reaction materials will result in different conversion rates of biodiesel in the transesterification process. Nonetheless, the conversion range achieved by AC is comparable to those of other catalysts. AC is a potential catalyst for transesterification due to its chemical stability, high specific surface area, and excellent catalytic activity.

11.3.3   Catalytic Cracking and Pyrolysis of Vegetable Oil

Catalytic cracking is another technology to convert plant-derived oil to green renewable fuel. The process is believed to be more cost-effective than transesterification since it does not necessitate the development of additional infrastructure. Many catalysts that have been used and tested in this process are zeolites, silica gel, resins, AC, graphene oxide, and magnetite nanoparticles. AC is used as a catalyst support for the catalytic cracking process due to its large specific surface area and highly internal pore structure.

As catalysts for catalytic cracking of hydrocarbon fuel, zeolites and complex oxides are commonly used. Due to the slight/weak acidity and extensive micropores, zeolite catalysts have predominantly been used for hydrocarbon catalytic cracking.[48] Nonetheless, carbon deposition clogging the pores in conventional ZSM-5 zeolites with strongly acidic sites and simple microporous structures is easily deactivated under supercritical reaction conditions.[49] As a consequence, using ZSM-5 for catalytic cracking gives a lower liquid product compared to AC.

According to Li *et al.*,[50] catalytic cracking with ZSM-5 yields 68.20% hydrocarbon biofuel, whereas Asikin-Mijan *et al.*[51] reported a 95% hydrocarbon yield with 83% n-$(C_{15} + C_{17})$ selectivity for catalytic cracking using 5 wt% AC loading. In addition, the oxygen-containing products were found to drop from 74% to 36% in the liquid product.[51] AC catalyst outperforms other catalysts due to the high stability of the carbon material. Further, AC yields a higher percentage of hydrocarbon as compared to the ZSM-5 catalyst. Recently, Naji *et al.*[52] observed that using coconut-AC with sulfuric acid pre-treatment has reached a high liquid yield of 88.67% with a hydrocarbon yield of 58.43%. Various AC-supported metal and bimetal oxides have also been investigated for catalytic cracking of vegetable oil. The synergistic effect of two metals for Co–Fe/AC catalysts was found to improve the liquid hydrocarbon yield (up to ~93%) and fatty acid conversion (up to 94%).[53] The development of AC-based catalysts for catalytic cracking is ongoing.

Another similar process is the pyrolysis of vegetable oil to generate bio-oil with gasoline, kerosene, and diesel boiling range components.[54] Pyrolysis is the thermal degradation of organic materials, such as biomass in an inert environment by cracking the chemical bonds at a temperature typically above 500 °C.[55] The type of catalyst is vital in the pyrolysis process and AC is one of the most widely tested catalysts in the pyrolysis of waste plant oil and animal fat.[56] AC with high surface area and acidity tends to produce more gases at the expense of the liquid yield in pyrolysis reactions.[57] Table 11.3 shows the ACs used as catalysts in the production of bio-oil *via* pyrolysis.

11.3.4 Conversion of Glucose to 5-Hydroxymethylfurfural

5-Hydroxymethylfurfural (HMF) is a heterocyclic furanic compound with hydroxide and aldehyde functionalities at the 2–5 positions. It has several applications in diverse sectors. Through various processes, HMF can be used as feedstock for polymer synthesis[62] and for pharmaceutical uses, and it can also be converted to fuel. Dehydration of C_6 carbohydrates, namely, glucose, sucrose, fructose, cellulose, starch, and raw biomass, is the conventional method to produce HMF.[63] In general, the most common route is the direct dehydration of hexoses *via* either an acyclic or cyclic intermediate into HMF, which is accomplished in a homogeneous catalysis system. With fructose, HMF conversion yields up to 99% have already been achieved.[64] However, there are issues with traditional homogeneous catalysts: it is difficult to recycle them, and they induce corrosion and impose considerable costs in separation and waste treatment processes.[65] This leads to the need for the development of heterogeneous catalysts that can be easily removed and reused from the liquid reaction mixture.[66]

Low-cost ACs with large surface areas, high porosity, thermal stability, and surface functionalities, such as oxygen-bearing surface groups with acid–base character, are potential catalyst candidates for the production of

Table 11.3   Pyrolysis of waste cooking oil and animal fat with AC as a catalyst.

AC material	Feed	Reaction conditions	Outcome	Ref.
Coconut husk	Waste cooking oil	400–550 °C; 1.5 hours reaction time	84 wt% liquid yield	58
Coconut husk	Waste cooking oil	800 °C; 20 minutes reaction time; 100 g catalyst	62 wt% liquid yield	59
Charcoal	Waste animal fats	420 °C; 1 hour reaction time	85 wt% liquid yield	60
Coconut shells	Waste cooking oil	430 °C; 1 hour reaction time	79.69 wt% liquid yield	61

HMF.[67] Several studies using AC-based catalysts to convert fructose to HMF have been reported. Deng *et al.*[68] used phosphoric acid functionalized sodium lignin sulfonate generated AC as a catalyst support, while Xiong *et al.*[69] used sulfuric acid sulfonated wood-based forestry charcoal for the reaction. These studies generated HMF yields of 55.6% and 42.3%, respectively. On the other hand, Villanueva and Marzialetti[70] reported a 10% HMF yield with carboxylated AC, and Zou *et al.*[71] observed a 5% HMF yield using sulfonated bagasse-based AC. Both provided moderate HMF yields.

The use of different types of AC-generating materials is critical in the conversion of fructose to HMF. Li *et al.*[72] recently explored the conversion of fructose to HMF using lignin-based AC as a catalyst, with an HMF yield of over 95%. A carbon-based solid derived from cellulose and lignin was found to effectively hydrolyze cellulose, indicating that lignin is a viable component for the synthesis of AC.[73] Therefore, the ultimate product yield will differ depending on the type of AC used.

11.3.5   Catalytic Oxidation of Nitric Oxide

Nitric oxide (NO) is one of the major air pollutants. Catalytic oxidation of NO ($2NO + O_2 \rightarrow 2NO_2$) using ACs, followed by NO_2 absorption as a more soluble NO_x, is considered to be a low-cost approach for NO_x elimination.[74] Using AC catalysts might be a multipollutant removal approach to air pollution. Other pollutants, in addition to NO, can be oxidized by AC to more soluble and readily removable forms. Several studies of NO catalytic oxidation on carbon catalysts in dry settings at varying temperatures and oxygen contents at atmospheric pressure have been carried out.[75] In terms of physical properties, microporosity, particularly the volume of smaller micropores, was found to have a vital role in NO catalytic oxidation. Besides, the availability and quantity of active catalytic sites, which are governed by the surface chemistry of ACs, affect the catalytic activity for the oxidation of NO.[74]

The catalytic oxidation of NO over microporous AC has attracted wide attention. Adding La_2O_3, CeO_2, MnO_x, V_2O_5, and other oxides to AC and AC fibres by impregnation or gas pretreatment to improve the denitrification performance has been extensively investigated.[76] Catalytic oxidation of NO to NO_2 is usually thought to be a critical step in the selective catalytic reduction (SCR) reaction. The pore structure of AC is generally accepted as a crucial element affecting NO catalytic oxidation. The influence of microporous width on NO catalytic oxidation has been studied.[77]

According to Zhang *et al.*,[78] the average micropore width has a substantial impact on NO conversion but without any direct association with surface area. The micropore width of 0.7 nm was adequate for the interaction of NO and O_2, with 0.317 nm and 0.346 nm molecular sizes, respectively.[77] On the other hand, O-containing functional groups have been found to be evolving during the NO catalytic oxidation and also influence the process itself.[79] After four cyclic NO catalytic oxidation runs, Atkinson *et al.*[80] reported that the O content increased from 2.27% to 8.61%. They also observed that increasing the O concentration accelerated the system to reach a steady state but had no effect on steady-state NO conversion. A high NO conversion of 51.58% over AC, which is proportional to the microporous surface area, was reported.[77]

11.3.6 Catalytic Wet Peroxide Oxidation

Catalytic wet peroxide oxidation (CWPO) is a reaction that uses hydrogen peroxide as an oxidant to oxidize organic pollutants in wastewater treatment. It is a heterogeneous Fenton process using a water-insoluble solid catalyst. ACs have been utilized as catalysts in CWPO[81] because they have donor–acceptor surface properties that allow the hydrogen peroxide to break down into radical species *via* an electron transfer mechanism similar to the Fenton reaction.[82] The Fenton reaction is commonly used in wastewater treatment and the degradation of organic pollutants in water. Accordingly, AC and AC^+ function as reduced and oxidized catalyst states, respectively, which lead to the generation of $^{\bullet}OH$ and $^{\bullet}OOH$ radicals. In the temperature range often used in CWPO, AC promotes parasitic reactions that consume $^{\bullet}OH$ and $^{\bullet}OOH$ radicals to form oxygen.

11.4 AC Modification

Heterogeneous catalysts include solid acid, base, acid–base bifunctional, nano, and biomass waste-based catalysts. Based on the required properties, the catalyst is prepared accordingly. Various strategies and modification techniques can be employed to functionalize the surfaces of ACs, depending on the starting materials or preparation processes used, as well as the materials' subsequent use.

11.4.1 Surface Area and Porosity

ACs' large specific surface area and well-developed porosity promote their usage as catalyst supports to provide a well-dispersed loading of active particles on the surface. The size of the pore is also essential. Active centres of highly porous ACs with small micropores can easily be blocked, hindering reactants' accessibility.[83] The surface area and porosity (pore size, pore volume, and pore shape) of ACs are affected by the preparation conditions.

Physical (CO_2 as an agent) and chemical ($HClO_4$ or $Mg(ClO_4)_2$ as an agent) activation methods were used to modify the porosity of anthracite-based AC.[84] The ultimate pore size distribution (mostly microporous or mesoporous) is affected by the activation agent, the treatment temperature and duration, and the starting textural qualities of the carbon source anthracite. Acidic treatments enhanced the porosity and surface area of ACs by removing inorganic impurities within. The treatment with hydrofluoric acid resulted in a significant increase in porosity and surface area.[85] Nonetheless, the porosity and surface area of AC materials are simply one factor influencing their use as catalyst supports. Another important feature is the surface chemistry of AC.[83]

11.4.2   Chemical Characteristics of AC

AC often contains trace amounts of chemically bound non-carbon heteroatoms such as nitrogen, oxygen, hydrogen, and phosphorus.[17] The nature, kind, and quantity of these components are determined by the raw material used and products formed during the activation stage. The heteroatoms near the periphery of the AC structure indicate surface functional groups. AC takes on its (acidic or basic) character based on the bonds formed between the AC and these atoms.[86] The basic structural unit of AC is similar to the graphite structure. AC is a disorganized form of graphite due to the presence of impurities.[87]

Surface functional groups can be identified *via* chemical and physical analyses.[86] Chemical analysis can be used to determine acidity, basicity, polarity, and surface charge. Physical analysis is used for determining the bond energy.[88] During the AC synthesis process, carbon-based materials are activated to form porous structures with defined shapes followed by surface modification.[89] The modifying agent and the modification process are the two main parameters to be considered during surface modification.

11.4.2.1   Acidic Surface

The presence of oxygen and its bond has the highest effect due to their high electronegativity and high tendency to modify *via* surface modification. There are a number of these functional groups in the AC structure. The oxygen functional groups are what give AC surfaces their acidic nature. Carboxylic, carbonyl, phenol, chromene, lactone, and pyrone groups are examples of oxygen-containing groups. Based on these groups' chemical properties, three types (acidic, basic, and neutral) have been established.

The main sources of acidity have been reported to be a carboxylic acid or carboxylic anhydride, phenolic hydroxyl, and lactone.[90]

Some activation and modification techniques work to change AC surface into highly acidic by adding carbon–oxygen surface groups. Chemical oxidation treatment is one of these methods which involves using oxidizing gases (O_2, O_3, CO_2, steam, *etc.*) or oxidizing solutions (HNO_3, H_2SO_4, H_2O_2, *etc.*) to enrich oxygen atoms.[91] Low-temperature oxidation produces strong carboxylic acid functional groups, whereas weakly acidic groups undergo phenolic oxidation at high temperatures. Besides, nitric acid oxidation, or a mixture of nitric and sulfuric acid oxidation, introduces a significant amount of carboxylic, lactone, and phenolic hydroxyl groups. In addition, oxidation treatment in the liquid phase leads to a high concentration of oxygen being deposited as carboxylic and phenolic hydroxyl groups on the surface of carbon.[92] The oxidation in the gas phase, on the other hand, promotes the formation of carbonyl and hydroxyl surface groups at a higher temperature.

Jaramillo *et al.*[93] investigated the effects of oxidising agents (O_2(air) or O_3 atmosphere, HNO_3, and H_2O_2 solutions) on the surface of AC prepared from cherry stones. Ozone and nitric acid were found to be the most efficient oxidizing agents for producing acidic oxygen surface groups, especially carboxylic acid groups. Hydrogen peroxide was a far less effective agent than nitric acid due to the less amount of oxygen surface groups produced. HNO_3 has been reported as one of the most effective compounds in promoting surface acidity. Further, HNO_3 enhanced mesoporosity and reduced microporosity.

Chemical sulfonation is another effective acidic modification that improves the number of acidic sites on the surface of AC. In this process, sulfonic acid groups are added to the carbon precursor, resulting in strong Brönsted acid sites. Large quantities of sulfonic acid groups can penetrate carbon bulk when the appropriate pore shapes and sizes are available. As a result, considerable covalent interaction between the sulfonic acid groups and carbon occurred, which improved the catalyst's hydrophilicity.[94] Shu *et al.*[95] evaluated the performance of AC produced by carbonized vegetable oil asphalt and sulfuric acid sulfonation. The catalyst's wide pore diameter of less than 2 nm allows the diffusion of large reagent molecules into the AC inner acidic sites.[96] The catalytic activity of sulfonated mesoporous carbon catalysts in the esterification reaction of oleic acid with ethanol is influenced by surface area, surface hydrophobicity, uniform pore sizes, thermal stability, mechanical properties, and acidic site concentration.[97]

Another highly recommended chemical for acidic modification of the AC surface is phosphotungstic acid (PWA). Ning *et al.*[98] produced biodiesel from oleic acid by using AC derived from bamboo and treated with PWA. The AC had a high acid density of 2.02 mmol g^{-1}, with a specific surface area of 576 m^2 g^{-1} and a pore volume of 0.52 cm^3 g^{-1}. It was prepared with 40% of PWA loaded capacity, at 600 °C activation temperature. The high loading capacity of PWA has a significant effect on the AC catalytic activity, improving the esterification efficiency from 56 to 94% while increasing the PWA loading capacity from 10 to 40%. However, with a PWA loaded capacity of 60%, it dropped to 88%. The superfluous PWA molecules tend to block the catalyst pores, resulting in hindering contact with reactants.

11.4.2.2   Basic Surface

The basicity of AC is connected to the presence of nitrogen functional groups, which may bind with protons and resonating π-electrons of carbon aromatic rings that attract protons. Surface modification with reagents such as ammonia and amines or activation of raw carbon with high nitrogen content are two ways to introduce nitrogen groups on the surface of AC.[99] The basic nature of ACs is mostly from delocalized electrons of graphene layers.[92] These electrons have been proposed to act as Lewis bases.[97] Protons from the mixture may be adsorbed onto the basic surface of carbon and non-carbon sites (oxygen sites).[100] Furthermore, these sites are found on the basic surface of carbon crystallites in π-electron-rich regions.[101] This showed that basic sites are Lewis bases that are associated with the carbon structure.

The effect of N-groups on the surface of AC was studied using ammonia.[102] By interacting with ammonia at high temperatures, the basic groups represented by nitrogen functional groups may be enhanced. During this process, acidic impurities are absorbed onto the functionalized carbon surface, resulting in a surface with more basic sites compared to pure AC. The amount of N_2 available on the surface is the main factor in the adsorption of acidic impurities.

N-enriched microporous AC may also be generated by ammoxidation.[103] Ammoxidation was carried out at 350 °C using a 1 : 3 combination of ammonia and air. Adding a considerable amount of nitrogen to the carbon has been shown to have a positive influence on the porous structure of the carbon during activation, yielding AC with specific surface areas of 2600–2800 $m^2 g^{-1}$ and pore volumes of 1.29–1.60 $cm^3 g^{-1}$. Stöhr *et al.*[104] modified an AC surface using ammonia treatment for oxidation reactions. According to their observations, adding N-functional groups on the AC surface improves AC catalytic activity at high temperatures (600–900 °C). In addition to ammonia treatment, heat treatment in an inert environment can also be used to remove selectively acidic groups from the surface of AC.[105] High temperatures (800 °C to 1000 °C) cause the breakdown of oxygen functional groups on the AC's surface, resulting in an increase in basic functional groups.

Strongly acidic groups such as carboxylic acids, anhydrides, and lactones degrade at high temperatures, whereas weakly acidic groups such as carboxyls, phenols, and quinones break down at low temperatures.[106] By eliminating the hydrophilic surface functions, the modification technique which operates under an inert atmosphere (nitrogen or helium) or hydrogen tends to enhance the hydrophobicity of carbon.[107] Hydrogen is more feasible in removing acidic groups and can stabilize the AC surface by deactivating the active sites *via* the introduction of stable C–H bonds and/or gasification of unstable carbon atoms.[106] In general, thermal treatment improves surface basicity by reducing the acidity-reducing oxygen content. Other than this, metal impregnation of AC to adjust the catalyst's basicity is another common practice used in catalysis.

11.5   Summary and Outlook

This chapter focused on the application of AC as a catalyst owing to its properties, as well as the methods used to improve its surface and catalytic properties. Several applications of ACs as catalysts were discussed. The major steps in the preparation of AC materials include precursor selection, precursor pre-treatment, and chemical and/or physical activation. The physicochemical properties of AC are derived directly from the raw material and the activation procedure. They are also closely associated with the catalytic performance of the AC in a particular reaction. Based on many recent studies, acidic treatment is recommended as a surface modification of AC due to its easy usage, the accessibility of strong oxidizing chemicals, and the carefully characterized oxidation reaction. For the application of AC as a catalyst, the catalyst performance is directly linked to its large surface area, well-developed porosity, and availability of specific surface functional groups.

In the case that AC properties need to be improved, the preparation conditions must be specifically chosen to tailor the microstructure, textural properties, performance, and surface chemistry of the obtained carbons including the carbon precursor, type of activation (chemical, physical, or mixed), activation temperature, reaction time, and impregnation ratio. To enhance the performance of AC as a catalyst, it is required to investigate and optimize the effects of these parameters.

Acknowledgements

This work was supported by the Ministry of Higher Education Malaysia under the FRGS grant scheme with the Project Code: FRGS/1/2019/TK02/USM/02/5 (A/C: 203. PJKIMIA. 6071445).

References

1. T. M. Alslaibi, I. Abustan, M. A. Ahmad and A. A. A. Foul, *J. Chem. Technol. Biotechnol.*, 2013, **88**, 1183.
2. G. S. Simate, N. Maledi, A. Ochieng, S. Ndlovu, J. Zhang and L. F. Walubita, *J. Environ. Chem. Eng.*, 2016, **4**, 2291.
3. M. Soleimani and T. Kaghazchi, *Chem. Eng. Technol.*, 2007, **30**, 649.
4. A. S. Al-Rahbi and P. T. Williams, *Waste Manage.*, 2016, **49**, 188.
5. N. Talreja, S. H. Jung, L. T. H. Yen and T. Y. Kim, *Chem. Eng. J.*, 2020, **379**, 122332.

6. T. Y. R. Oda, A. A. P. Rezende, R. de Cássia Superbi Sousa, C. M. Silva and A. C. Pereira, *J. Environ. Manage.*, 2021, **298**, 113477.

7. L. Niu, C. Shen, L. Yan, J. Zhang, Y. Lin, Y. Gong, C. Li, C. Q. Sun and S. Xu, *J. Colloid Interface Sci.*, 2019, **547**, 92.

8. A. Namane, A. Mekarzia, K. Benrachedi, N. Belhaneche-Bensemra and A. Hellal, *J. Hazard. Mater.*, 2005, **119**, 189.

9. M. Danish and T. A. Ahmad, *Renewable Sustainable Energy Rev.*, 2018, **87**, 1.

10. B. E. Narowska, M. Kułażyński and M. Łukaszewicz, *Catalysts*, 2020, **10**, 1049.

11. K. H. Büchel, H. H. Moretto and P. Woditsch, *CHEMKON*, 2001, **8**, 52.

12. F. Rodríguez-Reinoso, *Carbon*, 1998, **36**, 159.

13. S. Z. Naji and C. T. Tye, *Energy Convers. Manage.: X*, 2022, **13**, 100152.

14. M. Ilomuanya, B. Nashiru, N. Ifudu and C. Igwilo, *J. Microsc. Ultrastruct.*, 2017, **5**, 32.

15. M. J. Ahmed, *Process Saf. Environ. Prot.*, 2016, **102**, 168.

16. T. J. Bandosz, *Interface Sci. Technol.*, 2006, **7**, 571.

17. S. M. Yakout and G. S. El-Deen, *Arab. J. Chem.*, 2016, **9**, S1155.

18. D. R. Lobato-Peralta, E. Duque-Brito, A. Ayala-Cortes and D. M. Arias, *J. Environ. Chem. Eng.*, 2021, **9**, 105626.

19. A. Mohammad-Khah and R. Ansari, *Int. J. Chemtech Res. CODEN*, 2009, **1**, 859.

20. S. A. Sadeek, E. A. Mohammed, M. Shaban, M. T. H. Abou Kana and N. A. Negm, *J. Mol. Liq.*, 2020, **306**, 112749.

21. K. Reders, M. Schmidt and A. Schütze, Natural Gas, in *Handbook of Fuels*, Wiley-VCH GmbH, Boschstr, 2021, ch. 6, pp. 119–159.

22. P. Kaiser, R. B. Unde, C. Kern and A. Jess, *Chem. Ing. Tech.*, 2013, **85**, 489.

23. G. Zhang, Y. Dong, M. Feng, Y. Zhang, W. Zhao and H. Cao, *Chem. Eng. J.*, 2010, **156**, 519.

24. H. F. Abbas and W. M. A. W. Daud, *Int. J. Hydrogen Energy*, 2010, **35**, 1160.

25. L. Xu, Y. Liu, Y. Li and Z. Lin, *Appl. Catal., A*, 2014, **469**, 387.

26. B. Fidalgo and J. A. Menéndez, *Cuihua Xuebao Chinese J. Catal.*, 2011, **32**, 207.

27. R. Moliner, I. Suelves, M. J. Lázaro and O. Moreno, *Int. J. Hydrogen Energy*, 2005, **30**, 293.

28. G. Zhang, A. Su, Y. Du, J. Qu and Y. Xu, *J. Colloid Interface Sci.*, 2014, **433**, 149.

29. L. Li, J. Chen, Q. Zhang, Z. Yang, Y. Sun and G. Zou, *J. Clean Prod.*, 2020, **274**, 122256.

30. A. W. Budiman, S. H. Song, T. S. Chang, C. H. Shin and M. J. Choi, *Catal. Surv. Asia*, 2012, **16**, 183.

31. V. K. Gupta, A. Nayak, B. Bhushan and S. A. Agarwal, *Crit. Rev. Environ. Sci. Technol.*, 2015, **45**(6), 613.

32. H. Cui, S. Q. Turn and M. A. Reese, *Catal. Today*, 2009, **139**, 274.

33. M. M. V. M. Souza, D. A. G. Aranda and M. Schmal, *J. Catal.*, 2001, **204**, 498.
34. A. D. Ballarini, S. R. de Miguel, E. L. Jablonski, O. A. Scelza and A. A. Castro, *Catal. Today*, 2005, **107–108**, 481.
35. S. Li, J. Wang, G. Zhang, J. Liu, Y. Lv and Y. Zhang, *Fuel*, 2022, **311**, 122512.
36. Y. Sun, G. Zhang, J. Liu, P. Zhao, P. Hou, Y. Xu and R. Zhang, *Int. J. Hydrogen Energy*, 2018, **43**, 1497.
37. H. Noureddini and D. Zhu, Kinetics of Transesterification of Soybean Oil, *J. Am. Oil Chem. Soc.*, 1997, **74**, 1457.
38. B. Thangaraj, P. R. Solomon, B. Muniyandi, S. Ranganathan and L. Lin, *Clean Energy*, 2019, **3**, 2.
39. M. Zabeti, W. M. A. W. Daud and M. K. Aroua, *Appl. Catal., A*, 2009, **366**, 154.
40. S. Baroutian, M. K. Aroua, A. A. A. Raman and N. M. N. A. Sulaiman, *Bioresour. Technol.*, 2011, **102**, 1095.
41. S. Baroutian, M. K. Aroua, A. A. A. Raman and N. M. N. A. Sulaiman, *Fuel Process. Technol.*, 2010, **91**, 1378.
42. S. Gnanaserkhar, N. Asikin-Mijan, G. AbdulKareem-Alsultan, S. Seenivasagam, S. M. Izham and Y. H. Taufiq-Yap, *Biomass Bioenergy*, 2020, **141**, 105714.
43. D. A. Kamel, H. A. Farag, N. K. Amin and Y. O. Fouad, *Int. J. Environ. Sci. Technol.*, 2017, **14**, 785.
44. B. Narowska, M. Kułażyński, M. Łukaszewicz and E. Burchacka, *Renewable Energy*, 2019, **135**, 176.
45. A. Buasri, B. Ksapabutr, M. Panapoy and N. Chaiyut, *Korean J. Chem. Eng.*, 2012, **29**, 1708.
46. M. Kouzu, T. Kasuno, M. Tajika, Y. Sugimoto, S. Yamanaka and J. Hidaka, *Fuel*, 2008, **87**, 2798.
47. U. Rashid and F. Anwar, *Fuel*, 2008, **87**, 265.
48. A. A. Rownaghi, F. Rezaei and J. Hedlund, *Chem. Eng. J.*, 2012, **191**, 528.
49. Y. Ji, H. Yang and W. Yan, *Fuel*, 2019, **243**, 155.
50. C. Li, J. Ma, Z. Xiao, S. B. Hector, R. Liu and S. Zuo, *et al.*, *Fuel*, 2018, **218**, 59.
51. N. Asikin-Mijan, J. M. Ooi, G. Abdulkareem-Alsultan, H. V. Lee, M. S. Mastuli and N. Mansir, *et al.*, *J. Clean Prod.*, 2020, **249**, 119381.
52. S. Z. Naji, C. T. Tye and A. R. Mohamed, *Biomass Convers. Biorefin.*, 2022, DOI: 10.1007/s13399-022-03018-7.
53. T. Thangadurai and C. T. Tye, *Period. Polytech., Chem. Eng.*, 2021, **65**, 350.
54. T. K. Dada, A. Vuppaladadiyam, A. X. Duan, R. Kumar and E. Antunes, *Bioresour. Technol.*, 2022, **360**, 127515.
55. Z. X. Xu, P. Liu, G. S. Xu, Q. Liu, Z. X. He and Q. Wang, *Energy*, 2017, **133**, 666.
56. R. Idris, W. W. F. Chong, A. Ali, S. Idris, W. H. Tan and R. M. Salim, *et al.*, *Energy Convers. Manage.*, 2021, **244**, 114502.

57. R. Miandad, M. A. Barakat, A. S. Aburiazaiza, M. Rehan and A. S. Nizami, *Process Saf. Environ. Prot.*, 2016, **102**, 822.
58. S. S. Lam, W. A. W. Mahari, Y. S. Ok, W. Peng, C. T. Chong and N. L. Ma, *et al.*, *Renewable Sustainable Energy Rev.*, 2019, **115**, 109359.
59. W. A. W. Mahari, C. T. Chong, W. H. Lam, T. N. S. T. Anuar, N. L. Ma and M. D. Ibrahim, *et al.*, *Energy Convers. Manag.*, 2018, **171**, 1292.
60. T. Ito, Y. Sakurai, Y. Kakuta, M. Sugano and K. Hirano, *Fuel Process. Technol.*, 2012, **94**, 47.
61. P. Sommani, N. Mankong, T. Vitidsant and A. W. Lothongkum, *ASEAN Eng. J.*, 2015, **4**, 16.
62. A. Marshall, B. Jiang, R. M. Gauvin and C. M. Thomas, *Molecules*, 2022, **27**, 4071.
63. S. P. Teong, G. Yi and Y. Zhang, *Green Chem.*, 2014, **16**, 2015.
64. Y. Qu, C. Huang, J. Zhang and B. Chen, *Bioresour. Technol.*, 2012, **106**, 170.
65. M. Athar and S. Zaidi, *J. Environ. Chem. Eng.*, 2020, **8**, 104523.
66. M. Miceli, P. Frontera, A. Macario and A. Malara, *Catalysts*, 2021, **11**, 591.
67. E. Lam and J. H. T. Luong, *ACS Catal.*, 2014, **4**, 3393.
68. T. Deng, J. Li, Q. Yang, Y. Yang, G. Lv and Y. Yao, *et al.*, *RSC Adv.*, 2016, **6**, 30160.
69. X. Xiong, K. M. Y. Iris, S. S. Chen, D. C. W. Tsang, L. Cao and H. Song, *et al.*, *Catal. Today*, 2018, **314**, 52.
70. N. I. Villanueva and T. G. Marzialetti, *Catal. Today*, 2018, **302**, 100–107.
71. B. Zou, X. Chen, C. Zhou, X. Yu, H. Ma, J. Zhao and X. Bao, *Can. J. Chem. Eng.*, 2018, **96**, 1337.
72. M. Li, Q. Zhang, B. Luo, C. Chen, S. Wang and D. Min, *Ind. Crops Prod.*, 2020, **145**, 111920.
73. X. Li, P. Li, D. Ding, K. Chen, L. Zhang and Y. Xie, *BioRes.*, 2018, **13**(2), 4428.
74. S. A. Dastgheib, H. Salih, T. Ilangovan and J. Mock, *ACS Omega*, 2020, **5**, 21172.
75. Y. Shen, X. Ge and M. Chen, *RSC Adv.*, 2016, **6**, 8469–8482.
76. M. Pourkhalil, A. Z. Moghaddam, A. Rashidi, J. Towfighi and Y. Mortazavi, *Appl. Surf. Sci.*, 2013, **279**, 250.
77. X. Zhu, L. Zhang, T. Wang, J. Li, X. Zhou, C. Ma and Y. Dong, *Fuel*, 2022, **311**, 122627.
78. W. J. Zhang, S. Rabiei, A. Bagreev, M. S. Zhuang and F. Rasouli, *Appl. Catal., B*, 2008, **83**, 63.
79. H. L. Nicholas, I. Mabbett, H. Apsey and I. Robertson, *Gates Open Res.*, 2022, **6**, 96.
80. J. D. Atkinson, Z. Zhang, Z. Yan and M. J. Rood, *Carbon*, 2013, **54**, 444.
81. C. M. Domínguez, P. Ocón, A. Quintanilla, J. A. Casas and J. J. Rodriguez, *Appl. Catal., B*, 2013, **140–141**, 663.
82. S. Navalon, A. Dhakshinamoorthy, M. Alvaro and H. Garcia, *ChemSusChem*, 2011, **4**, 1712.

83. F. Rodríguez-Reinoso and A. Sepúlveda-Escribano, Carbon as Catalyst Support, in *Carbon Materials for Catalysis*, John Wiley & Sons, Inc., 2008, p. 131.
84. S. B. Lyubchik, R. Benoit and F. Béguin, *Carbon*, 2002, **40**, 1287.
85. S. Wang and G. Q. Lu, *Carbon*, 1998, **36**, 283.
86. M. V. Lopez-Ramon, F. Stoeckli, C. Moreno-Castilla and F. Carrasco-Marin, *Carbon*, 1999, **37**, 1215.
87. H. Chen, S. Wang, Y. Tang, F. Zeng, H. H. Schobert and X. Zhang, *Fuel*, 2021, **292**, 120373.
88. M. Pego, J. Carvalho and D. Guedes, *Surf. Rev. Lett.*, 2019, **26**(1), 1830006.
89. B. Hu, K. Wang, L. Wu, S. H. Yu, M. Antonietti and M. M. Titirici, *Adv. Mater.*, 2010, **22**, 813.
90. L. Li, P. A. Quinlivan and D. R. U. Knappe, *Carbon*, 2002, **40**, 2085.
91. L. J. Uranowski, C. H. Tessmer and R. D. Vidic, *Water Res.*, 1998, **32**, 1841.
92. M. A. Montes-Morán, D. Suárez, J. A. Menéndez and E. Fuente, *Carbon*, 2004, **42**, 1219.
93. J. Jaramillo, P. M. Álvarez and V. Gómez-Serrano, *Fuel Process. Technol.*, 2010, **91**, 1768.
94. H. H. Mardhiah, H. C. Ong, H. H. Masjuki, S. Lim and Y. L. Pang, *Energy Convers. Manag.*, 2017, **144**, 10.
95. Q. Shu, G. H. Xu, Z. Nawaz and Q. Zhang, *Fuel Process. Technol.*, 2009, **90**, 1002.
96. K. S. W. Sing, *Pure Appl. Chem.*, 1985, **57**, 603.
97. A. F. Pérez-Cadenas, C. Moreno-Castilla, F. J. Maldonado-Hódar and J. L. G. Fierro, *J. Catal.*, 2003, **217**, 30.
98. Y. Ning, S. Niu, K. Han and C. Lu, *Biomass Bioenergy*, 2020, **143**, 105873.
99. T. C. Drage, A. Arenillas, K. M. Smith, C. Pevida, S. Piippo and C. Snape, *Fuel*, 2007, **86**, 22.
100. C. A. L. Y. Leon, J. M. Solar, V. Calemma and L. R. Radovic, *Carbon*, 1992, **30**, 797.
101. N. Saeidi, F. D. Kopinke and A. Georgi, *Chem. Eng. J.*, 2020, **381**, 122689.
102. C. L. Mangun, K. R. Benak, J. Economy and K. L. Foster, *Carbon*, 2001, **39**, 1809.
103. R. Pietrzak, *Fuel*, 2009, **88**, 1871–1877.
104. B. Stöhr, H. P. Boehm and R. Schlögl, *Carbon*, 1991, **29**, 707.
105. M. S. Shafeeyan, W. M. A. W. Daud, A. Houshmand and A. A. Shamiri, *J. Anal. Appl. Pyrolysis*, 2010, **89**, 143.
106. S. A. Dastgheib and T. Karanfil, *J. Colloid Interface Sci.*, 2004, **274**, 1.
107. C. Pevida, M. G. Plaza, B. Arias, J. Fermoso, F. Rubiera and J. J. Pis, *Appl. Surf. Sci.*, 2008, **254**, 7165.

CHAPTER 12

Food Industry Applications of Activated Carbon

ISHRAT FATMA,[a] HUMIRA ASSAD,[a] ASHISH KUMAR*[b] AND CHAUDHERY M. HUSSAIN[c]

[a] Department of Chemistry, Faculty of Technology and Science, Lovely Professional University, Phagwara, Punjab, India; [b] NCE, Department of Science and Technology, Government of Bihar, India; [c] Department of Chemistry and Environmental Science, New Jersey Institute of Technology, Newark, NJ 07102, USA
*Email: drashishchemlpu@gmail.com

12.1 Introduction

The origin of activated carbon (AC) is linked with Ancient Egypt (1500 BC), where the Egyptians utilized its adsorbent-like feature for cleaning of H_2O as well as for medicinal purposes. Karl Wilhelm, a Swedish chemist, presented a report on the adsorption of various gases on charcoal. The production of AC started in Germany in the 20th century for a sugar refining industry. Later, various plants were developed to produce AC for wastewater treatment. In the late 1930s AC gained its attention as well as found application in numerous industrial areas. Between the years 1939 and 1945, great improvements occurred through which various chemically saturated carbons were developed for war as well as trapping of nerve gases.[1,2] In the year 2006, Marsh and Reinoso described AC (activated charcoal) as an amorphous carbon substance, which is generally characterized by its densely porous structure as well as its highly absorptive

Activated Carbon: Progress and Applications
Edited by Chandrabhan Verma and Mumtaz A. Quraishi
© The Royal Society of Chemistry 2023
Published by the Royal Society of Chemistry, www.rsc.org

character. It is a non-toxic black-colored substance with an extraordinary large surface area in the range of 500–2000 $m^2 g^{-1}$. It bears hexagonally arranged carbon atoms, which are covalently bonded with each other to form microcrystalline carbon layers due to which its surface appears to be non-polar and hydrophobic in nature, which in turn increases its interlayer distance as well as porosity. Furthermore, the bonding among functional groups, like –O and –H, at the border of the carbon stratum disturbs its surface characteristics and the dimensions of pores, and forms disordered layers; therefore, it is regarded as a non-graphite form.[3–5] AC is generally classified into three types: (i) powdered AC (PAC), (ii) granular AC (GAC), and (iii) AC fibers (ACFs).[6] PAC has particle size in the range of 0.015–0.025 mm. It is commonly used for wastewater treatment, for sugar decolorization, in the food industry, in the pharmaceutical industry, *etc.*[7–9] GAC shows certain advantages because of its renewable nature as well as proper distribution of pores on its surface.[10–12] ACFs are carbonized carbons which are mainly produced by employing saran and phenolic resins as precursors. They were first developed around 1970. They have a narrow pore size distribution and are generally utilized for the elimination of volatile organic substances.[13,14]

ACs represent a group of carbonaceous substances obtained from biomass, coal, and polymer scrap *via* thermochemical procedures as shown in Figure 12.1.

They are nanoporous adsorbents that play a significant role in gas and liquid phase separation methods. Features of AC are determined by the physical and chemical characteristics of the raw materials used. The functional moieties regulate the significant properties of the AC. Because of its versatile adsorption characteristics, AC has the ability to remove an extensive variety of materials. Utrilla and his team found that modified GAC could eliminate pathogenic bacteria like *Escherichia coli* from H_2O.[16]

Diban *et al.* (2008) verified that normal AC could recover an aroma complex from dispersed effluent.[17] Also, numerous investigators have stated its use in adsorbing colors, organic compounds, and several contaminants from the surroundings.[18–20] Because of its low price as well as non-toxic nature, AC is gaining importance rapidly in the food industry. Recently, AC has been introduced as an additive in numerous kinds of foodstuffs like baked goods, ice-creams, and beverages, so as to enhance their food texture, color, and health benefits.[21] Moreover, it has been used in several food-processing sectors. According to some researchers, AC could be used to obtain benzaldehyde and add coffee scent.[22] It may also be used to eliminate brown mixtures present in soya sauce.[23] Because of the upsurging customers' pressure concerning food safety and quality, companies try to satisfy customers' conditions and endure strong competition. They pay more attention to food supervision and approaches to conserve food safety, quality, and cost.

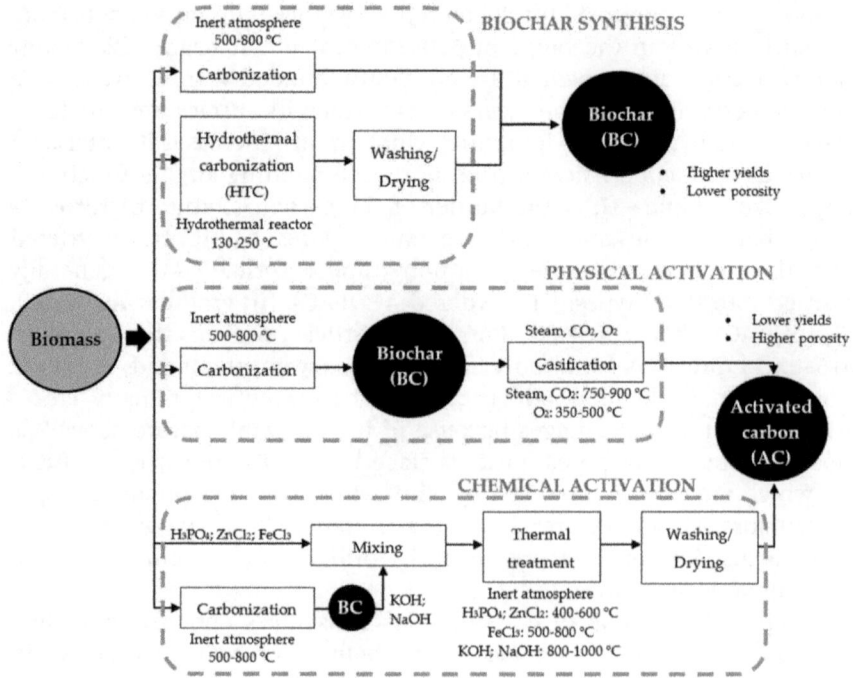

Figure 12.1 Synthesis routes to biochars and ACs from biomass. Reproduced from ref. 15, https://doi.org/10.3390/c4040063, under the terms of the CC BY 4.0 license https://creativecommons.org/licenses/by/4.0/.

12.2 Fabrication of AC

AC can be synthesized *via* direct activation using various raw materials or by means of a two-stage procedure such as carbonization followed by activation. During carbonization, the ingredients were subjected to very high temperature (<700 °C) so as to evaporate as well as to eliminate the various hydrocarbons present in it. In general, carbonization is hence a pyrolytic process and the product formed is called carbonized material or sometimes biochar.[24–26] Subsequently, several activation approaches are utilized to enhance the porosity as well as to generate structures which results in the formation of delicate solid cavities in AC. The pores produced on AC's surface are classified into macro-, meso-, and micropores.[27]

12.2.1 Carbonization/Pyrolysis

This involves the thermal breakdown of selected ingredients in a furnace in an inert atmosphere so as to eliminate volatile substances as well as moieties like N_2, O_2, and H, and to improve the fixed carbon content for the preparation of biochar.[28] In the course of devolatilization, fine pores of precursor-like structures start appearing that causes the accumulation of tarry moieties produced as the temperature upsurges.[29] In various cases, this deposition can result in collision of several tarry materials and the breakdown of the pore walls, which may cause hydrocracking as well as carbon deposition. The carbonization factors considerably influence this process as well as the quality of the resultant products; thus, a vigilant choice of factors is of great significance. The most important factors which affect this process include the carbonization temperature, heating rate, quantity of inert gas and its flow rate, and holding time. Usually, as the temperature increases, a large number of volatile moieties are liberated besides the escalation of fixed carbon as well as ash content. For example, Wang *et al.*[30] studied the impact of the carbonization temperature and found that as the carbonization temperature increased from 300 to 600 °C the adsorption ability of AC increased from 756.42 to 933.84 $mg\,g^{-1}$; however, its yield decreased from 41.23 to 32.79%. Conversely, from 400 °C to 600 °C, the adsorption ability intensely declined from 933.84 to 538.36 $mg\,g^{-1}$.

12.2.2 Physical Activation

Physical activation may occur in a single step as well as in dual steps.[31] Mainly, physical activation takes place in two steps, in which the carbonization of dried systems occurs at temperatures between 400 and 700 °C to form biochar, followed by activation in the presence of air, carbon dioxide, *etc.*, at temperatures of 800 to 1100 °C for some burn-off.[32] The biochar achieved *via* carbonization normally has a surface area of <300 $m^2\,g^{-1}$.[33] This small surface area as well as adsorption capability is attributed to the holes/pores sealed through tarry materials that may be eliminated through activation.[26] In a single-step process, carbonization as well as activation happens instantaneously in the temperature range of 600 to 800 °C.[34]

The same as that in the dual-step process, in the single-step process, the dehydrated substance is carbonized; however, biochar may be additionally heated for a short period in an oxidizing gas atmosphere. In spite of the processes being similar, the single-step process is highly practicable as it avoids the cooling effect just after carbonization which may lessen the physical exertion, electrical intake, price, and working time.[33]

In the course of physical activation, both pore widening and the formation of new pores take place simultaneously due to which the porosity as well as surface area of the carbonaceous material may be improved.[2] Conversely, AC produced by physical activation is of lower quality than that produced by chemical activation because of the greater activation temperature as well as lengthy activation time.[31] During physical activation, steam and carbon dioxide are mainly utilized as activating agents as they seem to provide largest BET surface areas and both are capable of combining with carbon to offer the largest surface area which may extend to 1000 $m^2\,g^{-1}$. Even though the processes are same, their procedures and reactions are marginally altered. Rafsanjani *et al.*[35] reported that, because of the lower size of water as compared to carbon dioxide, H_2O molecules are capable of dispersing inside the holes of biochar effortlessly and subsequently undergoing a faster reaction as compared to the carbon with carbon dioxide, and hence, additional holes are formed. At a near activation temperature, steam is found to respond towards the carbon more rapidly than to the carbon dioxide. Therefore, *via* steam activation, AC is capable of producing a very large surface area in a very short duration of time.[36] Carbon dioxide activation facilitates the formation of new pores rather than the broadening of small-sized pores, whereas steam activation supports the broadening of micropores to form mesopores as well as macropores from the start of the activation process. This confirms a broader pore distribution on the resulting ACs. In numerous cases, carbon dioxide activation is preferred because it is cleaner, portable and has control over the activation process even at very large temperatures due to its slow reaction rate.[2]

12.2.3 Chemical Activation

It may occur in a single-step as well as in a double-step procedure. Both procedures are comparable, excluding the carbonization stage.[2] In the single-step process, carbonization may not occur, but in the main step, the dehydrated moieties are activated *via* some interactions with some dehydrating agents such as sodium hydroxide and potassium hydroxide.[37] In the double-step process, just before chemical activation, the dried moiety is carbonized to form biochar in the temperature range of 400 to 600 °C.[31]

Chemical activation may be utilized in three forms, which include basic, acidic, and neutral activation.[38] Based on the kind of dehydrating agent utilized, various precursors combine differently to form a diversity of surface areas, pore sizes, and yields.[39] Generally, basic activation, which takes place in the presence of some alkaline metal hydroxides like potassium hydroxide and sodium hydroxide, yields ACs with an extraordinary surface area (2000 $m^2 g^{-1}$). On the other hand, the nature of precursors restricts the exploitation of alkaline metal hydroxides. Cao *et al.* presented that alkali metal hydroxides are mainly employed in the double-step process because they react more competently with biochar.[40] Char formed through carbonization of precursors has a definite number of holes, which supports the diffusion of dehydrating agents, and consequently, large surface area ACs with extraordinary porosity are formed.[2] Alkali metal hydroxides face some kind of difficulty while penetrating inside the precursors because of the uneven porosity of some raw materials used.[41] Only impregnation is ineffective in forming a porous active carbon, and thus, it may demand a double-step chemical activation. For example, Isoda *et al.* worked on rice husk in the presence of sodium hydroxide to produce some ACs with surface areas of 280 $m^2 g^{-1}$ and 660 $m^2 g^{-1}$.[42] It is als found that biochars are more liable to potassium hydroxide activation, and may form ACs with extraordinary surface characteristics.[43] Out of all the chemical activators, the most commonly employed acidic activator is phosphoric acid, which gives a better yield by increasing the porosity, reactivity, and surface area of the final product formed.[38,44]

However, alkali metal hydroxides as well as acids are less enviable because of their corrosive, poisonous, and menacing nature.[38,41] Hence, more common dehydrating agents like potassium carbonate are employed instead in the single-step activation process to produce ACs as well as to solve the problem of basic as well as acidic chemicals contributing to secondary waste disposal.[33] Conversely, chemical activation is preferred to physical activation in terms of surface area, porosity, yield, and economic stability because of the lower activation temperature as well as shorter activation duration.[38]

12.2.4   Physicochemical Activation

Other than physical and chemical activation, physicochemical activation may also be done,[45] as shown in Figure 12.2.

There are two ways by which ACs are produced through physicochemical activation: (1) chemical treatment prior to carbonization called precarbonization and (2) chemical treatment after carbonization called post-carbonization. The first approach has the precursors enduring carbonization that is monitored by impregnation of biochar, and after that thermal action by using oxidizing gas in an inert atmosphere and swapping to oxidizing gas for physical activation at the temperature of 600 to 850 °C,[33] whereas the second approach features precursors experiencing chemical treatment before thermal treatment as well as physical activation.[46] An investigation was done by Lee *et al.*,[47] in which they found that the order of chemical activation in this procedure does not have any influence on the quality as well as textural features of the resultant ACs. They also revealed that the chemically remedied pre-carbonized system exhibits

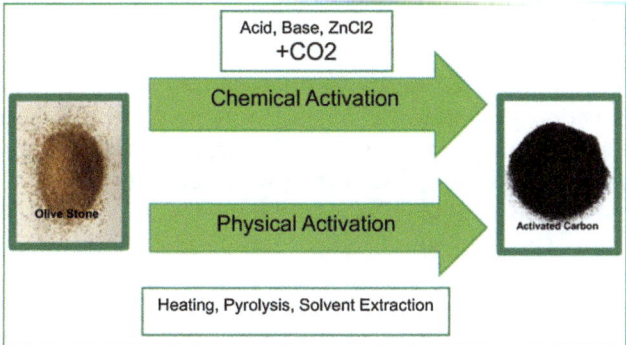

Figure 12.2   Physicochemical activation for AC production. Reproduced from ref. 45, https://doi.org/10.1007/s13399-019-00473-7, under the terms of the CC BY 4.0 license https://creativecommons.org/licenses/by/4.0/.

a surface area and a pore volume of around 990 $m^2 g^{-1}$ and 0.42 $cm^3 g^{-1}$, respectively, whereas the chemically remedied post-carbonized system exhibits a surface area and a pore volume of 680 $m^2 g^{-1}$ and 0.30 $cm^3 g^{-1}$, respectively. The lower surface area of the post-carbonized system is attributed to the pore-blocking effect of the dehydrating agent. Consequently, blockage is removed through physical activation. A comparison made between physicochemical and chemical activation processes showed that ACs produced through the dual-step process have excellent textural features as well as quality. Arami-Niya and his team specified that ACs obtained *via* physicochemical activation have well-built as well as uniformly scattered porous structures.[48] Tseng and co-workers[55] also revealed that when carbons are infused into potassium hydroxide, this resulted in the formation of surface micropore structures.[49] In physicochemical activation, when infused carbon is exposed to heat in the presence of carbon dioxide gas, the fraction of macropores to mesopores upsurges, which elucidates growth in pore size as well as in mass transfer for an enhanced adsorption ability.

12.2.5 Microwave-assisted Activation

This type of activation came out as an alternative to conservative methodologies for generating AC, because of its extraordinary characteristics like selectivity, speed, smooth and volumetric heating, and indirect interaction among the heat source and heated precursors. The chief functioning restraints for the microwave-sustained activation include the process configuration, microwave radiation strength, duration of activation, characteristics of precursors, interactions among microwaves as well as chemical moieties.[31] In microwave heating, the energy is simply deformed into heat within the particles through dipole orientation as well as *via* ionic transmission. When high-frequency voltages are present, the particles with induced/dipole moments are connected in a way that deviates from the direction of the applied force.[50] Hence, a huge temperature gradient grows from the inner side of the particle towards the cool external side, which makes microwave heating more effective and economical.[51]

Microwave-supported activation is the combination of physical and chemical activation that may yield better quality AC through single-[52] or double-step activation procedures.[53] Carbonization and activation are the two main processes for single-step microwave activation with a reactor.[54] The advantage of this activation is that a stable AC is obtained in a single-step procedure. On the other hand, double-step microwave activation encompasses carbonization and activation in the presence of microwave radiation.[55] In this process, the BET surface area relies on the source of biomass, pyrolysis temperature, radiation energy, and running duration. Microwave activation is proved to be an adaptable process for extraordinarily controllable features, as well as for producing ACs with excellent porosity as well as textural characteristics.[56]

12.3 Applications of AC in the Food Industry

ACs are utilized for various applications in the food and drinking H_2O industries.[57,58] They are also used for the adsorption of phenolic amalgams in the wastewater treatment of food plants. Various investigators explored the adsorption of olive wastewater phenolics. They utilized 8 g AC/100 mL of solution to achieve the maximum adsorption capability. Furthermore, the pseudo-second-order model was regarded as the most appropriate for kinetic outcomes. To summarize, ACs have the ability to capture polyphenols as well as carbohydrates of olive mill wastewater plants, which in turn are utilized to produce bioethanol and polyphenols significant in several sectors including the food industry,[59,60] as shown in Figure 12.3.

Lopez and co-workers established a new approach to produce AC. They extracted caffeine as well as chlorogenic acid from spent coffee grounds. The AC obtained through the carbonization process had maximum adsorption capacity for methylene blue (MB) between 411 and 813 mg g^{-1}.[61] In a further fascinating investigation, anaerobic assimilation of food waste in reactors was enhanced through the synergistic effect of ACs and trace elements. Thus, propionic acid was utilized faster, resulting in the formation of a large amount of methane. In addition, the addition of the AC improved the growth rate of some bacteria, which include archaea and syntrophic bacteria. Lastly, microbial investigations showed that hydrogenotrophic methanogens were the predominant strains, which established that the AC obtained from pecan shells had a surface area and a pore volume of 1500 m^2 g^{-1} and 0.7 cm^3 g^{-1}, respectively, besides exhibiting an excellent adsorption capacity for iron(II) at

Figure 12.3 Applications of AC in the food industry.

30 °C.[62] A pseudo-second-order model best described the adsorption isotherm.[63] Furthermore, Yangui and Abderrabba employed AC by using a film of milk proteins to obtain olive mill wastewater polyphenols. Remarkably, this eco-friendly approach improved the adsorption ability, as the efficacy was 75.4% for total phenols and 90.6% for hydroxytyrosol. The obtained polyphenols were then examined for radical scavenging power, followed by the estimation of antioxidant activity *via* DPPH assay.[64] Adsorption using ACs finds a large number of applications in various food industries in order to eliminate undesirable colors as well as to enhance the quality of the food. ACs are mostly utilized in the sugar industry to remove colors from sugar syrups, in the manufacturing of alcoholic beverages for eliminating unsolicited mixtures to develop the flavor and color besides other characteristics, and also in the fat and oil industry for eliminating unwanted colors (decolorization). The requirements of ACs in these industrial sectors are increasing day by day due to the endless growth of these industries.

12.3.1   Sugar and Sweetener Industry

Globally, ACs are utilized for decolorization as well as for decontamination of numerous sugar syrups. Bone charcoal which was used in the 19th century was replaced by activated carbons at the turn of the 20th century when their commercial production was started in Great Britain and Germany in 1911. At about the same time Norit also built an AC plant. ACs with superior decolorizing characteristics were developed *via* carbonization of wood as well as various carbonaceous raw elements by employing some activating agents like zinc chloride. Later on, there was a remarkable rise in the production of active carbons by employing various types of raw materials, production approaches, and altered production environments. Numerous adsorbents with ion-exchange characteristics and various artificial ion exchangers were made for removing the color from sugars, but most of them caused an upsurge in the pH of the sugar syrup, which delayed the crystallization process. The more significant coloring substances that need to be eliminated are (a) the melanoidin formed *via* the reactions among reducing sugars and amines, (b) the caramels that are nitrogen deficient dyes formed through partial thermal disintegration of sugars bearing phenols as well as quinoid moieties, and (c) phenolic complexes bearing Fe. These dyeing materials may act as both dissociated and associated particles. The dissociated charged forms are very common. Here, it is very important to note that materials having molecular weights from 8000 g mol^{-1} to 15 000 g mol^{-1} are present in the colloidal system. Scientifically, ACs enhance the color as well as the characteristics of both sugar syrup and the product formed. The elimination of colloidal materials improves the surface tension of the sugar solution but causes a decline in its viscosity. Both these features upgrade the crystallization rate of the sugar and enhance the separation of crystals. The metal salt contaminants of sugar solutions are also adsorbed through AC, by reducing the quantity of molasses formed. Refinement of sugar encompasses various steps in which decolorization *via* ACs is the last step, which is done before boiling the sugar solution in order to produce the mother liquor, crystallization of which results in the formation of white sugar. AC employed for removing the color of sugar solutions may be utilized in various forms, including PAC and GAC, and also in the form of an ion exchanger. Nevertheless, superior results may be achieved using mixtures of adsorbents in various steps of production. Decolorization through PAC may be done by employing the batch process and/or by the fixed-bed process. In the batch process, a specified quantity of AC is added to a measured volume of sugar syrup present in a vessel. The suspension is then placed at 80–90 °C for 30 minutes to achieve the adsorption equilibrium and then filtration is done in order to separate the syrup from AC. On the other hand, in the fixed-bed process, first the sugar solution is passed *via* a fixed bed containing AC. Since the decolorizing ability of the AC bed continuously decreases with passage of time, a system containing multiple filters is usually employed. This maintains the consistency of decolorization in syrup. When using GACs, decolorization may be done *via* the fixed-bed approach, the fluidized-bed approach, and the continuous countercurrent approach. In the case of the fixed-bed technique, a solution containing sugar is heated at 80 °C followed by its passage *via* an AC bed. This technique contains three stages: adsorptive decolorization of the system containing sugar, washing of the AC, and its regeneration. In the case of the fluidized-bed technique, a fixed quantity of AC is eliminated from the filtrate and a fresh quantity of AC is added to the bed. In the case of the countercurrent technique, the sugar solution moves in the upward direction *via* the AC layer which moves downwards towards the adsorption column. This causes the introduction of the whole layer of AC into the sugar solution. A specific plant consists of various adsorption columns that also consist of a column from which sugar solution is separated from the used carbon, an arrangement for eliminating H_2O from AC, as well as a revolving AC regeneration furnace. The regenerated AC is constantly reverted towards the decolorizing process. Out of all the approaches discussed earlier, the countercurrent technique is the most efficient since the sorption ability of AC is completely utilized and the requirements of space, buildings and the quantity of sweet H_2O are reduced to a minimum. It also lessens sugar wastage as well as the utilization of fuel. The decolorizing ability of AC relies on the pH of solution, its porous structure (which includes pore dimensions), the surface area of AC, and the nature of the AC surface, which is because of the variety and quantity of C–O surface moieties connected with the AC surface, besides the nature of the color present in the solution. The C–O surface moiety is responsible for the acidic or basic nature of the AC surface.[65–67] The sugar solution is subjected to sulfitation prior to treatment with AC; this enhances decolorization, as sulfitation reduces the alkalinity of the sugar solution. Active carbons are also effective for the decolorization of various other natural as well as synthetic sweeteners like glucose, fructose, syrups, dextrin, and vinegar. Decolorization of vinegar is significantly improved by upsurging the mesoporous structures of the AC.[68]

12.3.2   Oil and Fat Industry

ACs are a class of the most effective substances utilized for the decolorization of oils as well as fats. These are normally employed in the form of mixtures with some bleaching clays, as clays are less costly as well as their mixture may yield excellent outcomes. The composition of AC–clay mixtures relies on the quality of the triglycerides as well as coloring substance. A little quantity of AC may lessen the quantity of the clay utilized for decolorization of a certain quantity of oil as well as fat. The correct dose of the mixed adsorbent and the processing conditions rely on the color intensity of the oil. Decolorization is generally done through a batch method. During decolorization oxidation of oils as well as fats occurs because of the presence of acid in them. Thus oils as well as fats are first deacidified. After that, certain quantities of mixed adsorbent are added followed by decolorization inside a vacuum vessel at 100 to 120 °C. The content present inside the vacuum vessel is strongly blended. After that, the decolorized oil may be taken out from the AC by employing a filter press. The little quantity of left-out oil inside the filter cake may be obtained after washing the filtrate with solvents like petrol. In some cases decolorization is done at room temperature in order to protect some vitamins present in oils as well as fats, for which, first oil is thinned in the presence of some solvent prior to decolorization. Refined oil is obtained from crude oil after various stages of processing which include incubation, washing, removal of H_2O, decolorization, and deodorization. AC facilitates enhancement of the odor as well as color of the resultant oil.[69]

12.3.3   Alcoholic Beverage Industry

ACs have been used for numerous tasks in the production of alcoholic beverages. Handling alcoholic beverages with ACs enhances flavor, tint, and aroma, eradicates off-smells, interrupts oxidation, decontaminates the H_2O utilized, and supplies various minerals. The minerals existing in ACs upgrade the characteristics as well as nutritional value of beverages. ACs impregnated with silver as well as nonwoven fibers are generally utilized. ACs are employed in various steps of the fermentation of beer. Direct utilization of ACs during the brewing process alters the color and enhances the flavor and smell produced *via* phenols and by means of autolysis of yeast as well as infections. ACs also employed for the distillation of H_2O, air, and CO_2 utilized for brewing beer. The decontamination of beer is mainly done before bottling. However, if the beer attribute is somehow bad, then addition of about 2–2.5 g of AC in the cask for some days is done just before bottling. As universal adsorbents, ACs may also eliminate some beneficial constituents of beer. Thus, only smallest quantities of ACs should be employed. These quantities may be calculated *via* some laboratory tests.

Vodka is formed by combining alcohol, H_2O, and several kinds of alcoholic extracts obtained from seaweeds. About ten thousand liters of vodka prepared contain 100 to 200 liters of alcoholic extracts taken from

Undaria pinnatifida and *Costaria costata* and from certain other seaweeds. The seaweed extracts are taken from raw substances with 40% ethanol in H_2O. The subsequent combination is then cleaned by passing *via* AC filters followed by bottling. During the manufacture of vodka, the H_2O to be mixed with alcohol is first freed from organic acids *via* the addition of various suspensions including alkali and alkaline earth metals, oxides, and hydroxides followed by cooling and filtration *via* AC filters containing $KMnO_4$. Vodka is then mixed with AC impregnated with Ag. ACs may also eliminate fulvic acid residues in H_2O and may also lessen the quantity of oxides as well as peroxides. Vodka having huge nutrient value has been manufactured by combining certain proportions of vanillin, sucrose, soya bean extract, fructose, and citric acid. The substances are mixed carefully, decontaminated by passing over AC filters followed by bottling. The quantities of organic contaminants as well as unwanted materials in alcohol $+ H_2O$ systems utilized in the production of vodka may be calculated in terms of optical density. Addition of ACs causes some decrease in optical density. The best quality vodka produced utilizing distilled H_2O and treatment with AC has an optical density at 260 nm, which may be utilized as an index of its characteristics. Rectified spirit is manufactured through rectification of crude spirit in which three distinct fractions are accumulated, which include light, medium, and heavy fractions. The light fraction contains low-boiling RCHO including several O_2-containing organic moieties, whereas the heavy fraction contains only fusel oil. The medium fraction is a rectified spirit that always contains a little quantity of fusel oil which produces a displeasing taste as well as smell. This fusel oil may be separated by straining the rectified spirit *via* an AC bed, which subsequently may be employed for making alcoholic beverages. Impurities that brandies pick up during the manufacturing process like fusel oil, furfural, and tannins are responsible for producing an unpleasant taste as well as aroma. ACs may support eliminating these flavors by decreasing the quantity of these contaminants and some other constituents in the raw essence of the brandies. Addition of minor quantities of ACs (*i.e.*, 5 g L^{-1}) yields excellent quality brandies; still, to produce a unique taste as well as flavor, almost 30 g L^{-1} of AC has to be added to it. For use in the production of wines, ACs must meet some specific criteria. The AC must be capable of altering the color as well as eliminating unwanted constituents coming from corks, holders, yeast, *etc.* Sometimes ACs also eliminate the constituents of wine that endow it with some pleasing features. Various AC companies have established and advertised distinctive ACs which are more suitable for use in the production of particular wines. Comprehensive investigations into the exclusion of specific materials from wines utilizing ACs have been done by using numerous kinds of ACs.[70]

Thus, it is necessary to perform the laboratory investigation of wine by selecting an appropriate AC and its quantity that may provide excellent outcomes. Typically, the quantity of carbon varies from 0.5 to 1.0 g for every liter of wine; however, to prevent the loss of its quality, the minimum operational amount must be experimentally analyzed. For this, the AC is added in the form of suspension to a small quantity of wine in a container followed by stirring the constituents of the vessel. This process requires several hours; after this, the carbon is detached *via* sedimentation and is then filtered with the help of a filter press or *via* centrifuge. Even though ACs are ineffective in eliminating the wine constituents, which may possibly lower the quality of wine, the utilization of ACs in refining the taste as well as color of the wines is a significant phase in the production of wine. It is estimated that the development of innovative categories of ACs with additional specificity will cause additional developments in the manufacture of good quality wines.

12.4   Conclusion

After being identified as a potent and dependable adsorbent, AC is employed in a variety of applications. Numerous initiatives to develop natural, biodegradable food packaging and additives have been made in response to the growing concern over environmental sustainability. An intriguing option is AC, a substance with multiple uses. In addition to having extraordinary adsorption capability, it is non-toxic, naturally degradable, and inexpensive. Its characteristics are adaptable and could be applied to a variety of situations. As a result, both its surface chemistry and pore structure have a significant impact on its capabilities. Although AC has long been used for its influence on hydrophobic compounds, altering its pore size and surface characteristics may increase its affinity for hydrophilic substances. This chapter primarily focused on the production of AC and its subsequent uses for food additives.

References

1. T. F. Chyad, R. F. C. Al-Hamadani, Z. A. Hammood and G. Abd Ali, Removal of Zinc(ɪɪ) ions from industrial wastewater by adsorption on to activated carbon produced from pine cone, *Mater. Today: Proc.*, 2021, 1–6.
2. P. González-García, Activated carbon from lignocellulosics precursors: A review of the synthesis methods, characterization techniques and applications, *Renewable Sustainable Energy Rev.*, 2018, **82**, 1393–1414.
3. H. Marsh and F. R. Reinoso, *Activated carbon*, Elsevier, 2006.
4. L. Khezami, A. Chetouani, B. Taouk and R. Capart, Production and characterisation of activated carbon from wood components in powder: Cellulose, lignin, xylan, *Powder Technol.*, 2005, **157**(1–3), 48–56.
5. K. Le Van and T. T. L. Thi, Activated carbon derived from rice husk by NaOH activation and its application in supercapacitor, *Prog. Nat. Sci.: Mater. Int.*, 2014, **24**(3), 191–198.
6. S. Babel and T. A. Kurniawan, Cr(ᴠɪ) removal from synthetic wastewater using coconut shell charcoal and commercial activated carbon modified with oxidizing agents and/or chitosan, *Chemosphere*, 2004, **54**(7), 951–967.

7. D. Cook, G. Newcombe and P. Sztajnbok, The application of powdered activated carbon for MIB and geosmin removal: predicting PAC doses in four raw waters, *Water Res.*, 2001, **35**(5), 1325–1333.
8. M. P. Ormad, N. Miguel, A. Claver, J. M. Matesanz and J. Ovelleiro, Pesticides removal in the process of drinking water production, *Chemosphere*, 2008, **71**(1), 97–106.
9. K. Foo and B. Hameed, An overview of landfill leachate treatment via activated carbon adsorption process, *J. Hazard. Mater.*, 2009, **171**(1–3), 54–60.
10. P. J. Cerminara, G. A. Sorial, S. P. Papadimas, M. T. Suidan, M. A. Moteleb and T. F. Speth, Effect of influent oxygen concentration on the GAC adsorption of VOCs in the presence of BOM, *Water Res.*, 1995, **29**(2), 409–419.
11. F. I. Hai, K. Yamamoto, F. Nakajima and K. Fukushi, Bioaugmented membrane bioreactor (MBR) with a GAC-packed zone for high rate textile wastewater treatment, *Water Res.*, 2011, **45**(6), 2199–2206.
12. R. G. Scharf, R. W. Johnston, M. J. Semmens and R. M. Hozalski, Comparison of batch sorption tests, pilot studies, and modeling for estimating GAC bed life, *Water Res.*, 2010, **44**(3), 769–780.
13. J. Menéndez-Díaz and I. Martín-Gullón, Types of carbon adsorbents and their production, in *Interface science and technology*, Elsevier, 2006, vol. 7, pp. 1–47.
14. D. Das, V. Gaur and N. Verma, Removal of volatile organic compound by activated carbon fiber, *Carbon*, 2004, **42**(14), 2949–2962.
15. J. Bedia, M. Peñas-Garzón, A. Gómez-Avilés, J. J. Rodriguez and C. Belver, A review on the synthesis and characterization of biomass-derived carbons for adsorption of emerging contaminants from water, *C*, 2018, **4**(4), 63.
16. J. Rivera-Utrilla, I. Bautista-Toledo, M. A. Ferro-García and C. Moreno-Castilla, Activated carbon surface modifications by adsorption of bacteria and their effect on aqueous lead adsorption, *J. Chem. Technol. Biotechnol.*, 2001, **76**(12), 1209–1215.
17. N. Diban, G. Ruiz, A. Urtiaga and I. Ortiz, Recovery of the main pear aroma compound by adsorption/desorption onto commercial granular activated carbon: Equilibrium and kinetics, *J. Food Eng.*, 2008, **84**(1), 82–91.
18. H. D. Ozsoy and J. H. van Leeuwen, Removal of color from fruit candy waste by activated carbon adsorption, *J. Food Eng.*, 2010, **101**(1), 106–112.
19. S. M. M. Kamal, N. L. Mohamad, A. G. L. Abdullah and N. Abdullah, Detoxification of sago trunk hydrolysate using activated charcoal for xylitol production, *Procedia Food Sci.*, 2011, **1**, 908–913.
20. S. Deng, Y. Nie, Z. Du, Q. Huang, P. Meng and B. Wang, *et al.*, Enhanced adsorption of perfluorooctane sulfonate and perfluorooctanoate by bamboo-derived granular activated carbon, *J. Hazard. Mater.*, 2015, **282**, 150–157.
21. M. Valix, W. Cheung and G. McKay, Preparation of activated carbon using low temperature carbonisation and physical activation of high ash raw bagasse for acid dye adsorption, *Chemosphere*, 2004, **56**(5), 493–501.

22. J. Zhenchao, Z. Yuting, Y. Jiuming, L. Yedan, S. Yang and C. Jinyao, *et al.*, Safety assessment of dietary bamboo charcoal powder: A 90-day subchronic oral toxicity and mutagenicity studies, *Food Chem. Toxicol.*, 2015, **75**, 50–57.

23. A. Miyagi, H. Nabetani and M. Nakajima, Decolorization of Japanese soy sauce (shoyu) using adsorption, *J. Food Eng.*, 2013, **116**(3), 749–757.

24. N. Byamba-Ochir, W. G. Shim, M. Balathanigaimani and H. Moon, Highly porous activated carbons prepared from carbon rich Mongolian anthracite by direct NaOH activation, *Appl. Surf. Sci.*, 2016, **379**, 331–337.

25. F.-C. Huang, C.-K. Lee, Y.-L. Han, W.-C. Chao and H.-P. Chao, Preparation of activated carbon using micro-nano carbon spheres through chemical activation, *J. Taiwan Inst. Chem. Eng.*, 2014, **45**(5), 2805–2812.

26. M. A. Yahya, Z. Al-Qodah and C. Z. Ngah, Agricultural bio-waste materials as potential sustainable precursors used for activated carbon production: A review, *Renewable Sustainable Energy Rev.*, 2015, **46**, 218–235.

27. N. Radenahmad, A. T. Azad, M. Saghir, J. Taweekun, M. S. A. Bakar and M. S. Reza, *et al.*, A review on biomass derived syngas for SOFC based combined heat and power application, *Renewable Sustainable Energy Rev.*, 2020, **119**, 109560.

28. M. Alhinai, A. K. Azad, M. S. A. Bakar and N. Phusunti, Characterisation and thermochemical conversion of rice husk for biochar production, *Int. J. Renewable Energy Res.*, 2018, **8**(3), 1648–1656.

29. T. Odetoye, T. Afolabi, A. Bakar and J. Titiloye, Thermochemical characterization of Nigerian Jatropha curcas fruit and seed residues for biofuel production, *Energy, Ecol. Environ.*, 2018, **3**(6), 330–337.

30. J. Wang, F. Wu, M. Wang, N. Qiu, Y. Liang and S. Fang, *et al.*, Preparation of activated carbon from a renewable agricultural residue of pruning mulberry shoot, *Afr. J. Biotechnol.*, 2010, **9**(19), 2762–2767.

31. W. Ao, J. Fu, X. Mao, Q. Kang, C. Ran and Y. Liu, *et al.*, Microwave assisted preparation of activated carbon from biomass: A review, *Renewable Sustainable Energy Rev.*, 2018, **92**, 958–979.

32. E. Menya, P. Olupot, H. Storz, M. Lubwama and Y. Kiros, Production and performance of activated carbon from rice husks for removal of natural organic matter from water: a review, *Chem. Eng. Res. Des.*, 2018, **129**, 271–296.

33. N. A. Rashidi and S. Yusup, A review on recent technological advancement in the activated carbon production from oil palm wastes, *Chem. Eng. J.*, 2017, **314**, 277–290.

34. H. W. Lee, Y.-M. Kim, S. Kim, C. Ryu, S. H. Park and Y.-K. Park, Review of the use of activated biochar for energy and environmental applications, *Carbon Lett.*, 2018, **26**, 1–10.

35. H. H. Rafsanjani, H. Kamandari and H. Najjarzadeh, Study on pore and surface development of activated carbon produced from Iranian coal in a rotary kiln reactor, *Iran J. Chem. Eng.*, 2013, **10**, 27–38.

36. S. Wong, N. Ngadi, I. M. Inuwa and O. Hassan, Recent advances in applications of activated carbon from biowaste for wastewater treatment: a short review, *J. Cleaner Prod.*, 2018, **175**, 361–375.

37. A brief review on activated carbon derived from agriculture by-product, in *AIP conference proceedings*, ed. M. A. Yahya, M. H. Mansor, W. A. A. W. Zolkarnaini, N. S. Rusli, A. Aminuddin and K. Mohamad, *et al.*, AIP Publishing LLC, 2018.

38. M. I. Din, S. Ashraf and A. Intisar, Comparative study of different activation treatments for the preparation of activated carbon: a mini-review, *Sci. Prog.*, 2017, **100**(3), 299–312.

39. S. J. van den Toren, A. van Grieken, W. C. Mulder, Y. T. Vanneste, M. Lugtenberg and M. L. de Kroon, *et al.*, School absenteeism, Health-Related Quality of Life [HRQOL] and happiness among young adults aged 16–26 years, *Int. J. Environ. Res. Public Health*, 2019, **16**(18), 3321.

40. Q. Cao, K.-C. Xie, Y.-K. Lv and W.-R. Bao, Process effects on activated carbon with large specific surface area from corn cob, *Bioresour. Technol.*, 2006, **97**(1), 110–115.

41. Z. Z. Chowdhury, S. B. Abd Hamid, R. Das, M. R. Hasan, S. M. Zain and K. Khalid, *et al.*, Preparation of carbonaceous adsorbents from lignocellulosic biomass and their use in removal of contaminants from aqueous solution, *BioResources*, 2013, **8**(4), 6523–6555.

42. N. Isoda, R. Rodrigues, A. Silva, M. Gonçalves, D. Mandelli and F. C. A. Figueiredo, *et al.*, Optimization of preparation conditions of activated carbon from agriculture waste utilizing factorial design, *Powder Technol.*, 2014, **256**, 175–181.

43. M. A. A. Zaini and M. J. Kamaruddin, Critical issues in microwave-assisted activated carbon preparation, *J. Anal. Appl. Pyrolysis*, 2013, **101**, 238–241.

44. B. Caglar, B. Afsin, E. Koksal, A. Tabak and E. Eren, Characterization of Unye bentonite after treatment with sulfuric acid, *Quim. Nova*, 2013, **36**, 955–959.

45. J. Saleem, U. B. Shahid, M. Hijab, H. Mackey and G. McKay, Production and applications of activated carbons as adsorbents from olive stones, *Biomass Convers. Biorefin.*, 2019, **9**(4), 775–802.

46. A. T. M. Din, B. Hameed and A. L. Ahmad, Batch adsorption of phenol onto physiochemical-activated coconut shell, *J. Hazard. Mater.*, 2009, **161**(2–3), 1522–1529.

47. T. Lee, Z. A. Zubir, F. M. Jamil, A. Matsumoto and F.-Y. Yeoh, Combustion and pyrolysis of activated carbon fibre from oil palm empty fruit bunch fibre assisted through chemical activation with acid treatment, *J. Anal. Appl. Pyrolysis*, 2014, **110**, 408–418.

48. A. Arami-Niya, W. M. A. W. Daud, F. S. Mjalli, F. Abnisa and M. S. Shafeeyan, Production of microporous palm shell based activated carbon for methane adsorption: modeling and optimization using response surface methodology, *Chem. Eng. Res. Des.*, 2012, **90**(6), 776–784.

49. R.-L. Tseng, S.-K. Tseng and F.-C. Wu, Preparation of high surface area carbons from Corncob with KOH etching plus CO_2 gasification for the adsorption of dyes and phenols from water, *Colloids Surf., A*, 2006, **279**(1–3), 69–78.

50. T. M. Alslaibi, I. Abustan, M. A. Ahmad and A. A. Foul, A review: production of activated carbon from agricultural byproducts via conventional and microwave heating, *J. Chem. Technol. Biotechnol.*, 2013, **88**(7), 1183–1190.

51. P. D. Pathak and S. A. Mandavgane, Preparation and characterization of raw and carbon from banana peel by microwave activation: application in citric acid adsorption, *J. Environm. Chem. Eng.*, 2015, **3**(4), 2435–2447.

52. H. Deng, G. Li, H. Yang, J. Tang and J. Tang, Preparation of activated carbons from cotton stalk by microwave assisted KOH and K_2CO_3 activation, *Chem. Eng. J.*, 2010, **163**(3), 373–381.

53. A. F. Abbas and M. J. Ahmed, Mesoporous activated carbon from date stones (Phoenix dactylifera L.) by one-step microwave assisted K_2CO_3 pyrolysis, *J. Water Process Eng.*, 2016, **9**, 201–207.

54. J. Li, J. Dai, G. Liu, H. Zhang, Z. Gao and J. Fu, *et al.*, Biochar from microwave pyrolysis of biomass: A review, *Biomass Bioenergy*, 2016, **94**, 228–244.

55. W. Li, L.-B. Zhang, J.-H. Peng, N. Li and X.-Y. Zhu, Preparation of high surface area activated carbons from tobacco stems with K_2CO_3 activation using microwave radiation, *Ind. Crops Prod.*, 2008, **27**(3), 341–347.

56. D. Angın, E. Altintig and T. E. Köse, Influence of process parameters on the surface and chemical properties of activated carbon obtained from biochar by chemical activation, *Bioresour. Technol.*, 2013, **148**, 542–549.

57. M. E. Goher, A. M. Hassan, I. A. Abdel-Moniem, A. H. Fahmy, M. H. Abdo and S. M. El-sayed, Removal of aluminum, iron and manganese ions from industrial wastes using granular activated carbon and Amberlite IR-120H, *Egyptian J. Aquat. Res.*, 2015, **41**(2), 155–164.

58. P. McCleaf, S. Englund, A. Östlund, K. Lindegren, K. Wiberg and L. Ahrens, Removal efficiency of multiple poly- and perfluoroalkyl substances (PFASs) in drinking water using granular activated carbon (GAC) and anion exchange (AE) column tests, *Water Res.*, 2017, **120**, 77–87.

59. D. Zhang, P. Huo and W. Liu, Behavior of phenol adsorption on thermal modified activated carbon, *Chin. J. Chem. Eng.*, 2016, **24**(4), 446–452.

60. B. Aliakbarian, A. A. Casazza and P. Perego, Kinetic and isotherm modelling of the adsorption of phenolic compounds from olive mill wastewater onto activated carbon, *Food Technol. Biotechnol.*, 2015, **53**(2), 207–214.

61. M. Ramón-Gonçalves, L. Alcaraz, S. Pérez-Ferreras, M. E. León-González, N. Rosales-Conrado and F. A. López, Extraction of polyphenols and synthesis of new activated carbon from spent coffee grounds, *Sci. Rep.*, 2019, **9**, 17706.

62. G. Capson-Tojo, R. Moscoviz, D. Ruiz, G. Santa-Catalina, E. Trably and M. Rouez, *et al.*, Addition of granular activated carbon and trace elements to favor volatile fatty acid consumption during anaerobic digestion of food waste, *Bioresour. Technol.*, 2018, **260**, 157–168.

63. A. R. Kaveeshwar, S. K. Ponnusamy, E. D. Revellame, D. D. Gang, M. E. Zappi and R. Subramaniam, Pecan shell based activated carbon for removal of iron(II) from fracking wastewater: adsorption kinetics,

isotherm and thermodynamic studies, *Process Saf. Environ. Prot.*, 2018, **114**, 107–122.

64. A. Yangui and M. Abderrabba, Towards a high yield recovery of polyphenols from olive mill wastewater on activated carbon coated with milk proteins: Experimental design and antioxidant activity, *Food Chem.*, 2018, **262**, 102–109.

65. M. Smisek and S. Cerny, New books – Active carbon: Manufacture, properties, and applications, *Anal. Chem.*, 1970, **42**(14), 81A.

66. T. J. Bandosz and C. Ania, Surface chemistry of activated carbons and its characterization, in *Interface science and technology*, Elsevier, 2006, vol. 7, pp. 159–229.

67. R. C. Bansal and M. Goyal, *Activated carbon adsorption*, CRC Press, 2005.

68. F. López, F. Medina, M. Prodanov and C. Güell, Oxidation of activated carbon: application to vinegar decolorization, *J. Colloid Interface Sci.*, 2003, **257**(2), 173–178.

69. J. Przepiórski, Activated carbon filters and their industrial applications, in *Interface Science and Technology*, Elsevier, 2006, vol. 7, pp. 421–474.

70. J. M. Tascón, *Novel carbon adsorbents*, Elsevier, 2012.

Influence of Activated Carbon on Metallic Corrosion

R. ASLAM, M. MOBIN* AND LEI GUO*

Corrosion Research Laboratory, Department of Applied Chemistry, Faculty of Engineering and Technology, Aligarh Muslim University, Aligarh, 202002-India
*Emails: drmmobin@hotmail.com; cqglei@163.com

13.1 Introduction

13.1.1 Corrosion and Corrosion Mitigation

According to the National Association of Corrosion Engineers (NACE), corrosion is a serious issue in many industrial sectors and costs the world trillions of dollars annually, particularly in industrial processes where metal or metal alloys are exposed to various acidic environments that are highly corrosive.[1–3] An estimated 26.1 billion dollars are lost annually due to corrosion, a natural phenomenon that causes material loss all around the world. Chemically, it is defined as the deposition of an oxide, sulfide, or chloride layer on a material's surface. Corrosion was formerly exclusively studied in terms of the deterioration of metal surfaces, but today's studies cover the ageing or naturally occurring degradation of plastics and polymers. Wherever there is moisture, corrosion exists. In fact, in the absence of moisture, corrosion of steel that causes its cracking is observed. This is primarily due to exposure to di-hydrogen gas, which as a result releases methane by reacting with the carbon present in steel, and this is referred to as "dry corrosion." Following the chemistry of the redox reaction, one component serves as an anode and the other as a cathode.

Activated Carbon: Progress and Applications
Edited by Chandrabhan Verma and Mumtaz A. Quraishi
© The Royal Society of Chemistry 2023
Published by the Royal Society of Chemistry, www.rsc.org

Like a regular galvanic cell, degradation typically takes place at the anode where oxidation takes place, whereas oxidation product deposition is typically seen at the cathode where reduction happens. Chemical substances are employed to block corrosion or to postpone its progression since corrosion destroys material and causes mechanical failure. These chemicals are known as inhibitors. They typically contain elements with a lot of electrons, like aromatic rings, nitrogen, oxygen, and sulfur. Among the various known and established alternatives, the use of organic compounds is one of the most popular methods. These compounds possess electron-rich donor sites that can strongly coordinate with metallic surfaces. These sites are called adsorption sites or centers.[1–3]

13.1.2 Activated Carbon: Fundamentals, Properties, and Applications

Numerous studies have been conducted on the potential of activated carbon (AC). It has various characteristics such as high porosity, high surface area, and low cost.[4,5] The surface area of AC, also known as activated charcoal, can exceed $1000 \text{ m}^2 \text{ g}^{-1}$. Due to its properties, AC can have a surface area that is equivalent to that of a football field. It can be used in various industrial processes such as water treatment, energy storage, and pollution removal. It can also be used in the treatment of various hazardous and poisonous compounds. For instance, it can remove various harmful chemicals from the environment by absorbing my compounds including dye, water, and oxygen. Besides its properties, AC has also been known as an ideal material for supercapacitors due to its low cost and excellent conductivity.[6]

13.2 Types of AC

The end product of AC can vary significantly due to the various factors that affect its chemical and physical properties. This makes it difficult to predict which type of carbon will perform best in each application. Due to this, the different types of AC are highly specialized. Despite the varying characteristics of these products, there are still three main types of ACs (Figure 13.1):

POWDER   GRANULES   PELLETS

Figure 13.1   AC powder, granules, and pellets.

(1) *Powdered AC (PAC)*: PACs are generally categorized into the size range of 5 to 150 Å. They are commonly used in liquid-phase applications. Compared to other types of AC, PACs offer a lower processing cost and flexibility in operation.

(2) *Granular AC (GAC)*: Applications for GACs can be found in both the gas and liquid phases and typically range from 0.2 mm to 5 mm in particle size. Because they allow for clean handling and typically last longer than PACs, GACs are well-liked. They can also be recycled and reused and offer increased strength (hardness).

(3) *Extruded AC (EAC)*: EACs are cylindrical pellet products that can range from 1 to 5 mm in size. They are commonly used in gas-phase reactions. Due to their extrusion process, ACs are heavy-duty.

13.3 Production of AC

AC is produced through two main processes: carbonization and activation.

13.3.1 Carbonization

Carbonization is a process utilized in bioresource engineering and bioresource technology to remove solid carbon from biomass. It involves heating the biomass to leave behind its carbonaceous material, which then becomes solid. Coke is a byproduct of this process, and it can be produced at either high or low temperatures. The process of carbonization is carried out through the destruction of organic matter, such as dead animal remains and plants. It is a complex process that involves various reactions. Some of these reactions include dehydrogenation, isomerization, and hydrogen transfer. The raw material used for the process of carbonization is placed in an inert environment, which is below 800 °C. Through gasification, various elements, such as nitrogen, sulfur, oxygen, and hydrogen, are taken out of the biomass, and biochar is left behind as a solid product.

13.3.2 Activation

To completely create the pore structure, the carbonized material, or char, must now be activated. This is accomplished by heating the char to 800–900 °C and oxidizing it using air, carbon dioxide, or steam. Both thermal (physical/steam) activation and chemical activation can be used to produce AC, depending on the parent material. During the physical activation step, the substance is pyrolyzed, either with acid as a dehydrating agent or with Zn, Ni, or Cd compounds. Because the surface area of the AC produced by the chemical process is higher than that of the AC formed through pyrolysis and it has a functional group that can interact with the adsorbate, it has an advantage over that produced by the physical approach. The porosity of the material develops more effectively when a gaseous

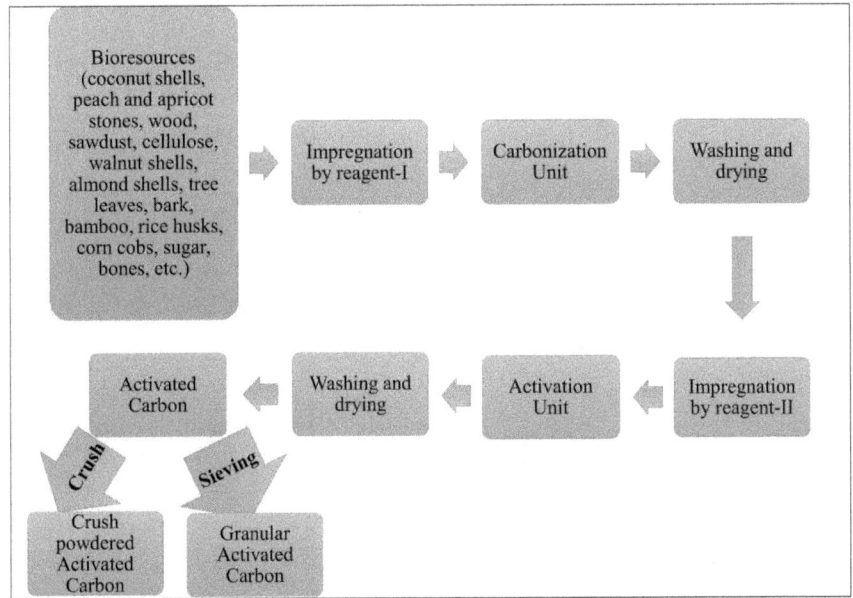

Figure 13.2 General flow diagram for manufacturing of AC from bioresource feedstock.

stream, such as air, nitrogen, or argon, is applied. Figure 13.2 is a general illustration of how AC can be made from bioresources.

13.4 Feedstocks for AC

AC is typically generated from expensive and finite materials such as coal, lignite, peat, petroleum waste, and wood (Figure 13.3). Recently, there has been increased interest in using AC generated from biowaste from forest, municipal, industrial, and agricultural sources. These bioresources are affordable, widely accessible, replenishable, and ecologically sound. Additionally, the carbon material that was extracted from the biomass demonstrated outstanding surface characteristics, including a high degree of porosity and a high specific surface area. Palm shells, fruit shells, groundnut shells, coconut shells, *Ricinus communis* seed shells, and other materials are examples of agricultural waste. The biowaste includes plant parts such as the root, stem, bark, flower, leaf, fruit peel, husk, shell, and stone. Woody and non-woody resources are the two categories of these wastes. The three primary components of woody resources are cellulose, hemicellulose, and lignin, as opposed to non-woody resources, which are made up of cellulose, hemicellulose, lignin, lipids, proteins, sugars, water, hydrocarbons, starches, and several functional groups. Finding lower-priced and more powerful substitutes for the current commercial AC is very popular these days. Investigating efficient and affordable AC may help to promote environmental sustainability and provide advantages for upcoming

Figure 13.3 Various biomass feedstocks.

commercial uses. Comparatively speaking, the AC made from bioresources is quite inexpensive. For the synthesis of AC, bioresources must therefore be given proper consideration and weight.

AC has been produced using banana fibers, argan seed shells, corn grains, *Camellia oleifera* shells, sugarcane bagasse, scrap waste tyres, and onion dry peel waste.[7–9]

Rice husk was used by Song *et al.*[10] to make ACs using the dry activation method. N_2 adsorption, X-ray diffraction, infrared spectroscopy, and scanning electron microscopy were used to analyze the materials. To clarify the mechanism of pore creation, the effects of the mass ratio (1–4) and the activation method (dry activation or impregnation) on the surface area were examined. With a mass ratio of 3, dry-ACs had a surface area of 2841 $m^2\,g^{-1}$ compared to 933 $m^2\,g^{-1}$ for impregnation. In order to manufacture inexpensive AC with exceptional surface area, the authors suggested dry activation as a practical and efficient method.

Wood apple fruit shells were used to make fine powdered activated charcoal.[11] After two hours of carbonization at 300 °C and 500 °C, the mixture was allowed to cool to ambient temperature. The produced charcoal was chemically activated when it was soaked in the activating reagent 1 N $ZnCl_2$ and heated for an hour at temperatures of 300 °C and 500 °C, respectively. The methods given by the American Water Works Association (AWWA) and the European Chemical Industry Council (CEFIC) and scanning electron microscopy were used to determine the properties of the ACs. The ACs' pH, conductance, ash content, moisture, methylene blue number, iodine value, and calorific value all differed noticeably from one another.

At a temperature of 400 °C for one hour, carbon derived from the plantain (*Musa paradisiaca*) stem was carbonized.[12] The carbonized carbon was split into two portions and they were individually activated. H_3PO_4 and $ZnCl_2$ were used in the chemical activation method to create the ACs CPPAC (carbonized plantain phosphoric acid AC) and CPZAC (carbonized plantain zinc chloride AC). This was done to characterize the pH, bulk density, moisture, ash, volatile matter, iodine number, and oxygen functional group. Except for the carboxylic group for CPPAC and the phenolic group for both CPPAC and CPZAC, there were substantial variations in all the surface attributes when compared to the untreated plantain carbon (UPC), indicating that a chemical transformation had taken place.

Using beet waste as the starting material and KOH as the activator, Zhao *et al.*[13] reported the preparation of several ACs. The final product had a large specific surface area and a dense pore network. The ratio of KOH to beet residue and the impact of the activation temperature on the characteristics of the AC were investigated.

13.5   AC: As an Anti-corrosive Material

Utilizing nanoparticle-modified AC (NMAC), mild steel industrial corrosion is reduced. AC and ZnO- or NiO nanoparticle-modified AC were both produced using the co-precipitation method (NMAC). SEM, EDS, FTIR, and TEM studies were used to determine the surface shape and size of the modified nanoparticle. The weight reduction method in a solution of 1 M HCl was used to determine the mild steel corrosion rate and inhibition effectiveness. The corrosion rate of mild steel was slowed down by the modified ACs, particularly that modified with NiO NPs. The mild steel inhibition efficiency in 1 M HCl at ambient temperature was 43.69% in NiO NPs and 38.59% in NiO-AC nanocomposites.[14]

The effect of immersion of mild steel specimens in 1 M HCl while using both regular and modified AC NPs was studied by the SEM method. The surface was highly damaged and corroded in the absence of AC NPs due to free acid attacks. However, significantly less damage was apparent in the micrographs in the presence of the AC and modified AC NPs than it was in the absence of the AC and modified ACs. On interaction with the AC, a strong promotion of spontaneous potential (E_{sp}) was seen, and this was higher than the repassivation potential for crevice corrosion (E_R, CREV). As a result, the presence of the AC significantly increased the E_{sp} of 316L SS, which boosted the galvanic effect's ability to cause crevice corrosion.[15]

A commercial AC fiber (ACF) was modified by Yang *et al.*[16] and used to stop iron corrosion in industrial water supply and circulation systems. The addition of the modified ACF to the water between pH 4 and 10 greatly reduced the rate of iron corrosion. Notably, the modified ACF became more effective as the pH increased, suggesting that the modified ACF performed better in basic settings than under acidic conditions. It was found that at pH 4 the corrosion of iron was slowed down more when the amount of the modified ACF added was increased from 50 mg to 100 mg.

13.6 Conclusion

Metal corrosion can destroy equipment and increase maintenance costs, perhaps reducing the device's lifespan and raising safety concerns. Due to its exceptional adsorption properties, AC is an incredibly versatile material that lends itself to thousands of uses. Due to its high surface area and its adsorptive capacity, AC is an important component in many industries. A useful tool for recovering precious metals like gold and silver is AC. It is frequently used to decaffeinate coffee and remove unwanted flavours, colours, and aromas from food and beverages. It can be used to purify air and treat a range of illnesses and poisonings. Additionally, metals can be protected from corrosive chemical species by using AC in corrosion prevention. Functionalization is necessary for this application, though. However, the production of such materials typically calls for extremely difficult conditions and has several restrictions, such as the initial high temperatures of the carbonization process (up to >800 °C), which is followed by chemical or physical activation to turn carbon materials into AC. In addition, it is crucial to investigate cost-effective and environmentally friendly alternatives to crude oil and natural gas as raw materials to produce carbon, which will lead to a reexamination of this area.

References

1. C. Verma, E. E. Ebenso, M. A. Quraishi and C. M. Hussain, *Mater. Adv.*, 2021, **2**, 3806–3850.
2. M. Benarioua, A. Mihi, N. Bouzeghaia and M. Naoun, *Egypt. J. Pet.*, 2019, **28**, 155–159.
3. A. Gruca and M. Greczek-Stachura, *Inżynieria Materiałowa*, 2019, **1**, 19–24.
4. K. Nutan, C. Alok and K. Rajesh, Comparative study of corrosion inhibition efficiency of naturally occurring eco-friendly varieties of holy Basil (Tulsi) for Tin in HNO₃ solution, *Sci. Res.*, 2012, **2**, 68–73.
5. A. Singh, K. R. Ansari, J. Haque, P. Dohare, H. Lgaz, R. Salghi and M. A. Quraishi, Effect of electron donating functional groups on corrosion inhibition of mild steel in hydrochloric acid: experimental and quantum chemical study, *J. Taiwan Inst. Chem. Eng.*, 2018, **82**, 233–251.
6. W. Deng, Potassium hydroxide activated and nitrogen doped graphene with enhanced supercapacitive behavior, *Sci. Adv. Mater.*, 2018, **10**, 937949.
7. G. A. Ali, O. Abed Habeeb, H. Algarni and K. F. Chong, CaO impregnated highly porous honeycomb activated carbon from agriculture waste: symmetrical supercapacitor study, *J. Mater. Sci.*, 2019, **54**, 683–692.

8. G. Ali, S. Supriya, K. F. Chong, E. R. Shaaban, H. Algarni, T. Maiyalagan and G. Hegde, Superior supercapacitance behavior of oxygen self-doped carbon nanospheres: a conversion of Allium cepa peel to energy storage system, *Biomass Convers. Biorefin.*, 2019, **10**, 319–399.

9. J. H. Yun, D. K. Choi and S. H. Kim, Adsorption equilibria of chlorinated organic solvents onto activated carbon, *Ind. Eng. Chem. Res.*, 1998, **37**, 1422–1427.

10. S. Xiaolan, Z. Ying and C. Caimin, Novel method for preparing activated carbons with high specific surface area from rice husk, *Ind. Eng. Chem. Res.*, 2012, **51**, 15075–15081.

11. P. D. AshtaPutrey and S. D. Ashtaputrey, Preparation and characterization of activated charcoal derived from wood apple fruit shell, *J. Sci. Res.*, 2020, **64**, 236–240.

12. O. A. Ekpete, A. C. Marcus and V. Osi, Preparation and characterization of activated carbon obtained from plantain (*Musa paradisiaca*) fruit stem, *J. Chem.*, 2017, 8635615.

13. J. Zhao, L. Yu, F. Zhou, H. M. K. Yanga and G. Wu, Synthesis and characterization of activated carbon from sugar beet residue for the adsorption of hexavalent chromium in aqueous solutions, *RSC Adv.*, 2021, **11**, 8025–8032.

14. H. M. El Refay, A. M. Hyba and G. A. Gaber, Fabrication, characterization and corrosion feature evaluation of mild steel in 1 M HCl by nanoparticle-modified activated carbon, *Chem. Pap.*, 2022, **76**, 813–825.

15. Y. M. Hiroshi and A. Y. Watanabe, Influence of activated carbon on crevice corrosion in adsorption tower of advanced liquid processing system, *Corros. Eng., Sci. Technol.*, 2018, **53**, 39–43.

16. J. Yang, J. Peng, Z. Shen, J. Jia and F. Zhang, Corrosion protection of iron in water by activated carbon fiber (ACF), *Carbon*, 2006, **44**, 19–26.

Subject Index